ELECTRICAL ENGINEERING
Theory and Practice

NEW YORK · JOHN WILEY & SONS, INC.
London · Chapman & Hall, Limited

2nd Edition

ELECTRICAL ENGINEERING
Theory and Practice

William H. Erickson
Professor of Electrical Engineering

Nelson H. Bryant
Associate Professor of Electrical Engineering
Cornell University

Copyright 1952 © 1959 by John Wiley & Sons, Inc.
All Rights Reserved

This book or any part thereof must not be reproduced in any form without the written permission of the publisher.

Library of Congress Catalog Card Number: 59-5879
Printed in the United States of America

■ Preface to the Second Edition

Because the use of transistors has developed so rapidly since the first publication of the book, considerable material on transistors has been added in this revision. This addition is the principal change in the revision. An extensive reorganization of the electronics section was necessary to present the transistor on an equal footing with the electron tube. The transistor is introduced as a device operating on principles quite different from a tube but capable of all the familiar functions of the tube. Therefore, methods developed for the analysis of vacuum-tube circuits are applied whenever possible to problems of transistor circuit analysis.

In addition, the number of problems at the end of each chapter has been increased, the discussion of the application of complex algebra to a-c circuit problems has been transferred from the appendix to the chapter on a-c circuits, and material on magnetic amplifiers has been included.

WILLIAM H. ERICKSON
NELSON H. BRYANT

October 1958

Preface to the Second Edition

■ *Preface to the First Edition*

This book has been developed from mimeographed material used for several years as the textbook for a sequence of courses given to mechanical, civil, and chemical engineering students. Although the sequence now consists of three 3-semester-hour courses, the book has also been successfully tested in a sequence of two 4-hour courses.

The topics covered may be classified into three general categories—circuits, machines, and electronics—according to the order of presentation. Thus, both d-c and a-c circuits are discussed before any consideration is given to machine details. Wherever possible, basic relations are introduced in a general form before applying them to either d-c or a-c circuits. In addition to the ordinary d-c and a-c machinery, the general theory and application of synchro units and rotating amplifiers are included. The chapters relating to electronics are intended to serve as an introduction to industrial electronics. A short chapter on communications is included since communication circuits incorporate many components considered elsewhere in the book.

In presenting the theory, emphasis is placed upon those factors that influence intelligent application or selection of equipment; little stress is placed upon design details. Although mathematical analysis of the operation of circuits and machines is not avoided, physical analysis is stressed. Thus, certain methods of treatment, such as the equivalent-circuit concept for a-c machines, have been omitted because it is felt that these methods do not contribute to the students' understanding in proportion to the time necessary to gain facility with such methods. Similarly, a-c circuit theory is developed without complex notation;

however, an appendix on complex notation is included for those who prefer this approach.

The theory is presented in such a way that the system of units employed is unimportant. Although the desirability of a universal system of units is recognized, the mks system of units is not employed in some sections of the book, for example, magnetics; in these sections, the units are those in ordinary usage.

The problems at the ends of the chapters are, for the most part, of a practical nature and require analysis rather than the mere substitution of numbers in formulas. It is felt that illustrative problems encourage memorization; therefore illustrative problems in the text are kept to the minimum and are used only for analysis.

The authors are greatly indebted to the students who have helped to develop this book by their patience, suggestions, and encouragement.

WILLIAM H. ERICKSON
NELSON H. BRYANT

Ithaca, New York
May 1952

■ Contents

1 INTRODUCTORY CONCEPTS 1

The Electron Theory. Electromotive Force and Potential Difference. Current. Resistance. Ohm's Law. Direct and Alternating Current. Power and Energy. Some Circuit Concepts.

2 RESISTANCE 15

Definition. Physical Aspects of Resistance. Resistivity. Conductivity. The Circular Mil. Wire Gages and Conductors. Current-Carrying Capacity. Effect of Temperature. Resistance Thermometry. Temperature versus Time. Rheostats. Rating of Resistors. Strain Gages.

3 ELECTRIC CIRCUIT LAWS AND D-C CIRCUITS . . . 31

Introduction. Fundamental Circuit Laws. Procedure in D-C Circuit Solutions. Equivalent Circuits. Thévenin's Theorem. Nonlinear D-C Circuits.

4 MAGNETICS AND MAGNETIC CIRCUITS 52

Magnetics. Lines of Force. Flux and Flux Density. Production of Flux. Magnetomotive Force. Reluctance. Electric Circuit Analogies. Magnetic Relations. The Magnetization Curve. Permeability of Magnetic Materials. Magnetic Circuits. The Solution of Magnetic Circuit Problems. Hysteresis. Electromagnetic Forces.

5 ELECTROMAGNETIC INDUCTION 83

Electromagnetic Induction. Flux Linkages. Magnitude of Induced Voltage. Polarity of Induced Voltage. Flux Cutting. Eddy Currents. Self-Induced Voltages.

6 INDUCTANCE AND CAPACITANCE 100
Introduction. Coefficient of Self-Inductance. Physical Aspects of Inductance. Effects of Self-Inductance. A Mechanical Analogy. Energy Relations. Mutual Inductance. Effects of Mutual Inductance. Capacitance. Effects of Capacitance. Connection of Capacitors.

7 ALTERNATING CURRENTS 122
Introduction. Double-Subscript Notation. Definitions. Production of Sine Waves. Effective and Average Values of Sine Waves. Phase Relations. Addition of Sine Waves. Vectors. Addition of Vectors.

8 SINGLE-PHASE A-C CIRCUITS 139
Introduction. Pure Resistive Circuits. Pure Inductive Circuits. Pure Capacitive Circuits. Practical Considerations. Nonlinear Elements in A-C Circuits. Series Combinations of Resistance, Inductance, and Capacitance. Impedance. Apparent Power and Power Factor. Reactive Power. Parallel Combinations of Resistance, Inductance, and Capacitance. Series Resonance. Parallel Resonance. Complex Notation. Application of Complex Notation. Harmonics.

9 POLYPHASE A-C CIRCUITS 174
Introduction. Three-Phase Four-Wire Circuit. Balanced Wye-Connected Loads. Polyphase Vector Diagrams. Unbalanced Wye-Connected Loads. Delta-Connected Loads. Three-Phase Ratings.

10 ELECTRICAL INSTRUMENTS AND MEASUREMENTS . . 189
Introduction. The Permanent-Magnet Moving-Coil Mechanism. The Electrodynamometer Mechanism. The Moving-Iron Mechanism. D-C Voltmeters. D-C Ammeters. A-C Voltmeters. A-C Ammeters. Wattmeters. Measurement of Power in Three-Phase Circuits. Ohmmeters. Watthour Meters. Measurement of Non-electrical Quantities.

11 THE D-C MACHINE 220
Introduction. D-C Machine Classification. Generator Action. A-C to D-C Conversion. Motor Action. Machine Construction. Armature Windings. Armature Resistance. Commutation. Field Windings. Commutating Windings. Machine Ratings.

12 D-C GENERATORS 240
D-C Generator Rating. Generated Voltage. Factors Affecting Gen-

erated Voltage. The Saturation Curve. Shunt Generator Build-up. Generator Operating Characteristics. Voltage Regulation. Generator Prime Movers.

13 D-C MOTORS 256
Motor Rating. Motor Torque. Starting a D-C Motor. Motor Operation. Motor Power Requirements. Speed–Load Characteristics. Torque–Current Characteristics.

14 D-C MOTOR CONTROL 268
Introduction. Fundamentals of Motor Starting. Rating of Starters. Manual Starters. Automatic-Starter Components. Definite-Time Automatic Starters. Current-Limit Automatic Starters. Speed Control. Motor Reversal. Motor Braking.

15 TRANSFORMERS 286
Definitions. No-Load Conditions—Ideal Transformer. Load Conditions —Ideal Transformer. The Actual Transformer. Transformer Rating. Transformer Losses. Single-Phase Transformer Connections. Autotransformers. Polyphase Connection of Transformers. Instrument Transformers.

16 THE THREE-PHASE INDUCTION MOTOR 312
Definitions. The Stator. The Rotor. The Two-Pole Rotating Field. The Four-Pole Rotating Field. Synchronous Speed. Induction-Motor Torque at Standstill. Slip. Effects of Slip. The Torque–Slip Curve. Induction Motor Operation. Induction Motor Rating. Induction Motor Current. Starting Current and Torque. Starting Methods. Speed Control.

17 THREE-PHASE SYNCHRONOUS MACHINES 336
Introduction. Rotor Construction. Stator Construction. The Synchronous Generator. Synchronous Generator Operation. Generator-to-Motor Conversion. Synchronous Machine Torque. A Mechanical Analogy. Synchronous Motor Operation. The Synchronous Condenser. Starting Synchronous Motors. Synchronizing Methods.

18 SINGLE-PHASE MOTORS 359
Introduction. Induction-Type Single-Phase Motors. Theory of Operation. Starting. Split-Phase Motors. Shaded-Pole Motors. Commutator Single-Phase Motors. Series Motors. Repulsion Motors.

19 CONTROL EQUIPMENT 373
Introduction. Synchro Units. Synchro Voltages. Synchro Generators

and Motors. Differential Synchro Units. Synchro Control Transformers. Some Synchro Applications. Rotating Amplifiers. The Amplidyne. The Rototrol and the Regulex. Some Rotating-Amplifier Applications. Magnetic Amplifiers. Self-Saturating Magnetic Amplifiers. Magnetic Amplifier Connections.

20 HIGH-VACUUM TUBES 407

Introduction. Electron Tubes. Thermionic Emission. Construction of Thermionic Tubes. Characteristics of High-Vacuum Thermionic Diodes. The High-Vacuum Diode as a Circuit Element. Photoelectric Emission. Secondary Emission. High-Field Emission. The Electric Field. Power Loss in a Diode. Triodes. Tetrodes. Pentodes. Beam-Power Tubes. The Triode as a Circuit Element.

21 GAS-DISCHARGE TUBES 431

Introduction. Starting a Gas Discharge. Glow Discharge. Arc Discharge. Gas-Filled Phototubes. Glow Tubes. Arc-Discharge Lamps. Phanotrons. The Phanotron as a Circuit Element. Thyratrons. Thyratron Control Characteristics. Thyratron as a Circuit Element. Cold-Cathode Trigger Tube.

22 SEMICONDUCTOR DEVICES 452

Introduction. Conduction in Pure Germanium and Silicon. Effect of an Impurity in Semiconductors. *P-N* Junctions. Electrical Characteristics of a *P-N* Junction. Other Solid-State Junctions. Application of Semiconductor Diodes. Transistors. Transistor Characteristics. Transistor as a Circuit Element.

23 SINGLE-PHASE RECTIFIERS 467

Introduction. Ratings for Rectifier Elements. Half-Wave Rectifier. Half-Wave Rectifier with Capacitor Filter. Voltage Doubler. Full-Wave Rectifier. Bridge Rectifier. Rectifier Filters. Vibrator Rectifiers. Grid-Controlled Rectifiers. Phase-Shift Control. Bias-Phase Control.

24 POWER RECTIFIERS 490

Introduction. Mercury-Pool Cathodes. Ignitrons. Multianode Tank Rectifiers. Three-Phase Half-Wave Rectifiers. Zig-zag Connection. Six-Phase Half-Wave Rectifier. Three-Phase Full-Wave Rectifier. Double-Wye Connection. Rectification Efficiency. Transformer Utilization.

25 ELECTRONIC AMPLIFIERS 508

Introduction. D-C Amplifiers. A-C Amplifiers. Biasing. Resistance–Capacitance Coupling. Transformer Coupling. Push-Pull Amplifier.

Efficiency of the Class A Amplifier. Push—Pull Class B Amplifier. Tuned Amplifiers. Oscillators.

26 ANALYSIS OF LINEAR ELECTRONIC CIRCUITS   533

Introduction. A-C Equivalent Circuit for the Vacuum Tube. Voltage Gain of the Basic Vacuum-Tube Amplifier. Gain of an R–C Coupled Amplifier. Distortion in Amplifiers. A-C Equivalent Circuit of a Transistor. Gain of a Transistor Amplifier.

27 ELECTRONICS IN COMMUNICATIONS   552

Modulation. Amplitude Modulation. Detection of Amplitude-Modulated Signals. Amplitude-Modulation Receivers. Frequency Conversion. Frequency Modulation. Frequency-Modulation Receiver. Other Applications of Modulation.

28 ELECTRONIC CONTROL   565

Introduction. Voltage Regulator. Generator Voltage Regulator. Speed Regulation of a D-C Motor. Anti-Hunting. Electric Response from Nonelectrical Quantities. Photo-Controlled Relay. Timing Circuits. Resistance-Welding Circuits.

29 MISCELLANEOUS ELECTRONIC CIRCUITS   580

Introduction. Vacuum-Tube Voltmeters. Cathode-Ray Oscilloscope. Thyratron Relaxation Oscillator. Diode Wave Shaping. Multivibrators. Monostable and Bistable Multivibrators. Analog Computers.

■1 Introductory Concepts

1-1. THE ELECTRON THEORY

If the student were to undertake research into the nature of electricity, he would eventually have to consider the characteristics of the materials that are important in electrical applications. This research would inevitably lead to a study of the **atom,** a constituent of all matter. Some consideration of the atom is necessary, therefore, if he is to "understand" the nature of electricity and to "explain" the reaction of electric components. The consideration here is a general one, however, and is limited to solids. The atomic behavior of gases is considered later in the book.

The atom is analogous to a miniature solar system, with a **nucleus** as the "sun" and **electrons** as the "planets" revolving in orbits around the nucleus. Like the solar system, the atom is essentially free space.

The electrons are held in their orbits by the attraction of the nucleus. This force of attraction is presumed to exist because the nucleus and the electrons carry a **charge.** The characteristics of charged bodies are such that bodies of similar charge exert a force of repulsion upon each other, whereas bodies of opposite charge experience a force of attraction. The nucleus is considered to have a **positive** charge, an electron a **negative** charge.

In most atoms, the amount of positive charge equals the amount of negative charge, and therefore the atom exhibits no charge externally. If an electron is removed from the atom, however, there is an excess of positive charge, and the atom exhibits a positive charge. Similarly, the addition of an electron makes the atom negative. Atoms that have lost or gained one or more electrons are called **ions**.

Some materials are so constituted that the electrons farthest from the nucleus are loosely bound to the atom and can leave the atom upon the application of a small force, thus becoming **free electrons** within the material. An individual free electron does not remain free very long, however, because it soon encounters an atom lacking an electron, and an atomic reunion occurs. This continual interchange of electrons occurs throughout the material. Other materials have their electrons tightly bound to the atoms; in these materials it is exceedingly difficult to free any electrons.

Materials having free electrons are classified as **conductors;** those having tightly bound electrons are classified as **insulators.** In general, those materials that are good heat conductors are also good electric conductors, and their relative abilities to conduct electricity and heat are about the same.

Under certain conditions an insulator may become a conductor. An example is glass, which is a good insulator at normal temperatures but which becomes a conductor when heated to a dull red. This characteristic of glass is important to glass manufacturers who weld glass products by electrical means.

1-2. ELECTROMOTIVE FORCE AND POTENTIAL DIFFERENCE

If a group of negatively charged atoms is separated from a group of positively charged atoms, there is a difference in potential between the groups. The creation of a difference in potential necessitates an expenditure of energy, because a force of attraction exists between bodies of opposite charge, and work is required to separate the bodies. Thus, there must be a conversion of energy when either electrons or positively charged atoms are moved relative to each other through a body. The property or state that enables a device to effect a difference in potential between two points is called its **electromotive force** (abbreviated **emf**). The term electromotive force is most often used in reference to potential differences created by the conversion of some other form of energy to electric energy. The term may also be applied to similar potential differences effected when such conversions can be, and are, reversed.*

* The concept of potential difference is not restricted to an electric source. The potential difference between any two points, no matter where located, is the work required to move a unit positive charge from one point to the other. If work must be done on the charge to move it, the starting point is negative with respect to, or at a lower potential than, the other point.

CURRENT

Since a potential difference is created by the separation of a group of positively charged atoms and a group of negatively charged atoms, the terminals, or connecting points, of the device in which the separation takes place are therefore marked + and −. The positive terminal is said to be at a higher potential than the negative terminal.

Examples of devices having electromotive force are:

(*a*) Storage batteries: chemical energy is converted into electrical energy in maintaining a difference in potential between the battery terminals.

(*b*) Electric generators: mechanical energy is converted into electric energy. The mechanical energy is supplied by the prime mover of the generator: that is, an electric motor, a Diesel engine, a steam turbine, or another form of driver.

(*c*) Thermocouples: thermal energy is converted into electric energy.

The unit of potential difference is the **volt**. (The term **voltage** is often used in place of difference in potential.) It is frequently more convenient to use the **kilovolt** (abbreviated **kv**), which is 1000 volts, or the **millivolt** (abbreviated **mv**), which is 0.001 volt, as the unit of potential difference.

1-3. CURRENT

If a bar of material having free electrons is connected between the terminals of a battery having electromotive force, as shown in Fig. 1-1, the electrons in the bar, being negatively charged, are continuously attracted to the positive terminal of the battery, and repelled or expelled at the negative terminal. This flow of electrons constitutes an electric **current**. (Note that an electric current is a flow of electrons, and therefore that the expression "current flows" is redundant. However, this expression is a convenient one that is widely used.)

Although electricity is said to travel with the speed of light (186,000

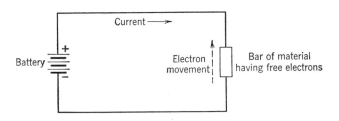

Fig. 1-1. Connection of battery and bar of material.

miles per second), the movement of electrons toward the positive terminal of the battery is a relatively slow drift. (The apparent speed of electrical effects is the speed of the impulse that is transmitted when the battery and the bar of material, in this example, are first connected.)

The rate at which electrons pass a given point determines the magnitude of the current, and an increase in the rate of flow constitutes an increase in the current. The unit of current is the **ampere**. For small currents, the **milliampere** (abbreviated **ma**), which is 0.001 ampere, and the **microampere** (abbreviated μa), which is 0.000001 ampere, are often used as the units.

As shown in Fig. 1–1, the drift of electrons in the bar is toward the positive terminal of the battery, whereas the direction of current is taken as being opposite to the electron movement. Established by experimenters before very much was known about the nature of electricity, this direction of current has almost unanimous acceptance and is known as the conventional current. The term electron current is used by those who have deviated from this convention and consider that electron movement and current direction are the same. The direction of conventional current still prevails overwhelmingly, however, and is used throughout this book except where specific designation of electron current may be indicated.

The connection shown in Fig. 1–1 is called an **electric circuit**. Since the word circuit usually denotes continuity or completeness, the expression open circuit, which is used in electrical terminology, would seem to be inconsistent. It would seem that, if a circuit were broken, what remained could not still be a circuit. However, the word circuit is defined for electrical purposes as "a conducting part, or a system of conducting parts, through which an electric current is intended to flow." Thus, should a circuit be broken, deliberately or otherwise, the original intention, that there would be a current, is not eliminated.

Figure 1–1 shows a **closed circuit**. If an open switch were placed in one of the connecting wires between the battery and the bar of material, Fig. 1–1 would be an **open circuit**. When electric connections are more complex than the connection shown in Fig. 1–1, both open circuits and closed circuits may be involved, and the whole arrangement is referred to as a circuit or a network.

1–4. RESISTANCE

If a bar of a different material is substituted for the original bar in the circuit shown in Fig. 1–1 and the subsequent electron-flow rate is measured, the currents for the two connections may differ. The bar with the smaller current offers the greater impedance to electron flow and, therefore, is said to have the higher electrical **resistance**. Similarly, the bar with the greater current is said to have the lower resistance. A device that introduces resistance into an electric circuit is called a **resistor**.

The unit of resistance is the **ohm**. A more convenient resistance unit for high-resistance devices, such as insulators, is the **megohm** (1,000,000 ohms).

(The reciprocal of resistance is called **conductance**. Thus, the bar having the lower resistance has the higher conductance and therefore the greater current. The unit of conductance is the **mho**.)

1-5. OHM'S LAW

Preceding articles indicate that the current in an electric connection is dependent upon the electromotive force and the resistance in the connection. The expression relating these three factors—current, voltage, and resistance—is known as Ohm's law. *When a simple conductor carries an unvarying current, the potential difference between any two points on that conductor is directly proportional to the current and to the resistance of the current path between the points.*

The mathematical expression for Ohm's law is

$$E = IR \tag{1-1}$$

where E is the symbol for an unvarying voltage, I the symbol for an unvarying current, and R the symbol for a constant resistance. When the unit of current is the ampere and the unit of resistance the ohm, the potential difference is in volts.

Although apparently well known even by some who have never formally studied electrical phenomena, Eq. 1-1 can be easily misapplied if the statement of Ohm's law as given is not fully understood. The important parts of the statement are "simple conductor" and "unvarying current." Thus, the equation cannot be applied if, between the two points of interest, there is an emf or anything else that prevents the circuit from being a simple conductor. Similarly, a current that is not steady may produce effects, as discussed later in this book, that make the equation inapplicable. When the effects of current variation are negligible, Ohm's law can be applied to instantaneous values of voltage, current, and resistance.

1-6. DIRECT AND ALTERNATING CURRENT

The direction of the current in the circuit of Fig. 1–1 is always the same. Variation of the magnitude of the emf or the resistance changes the magnitude of the current but does not change its direction. The current is unidirectional because the polarities of the battery terminals remain fixed; that is, one terminal of the battery remains positive with respect to the other.

If the emf of the battery and the resistance of the circuit remain constant, resulting in a unidirectional current of constant magnitude, the current is called a **direct current** and the circuit is called a direct-current, or d-c, circuit. Circuits of this type commonly experience slight changes in emf or resistance, with the result that the current is unidirectional but is not constant. If these variations are small and of a haphazard nature, the circuit is still called a d-c circuit, even though the strict definition of a direct current has been violated.

By definition, a pulsating current is one that is unidirectional but that has regular changes in magnitude; that is, these changes occur in definite time intervals and are repetitive. The ordinary d-c generator produces a voltage that has small regular changes in magnitude, but these changes are usually too small to produce a current that pulsates appreciably.

If a device having emf has regular reversals of the polarities of its terminals, the resulting current, in a constant-resistance circuit containing no other emf's, changes direction with each reversal of polarity. Therefore, if a positive value is assigned to a given direction of current, there are, alternately, positive and negative values of current in the circuit. This current is called an **alternating current,** the emf is an alternating emf, and the circuit is an alternating-current, or a-c, circuit. The characteristics of an alternating current are regular changes in magnitude and direction. The average algebraic value of a pure alternating current over a period, that is, over the time for one complete reversal, is zero.

Plots of current versus time to illustrate the various types of current are shown as Fig. 1–2. Figure 1–2a represents the plot of a true direct current, as obtained with constant emf and constant resistance in the circuit of Fig. 1–1. If the battery in the circuit of Fig. 1–1 has its terminals instantaneously and periodically reversed, a form of alternating current results, as shown in Fig. 1–2b. The alternating current produced by the alternating emf of an a-c generator is shown as Fig. 1–2c. If the negative loop of the alternating current of Fig. 1–2c is reversed, which is

POWER AND ENERGY

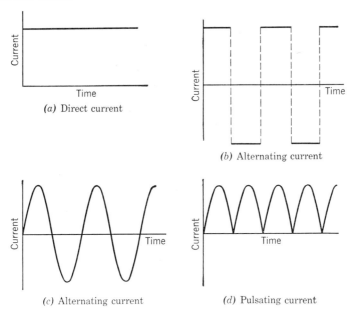

Fig. 1–2. Current versus time diagrams.

possible, a form of pulsating current results, as shown in Fig. 1–2d. Pulsating currents in general, however, need not have zero values.

1–7. POWER AND ENERGY

Some form of energy must be converted to electric energy if there is to be a current in an electric circuit. The electric energy is subsequently converted into other forms. Thus, in the circuit of Fig. 1–1, chemical energy in the battery is converted into electric energy, which in turn is converted into heat in the resistance of the circuit. Although engineers are interested in the amount of energy dissipated in the processes or devices under their control, they are often more interested in the rate at which the energy is dissipated or converted. The time rate of energy dissipation is known as the **power**.

In electric circuits, the unit of power is the **watt**. The instantaneous power in watts is the product of the instantaneous voltage in volts and the instantaneous current in amperes. Thus, using the symbol p for instantaneous power,

$$p = ei \tag{1-2}$$

(Lower-case letters indicate instantaneous values.) When e is the voltage across a resistor having constant resistance, and the effects of current variation are negligible, Eq. 1–2 can be modified by substituting for the voltage value in terms of the current and the resistance. Thus,

$$p = ei = (iR)i = i^2 R \qquad (1\text{--}3)$$

or, the power dissipated in a resistor is proportional to the resistance of the resistor and the square of the current through the resistor.

The average power, rather than the instantaneous, is usually of greater interest to the engineer. When the voltage and current are constant over a given time interval, the average power is, of course, the same as the instantaneous. If, however, either the voltage or the current, or both, are not constant, the average power must be determined by integration. Thus,

$$P = \frac{1}{T} \int_0^T ei \, dt \qquad (1\text{--}4)$$

where P is the symbol for average power. When the integration indicated is applied to an a-c circuit, the time interval T over which the integration is taken must be a complete period: that is, the time for a complete reversal of the alternating current. The use of a time interval shorter than this may result in an erroneous conclusion as to the average power over a long period.

The student should keep in mind that the identity of the quantities substituted in the power relations identifies the power quantity. As examples: using circuit voltage and circuit current gives the circuit power input or output; using resistor voltage and current or resistor current and resistance gives resistor power. If the quantities to be inserted are carefully examined, there should be no doubt as to the identity of the result.

The common mechanical unit for power is the horsepower, which is equivalent to an energy dissipation of 550 foot-pounds per second, or 33,000 foot-pounds per minute. The conversion of mechanical power units to those of the electric system, or vice versa, may be accomplished by means of the relation

$$1 \text{ horsepower} = 746 \text{ watts} \qquad (1\text{--}5)$$

Factors that may be applied to convert electric or mechanical power units to those of some other system are available in most handbooks.

Since power is the time rate of energy dissipation or conversion, the unit of electric energy may be obtained directly. Thus, if the voltage and current are constant, the electric energy may be expressed by the relation

$$W = EIt \qquad (1\text{--}6)$$

SOME CIRCUIT CONCEPTS

where W is the energy in watt-seconds when E is in volts, I is in amperes, and t is the time in seconds. If the energy dissipated in a resistor is desired, and the current is constant, Eq. 1–6 is convertible to

$$W = I^2 R t \qquad (1\text{--}7)$$

If either the current or voltage, or both, are not constant during the time under consideration, the total energy dissipated or converted must be found by the summation of the incremental energies. Thus,

$$W_{(t)} = \int_0^t ei\,dt \qquad (1\text{--}8)$$

where $W_{(t)}$ represents the energy for the time interval t.

The watt-second represents little energy, and a larger unit, the **kilowatt-hour** (abbreviated **kw-hr**), is more often encountered. The **kilowatt** (abbreviated **kw**) is 1000 watts.

1–8. SOME CIRCUIT CONCEPTS

For purposes of the discussion to follow, the circuit of Fig. 1–1 is redrawn as Fig. 1–3.

The distance between the terminals of the battery and the terminals of the resistor in Fig. 1–3 is assumed to be relatively small; that is, the terminals of the battery are assumed to be adjacent to those of the resistor. The connection points are indicated as a and b.

For the battery alone, terminal a is positive with respect to terminal b, and there is a definite potential difference between the terminals. This is true regardless of the path one takes to get from a to b because there can be only one value of voltage between any two points. Therefore, since terminals a and b are also the terminals of the resistor, the voltage across the resistor is the same as the voltage across the battery.

The electric energy of the circuit of Fig. 1–3 is supplied by the con-

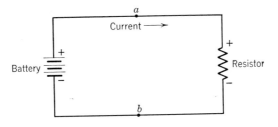

Fig. 1–3. Elementary electric circuit.

version of chemical energy in the battery. The battery is therefore considered to be the **source** of the electric energy. The fact that the current direction within the battery is from the − terminal to the + terminal illustrates the general rule that *current flows from negative to positive through a source*.

The electric energy produced by the battery is dissipated in the resistor as heat. Electrically, the resistor is therefore considered to be a **receiver** or **load**. (These names can be applied to any device in which electric energy is converted to some other form.) The direction of current within the resistor illustrates the general rule that *current flows from positive to negative in a receiver*.

If another resistor is added to the circuit of Fig. 1–3, the resistor can be connected in either of two ways: in **series** or in **parallel**. A series connection is one in which the elements of the circuit are connected end to end so as to form a single path for the current. Such a connection is shown as Fig. 1–4a. A parallel connection is one in which the elements of the circuit are connected to common terminals, as shown in Fig. 1–4b.

Since each of the resistor elements in Fig. 1–4a is a receiver, the polarities of the terminals are as indicated. Terminals c and d, which are assumed to be a negligible distance apart, are at the same potential but are labeled with different polarity markings in order to indicate the relative polarities of these terminals with respect to the other terminals of the resistors with which they are associated. Thus, terminal c is of negative polarity with respect to terminal a, and terminal d is positive with respect to terminal b. Since c and d are at the same potential, c is also positive with respect to b. Similarly, d is negative with respect to a. Since terminal a is at a higher potential than terminal c, which in turn is at a higher potential than b, the voltage across the two resistors in series is the summation of the individual voltages and must be equal to the voltage of the battery.

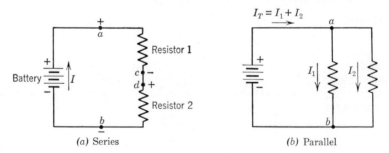

Fig. 1–4. Series and parallel connections.

SOME CIRCUIT CONCEPTS

If the circuit of Fig. 1–4a is traversed in the direction of the current, the voltage across each of the resistors may be considered to be a **voltage drop**. Thus, a drop in potential is experienced in going from a to c because the traverse is from a terminal of higher potential. Similarly, traverse from the negative terminal to the positive terminal of the battery constitutes a **voltage rise**. Students often have difficulty with the concepts of voltage rises and drops because they associate voltage rises only with voltage sources. Whether a potential difference is a voltage rise or a voltage drop depends not only upon the circuit element but also upon the direction in which the circuit is traversed. Thus, if the circuit of Fig. 1–4a is traversed in a counterclockwise direction, a drop in voltage, rather than a rise, is experienced in going from the positive to the negative terminal of the battery.

The resistors in the parallel circuit of Fig. 1–4b are connected to common supply terminals; therefore, the same voltage is applied to each. The resulting current in each resistor is dependent upon this voltage and the resistance of the individual resistors. These currents are appropriately labeled in Fig. 1–4b. Since the total flow of electrons must pass through the battery, the current in the battery is the sum of the currents in the individual resistors.

A common expression in electrical terminology is **short circuit**. As defined in the American Standard Definitions of Electrical Terms, a short circuit is an abnormal connection of relatively low resistance, whether made accidentally or intentionally, between two points of different potential in a circuit. In this book a short circuit is considered to have zero resistance.

The circuit concepts considered in this article are introductory and are important to an understanding of the discussions to follow. A more detailed consideration of circuits and circuit fundamentals is undertaken in Chapter 3.

PROBLEMS

1–1. A d-c voltmeter takes a current of 8 ma when connected to a 120-volt d-c supply. What is the resistance of the voltmeter?   *Ans.* 15,000 ohms.

1–2. The voltage between the terminals of a d-c ammeter is found to be 45 mv; the current through the ammeter is 9 amperes. What is the resistance of the ammeter?   *Ans.* 0.005 ohm.

1–3. How many kilovolts are required to pass a current of 3 ma through a 5-megohm resistor?   *Ans.* 15 kv.

1–4. Electric power loss in a resistor may be expressed as EI or I^2R. Derive still another electrical expression for the power loss in this same resistor.

1-5. The rated output of a given d-c generator is 10 kw at 125 volts. What is the rated current of the generator? *Ans.* 80 amperes.

1-6. (*a*) Neglecting losses, at what horsepower rate could energy be obtained from Niagara Falls, which has an average height of 168 ft and over which the water flows at a rate of 500,000 tons per minute?
Ans. 5,090,000 hp.

(*b*) If the over-all efficiency of conversion were 25%, how many 100-watt light bulbs could Niagara Falls supply? *Ans.* 9,500,000.

1-7. It is desired to raise the temperature of 1 pt (1.04 lb) of water from 20° C to 90° C in an electric heater which takes 3 amperes at 110 volts and has an efficiency of 80%. Determine:

(*a*) The resistance of the heater.
(*b*) The total electric energy to be supplied from the line.
(*c*) The time rate at which this energy is supplied.
(*d*) The time required to bring the water to 90° C.
(*e*) The cost of heating the water at 3 cents per kw-hr.

Ans. (*a*) 36.7 ohms. (*b*) 48 watt-hr. (*c*) 330 watts. (*d*) 8.72 min. (*e*) 0.144 cent.

1-8. A certain industrial process requires 48 gal of water per min at a temperature of 50° C (1 gal of water weighs 8.33 lb). The available water is at 20° C and must be raised 110 ft. A motor-driven direct-connected pump with an electric heating coil inserted in the discharge pipe of the pump is chosen to satisfy the requirements. For this load, the pump efficiency is 67%, and the motor efficiency is 80%. The power company's rate for electric energy is 1.5 cents per kilowatt-hour. Neglecting friction and heat losses in pipes and wiring, determine:

(*a*) Electric power required to pump the water.
(*b*) Electric power required to heat the water.
(*c*) Total cost of electric energy per gallon of water.

Ans. (*a*) 1865 watts. (*b*) 380,000 watts. (*c*) 0.2 cent.

1-9. What is the average power dissipated in a 6-ohm resistor that carries a current of the form shown in Fig. 1-5? *Ans.* 200 watts.

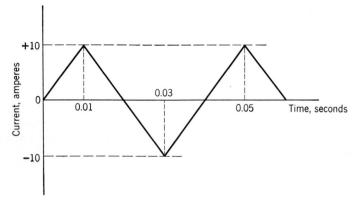

Fig. 1-5.

PROBLEMS

1-10. A voltage having the waveform shown in Fig. 1-6 is applied to a 5-ohm resistor. What is the value of average power dissipated in the resistor?
Ans. 33.3 watts.

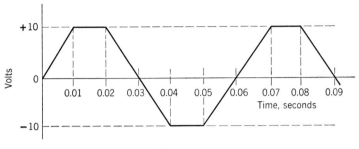

Fig. 1-6.

1-11. The instantaneous current and the voltage in a given a-c circuit are expressed as follows:

$$i = I_m \sin \omega t$$

$$e = E_m \sin (\omega t - \theta)$$

(a) Calculate the average power for this circuit. *Ans.* $(E_m I_m/2) \cos \theta$.

(b) Plot power versus time for this circuit for the conditions of $\theta = 0$ degree and $\theta = 90$ degrees.

1-12. Figure 1-7 is a plot of voltage versus time for the combination of an a-c supply connected in series with a d-c supply.

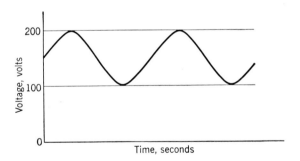

Fig. 1-7.

(a) What is the maximum voltage of the a-c supply and the voltage of the d-c supply? *Ans.* 50 volts; 150 volts.

(b) This combination voltage supply is connected to a resistor having constant resistance. Plot current versus time for this condition.

1-13. Identify the questioned elements in the circuit of Fig. 1-8 as electric sources or receivers.

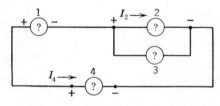

Fig. 1-8.

1-14. Assume that you have a radio, a desk lamp, and an electric fan in your room. How are these elements connected: in series or in parallel? Why?

1-15. An electric phenomenon that is usually brought to one's attention at an early age is the "blowing" of a fuse.
(a) What happens after a fuse "blows"?
(b) What causes a fuse to "blow"?
(c) Why might it not be good practice to replace a 25-ampere fuse with a 40-ampere fuse?

1-16. The doorbell circuit in a home has the bell and a push-button switch connected in series to the source. The push button has a light on it so that it is visible at night but the light goes out when the button is pressed. How is the light connected? What must be a characteristic of the light?

1-17. Assume that it is required to connect a lighting fixture so that it may be energized or de-energized from either of two separate switch locations. Draw a diagram showing how two single-pole double-throw switches should be connected to satisfy the requirement. (A single-pole double-throw switch has three terminals; the blade of the switch is connected to the center terminal and may be closed to complete the circuit to either of the other two terminals, as shown schematically in Fig. 1-9. The switches normally used in the application suggested by this problem are so constructed that the switch is always closed in one position or the other.)

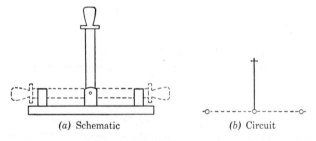

(a) Schematic   (b) Circuit

Fig. 1-9.

2 Resistance

2-1. DEFINITION

In Art. 1–4 the property of resistance is indicated to be the impedance to the directed movement of electrons within a resistor when an unvarying voltage is applied to the terminals of the resistor. A more general definition of resistance, as given by the American Standard Definitions of Electrical Terms, states that "resistance is the property of an electric circuit . . . which determines for a given current the rate at which electric energy is converted into heat . . . and which has a value such that the product of the resistance and the square of the current gives the rate of conversion of energy." Further, "In the general case, resistance is a function of the current, but the term is most commonly used in connection with circuits where the resistance is independent of current."

The general definition of resistance applies to both direct and periodic currents. When only direct currents are involved, the resistance of the circuit is a function of the dimensions, the material, and the temperature of the conducting medium and is called the d-c resistance.

Since all circuits have the property of resistance to some degree, it is the purpose of this chapter to consider some of the aspects of resistance as it is encountered in electric circuits. The consideration is limited, in general, to the d-c resistance; the resistance to periodic currents is discussed in Chapter 8.

2-2. PHYSICAL ASPECTS OF RESISTANCE

When a resistor having a resistance of 1 ohm is connected to a battery having an emf of 6 volts, as shown in Fig. 2–1a, the resulting current in the circuit is 6 amperes if any other resistance in the circuit (such as that in the connecting leads from the battery to the resistor and that within the battery) is negligibly small.

If another resistor, of equal resistance and of the same material and dimensions, is connected in parallel with the first, as shown in Fig. 2–1b, the current in each resistor is 6 amperes, since the voltage across each resistor is the same as it was for the single resistor in the original connection. The current in the battery is the sum of the individual currents and is therefore equal to 12 amperes, or double what it was originally. The parallel connection of the two resistors is equivalent to the connection of a single resistor having a resistance of 0.5 ohm, or one-half that of one of the paralleled resistors. The connection of an identical resistor in parallel with the original resistor doubles the area of the current path perpendicular to current flow and results in a total resistance that is one half of the original resistance. Further analysis of changes in area would show that the resistance of a portion of a current path is inversely proportional to the area of the path perpendicular to current flow.

If the second resistor is connected in series with the first, as shown in Fig. 2–1c, the voltage across each resistor is one half of the battery voltage, since the resistors are identical. The current in the circuit is therefore 3 amperes, or one half of the current in the original connection. The series connection of the two resistors is therefore equivalent to the connection of a single resistor having a resistance of 2 ohms, or double that of the original resistor. In addition, the series connection of the resistors doubles the length of the current path under consideration with-

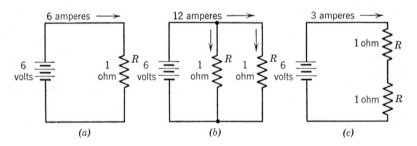

Fig. 2–1. Resistor connections.

out affecting the area of the path. Similar analyses for other changes in length lead to the conclusion that the resistance of a portion of a uniform current path is directly proportional to the length of the portion.

The expression for resistance as related to the dimensions of the resistor may be written

$$R = \rho \frac{l}{A} \qquad (2\text{--}1)$$

where R is the resistance of the portion under consideration, l is the length of the current path through the portion, A is the area of the portion perpendicular to current flow, and ρ is a constant of proportionality.

If a resistor of a different material but with the same dimensions is substituted for the original resistor in the circuit of Fig. 2–1a, the current of the circuit would, in general, have a value different from that in the original connection. The change in the value of current indicates that resistance is dependent upon the material of the circuit, as well as upon the dimensions. The value of the factor ρ of Eq. 2–1 is therefore dependent upon the material of the portion of the current path under consideration. The factor ρ is called the **resistivity** of the material.

2–3. RESISTIVITY

An analysis of Eq. 2–1 reveals that the unit for resistivity must be dimensionally expressed in terms of

$$\frac{\text{Resistance} \times \text{Area}}{\text{Length}} \qquad (2\text{--}2)$$

Thus, the unit might be expressed as the ohm-square inch per inch, the ohm-square centimeter per centimeter, or something similar.

The values of resistivity for the materials commonly applied in electric circuits may be found in most of the standard handbooks, but the units in which the resistivity is expressed might appear to be contrary to what has been previously stated. For example, the resistivity of copper is often expressed as 0.679×10^{-6} ohm-inch. When this value of resistivity is employed in the calculation of resistance, the length must be expressed in inches and the area in square inches for the calculated resistance to be in ohms. This unit is therefore a shortened version of the ohm-square inch per inch. Resistivity in other units may be expressed in a fashion similar to this, such as ohm-centimeters or microhm-centimeters (1 microhm = 10^{-6} ohm).

Many handbooks and textbooks list the resistivity of copper as 10.37

ohms per circular mil-foot. Because the circular mil is a unit of area, this method of expressing resistivity often leads students to believe that resistance is a function of volume. However, careful inspection of Eq. 2–1 indicates that one could have almost any volume he wished in order to produce a given resistance using a given material. Therefore the unit, ohms per circular mil-foot, is better expressed as ohm-circular mils per foot.

The values of resistivity for some of the materials used in electric circuits are indicated in Table 2–1. The significance of indicating the temperature at which the values of resistivity apply is discussed later.

TABLE 2–1
PROPERTIES OF MISCELLANEOUS MATERIALS
Temperature = 20° C

	Resistivity, microhm-cm	Conductivity, %	Specific Gravity
Aluminum	2.828	61	2.7
Copper	1.724	100	8.92
Iron	9.78	17.65	7.9
Lead	22	7.85	11.38
Nickel	7.2	24	8.8
Nichrome (Ni-Fe-Cr)	110	1.57	8.24
Silver	1.64	105	10.51
Tungsten	5.51	31.3	19.3

2–4. CONDUCTIVITY

Although values of resistivity are applicable for the comparison of materials with respect to their relative merits as conductors, values of **conductivity**, the reciprocal of resistivity, are more often used for this purpose. Thus, when resistivity is the basis for comparison, the larger resistivity is associated with the poorer conductor. However, when conductivity is the criterion, the larger conductivity denotes the better conductor.

The conductivity of a material is usually specified as a percentage, based on pure annealed copper at a temperature of 20° C as the standard having 100 per cent conductivity. Thus, taking the resistivity of pure annealed copper at 20° C to be 10.37 ohm-circular mils per foot, a type of steel having a conductivity of 8 per cent would have a resistivity given by 10.37 divided by 0.08, or approximately 129.6 ohm-circular mils per foot.

WIRE GAGES AND CONDUCTORS

Relative conductivities of some of the materials used in electric circuits are included in Table 2–1. (Although silver is a better electric conductor than copper, it is too expensive for most practical applications.)

2–5. THE CIRCULAR MIL

Any unit of resistivity that requires the area of the current path to be expressed in square measure, that is, square inches, square centimeters, etc., also requires the application of the cumbersome $\pi/4$ factor for round conductors. Because most electric conductors are round, it is logical that a unit of resistivity be established that permits the elimination of $\pi/4$ when calculating the resistance of such conductors. The unit created for this purpose is the ohm-circular mil per foot; the circular mil is the unit of the area.

A **circular mil** is the area of a circle having a diameter of 1 mil. Therefore, the cross-sectional area of a round conductor in circular mils is equal to the square of its diameter in mils. Thus, a round conductor having a diameter of 0.2 inch (200 mils) has a cross-sectional area of 40,000 circular mils.

2–6. WIRE GAGES AND CONDUCTORS

Rather than make round conductors of every imaginable diameter, wire manufacturers produce only certain sizes. The gage, or group of sizes, that is standard for copper and aluminum conductors in America is the American Wire Gage (AWG), also known as the Brown and Sharpe, or B&S, gage.

The gage numbers of the American Wire Gage run from 0000 to 40 in the sequence: 0000, 000, 00, 0, 1, 2, . . . , 40. The significance of the gage numbers becomes apparent when one considers how a round conductor is produced. A round bar of metal is reduced in size by being drawn through a die having a diameter smaller than the metal. A further reduction in the metal diameter requires a die smaller than the first. Thus, the material is drawn through a number of dies before reaching the desired size. The dies are numbered in the order in which they are used. Thus, die number 39 is smaller than die number 26; therefore, number 39 wire is smaller than number 26.

Conductors larger than 0000 are specified by their circular mil area, and these sizes are also standardized. Instead of being solid conductors, however, these large conductors usually consist of a number of smaller conductors stranded together to provide flexibility. Stranding is also used for many small-size conductors for the same reason. The resistance of a stranded conductor is about 2 per cent greater than that of a solid conductor of the same length and gage number because of the increased length of the strands due to spiraling. Although having the same conducting area, a stranded conductor occupies a larger space than a solid one of the same gage number, because of the voids between the strands.

Although copper is the predominant electric conductor, aluminum is applied extensively because of its lighter weight. The weight advantage is somewhat offset, however, by the lower tensile strength of aluminum. Also, because of its lower conductivity, an aluminum conductor must be larger in area to have the same resistance as an equivalent length of copper conductor. Stranded aluminum conductors with cores of stranded steel wires are often installed on cross-country transmission lines where weight is an important factor in the design of the towers that support the conductors, but where a high tensile strength is required.

A conductor that combines high strength and relatively high conductivity has the trade name Copperweld. A Copperweld conductor is made by welding a copper coating on the outside of a steel core, hot-rolling this combination into rods, and then cold-drawing the rods into wire. This procedure gives a solid conductor having a steel core and a copper shell.

2–7. CURRENT-CARRYING CAPACITY

In determining the size of conductor for a given installation, one must consider not only the amount of resistance required in the circuit but also the ability of the chosen conductor to carry the current of the circuit; the conductor might melt when carrying the required current, because of the power loss in the conductor and the resulting heat produced. Even though melting does not occur, the temperature of the conductor may become so high as to constitute a fire hazard.

Each size and type of conductor has a definite current-carrying capacity. This capacity is determined by the allowable ultimate temperature of the conductor for the given installation conditions. If the conductor is bare, the value of the ultimate temperature is dependent upon (*a*) the ability of the conductor to dissipate heat, (*b*) the ambient temperature, (*c*) the annealing point of the conductor, and (*d*) the nature of the surroundings. The type of insulation must also be considered if the conductor is insulated. A given size of conductor, therefore, may have a wide range of current-carrying capacities, and no fixed

EFFECT OF TEMPERATURE

rule can be formulated regarding the amount of current that a given conductor size can carry safely. However, several empirical relations applicable to fixed conditions of insulation and temperature rise have been developed and may be used to determine the size of conductor that will satisfy the current requirements of a circuit.

Tables such as Table 3 in the appendix, of the current-carrying capacities of various sizes and types of conductor are available in most standard handbooks. For many installations, such as building wiring, the size and type of conductor are established by law, for the rules of the National Electric Code, as adopted by the National Board of Fire Underwriters, are incorporated into the ordinances passed by most cities and towns throughout the United States.

2-8. EFFECT OF TEMPERATURE

The resistivity of a material is affected by temperature; it is not a constant factor. Thus, since resistance varies directly as the resistivity, the resistance is also affected by temperature.

If the resistance of a sample of pure annealed copper is measured under various conditions of temperature between 0 and 100° C, a plot of resistance versus temperature appears as a straight line with a positive slope, as shown in Fig. 2-2.

All pure metals and most conductor materials have a positive resistance–temperature coefficient; that is, the resistance increases with an increase in temperature. An exception is carbon, which has a negative resistance–temperature coefficient; that is, the resistance of carbon decreases when the temperatures increases. Insulating materials have

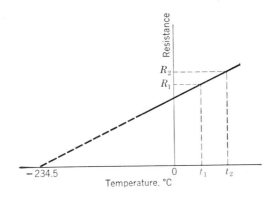

Fig. 2-2. Resistance versus temperature for pure annealed copper.

resistance–temperature characteristics similar to that of carbon. Some alloys, particularly Manganin, the trade name for a copper-manganese-nickel alloy, exhibit little or no change in resistance over relatively wide temperature ranges and are therefore said to have a zero resistance–temperature coefficient.

An extension of the straight line shown in Fig. 2–2 intercepts the temperature axis at $-234.5°$ C. This value of intercept is always obtained when a plot of resistance versus temperature is made for pure annealed copper, regardless of the values of resistance. It should not be inferred, however, that copper has zero resistance at this value of temperature. Exact measurements near $-234.5°$ C indicate larger values of resistance than those obtained from the graph extension. The only reason for the extension is to permit the derivation of a resistance–temperature relation that has practical application.

By the law of similar triangles, the relationship between two values of resistance, R_1 and R_2, at temperatures t_1 and t_2 is given by

$$\frac{R_1}{R_2} = \frac{234.5 + t_1}{234.5 + t_2} \qquad (2\text{–}3)$$

This relation is used to determine changes in the temperature of copper conductors as well as the change in resistance that may be expected when the temperature is changed.

The resistance–temperature relationship can be extended to other materials by writing Eq. 2–3 in the form

$$\frac{R_1}{R_2} = \frac{T + t_1}{T + t_2} \qquad (2\text{–}4)$$

where T is determined by the interception, with the temperature axis, of that portion of the plot of resistance versus temperature that is of interest. The values of T for materials other than copper are not ordinarily obtainable in handbooks but may be computed from values that are available.

The relationship between temperature and resistance may also be expressed in the form

$$R_2 = R_1[1 + \alpha_1(t_2 - t_1)] \qquad (2\text{–}5)$$

The quantity α_1 is known as the **temperature coefficient of resistance** and is dependent upon the base temperature t_1. The dependence of α upon t may be shown by revising Eq. 2–5 to the form

$$\frac{R_2 - R_1}{R_1} = \alpha_1(t_2 - t_1) \qquad (2\text{–}6)$$

and recognizing that α_1 represents the percentage change in resistance per degree change in temperature. For the linear portion of the resistance–temperature curve for copper, a temperature t_3 so selected that $(t_3 - t_2)$ is equal to $(t_2 - t_1)$ results in a resistance R_3 of such magnitude that $(R_3 - R_2)$ is equal to $(R_2 - R_1)$. It then becomes apparent that

$$\frac{R_3 - R_2}{R_2} \neq \alpha_1(t_3 - t_2) \tag{2-7}$$

but that

$$\frac{R_3 - R_1}{R_1} = \alpha_1(t_3 - t_1) \tag{2-8}$$

The fact that the temperature coefficient must be changed when the base temperature is changed makes Eq. 2–4 superior for some applications. Values of α are obtainable in handbooks, however.

The value of T in Eq. 2–4 may be found in terms of α from Eqs. 2–4 and 2–5. Thus,

$$T = \frac{1}{\alpha_1} - t_1 \tag{2-9}$$

where α_1 corresponds to the base temperature t_1. The value of T obtained by means of Eq. 2–9, however, applies only for the linear portion of the plot of resistance versus temperature including temperature t_1. If the plot is such that it is linear over relatively short ranges of temperature, then several different values of T are necessary, each value being applied to a linear section of the plot.

2–9. RESISTANCE THERMOMETRY

Equation 2–3 finds wide application in the determination of the temperature of equipment having copper conductors. Where a temperature rating is incorporated as a part of the nameplate data of electric equipment, the measurement of temperature is often made by noting the change in resistance, from "cold" to operating conditions, of some part of the electric circuit of the equipment. The resistance of that portion of the circuit is usually recorded for some known temperature so that the operating temperature may be quickly and easily determined. The temperature thus calculated is not that of the hottest spot in the equipment but is an average value. Equation 2–4 can be similarly applied for temperature measurement when the material of the circuit is not copper.

Temperature may also be measured by noting the change in resistance of a small resistor embedded in the location in which the temperature is of interest. Often, in this method, the instruments and circuits are so connected that the temperature may be read directly. Resistance thermometers, in which the resistor is placed in a porcelain "bulb," employ the principle of the embedded resistor. The common resistor materials are platinum and, in the less expensive thermometers, nickel. The accuracy of such thermometers is high, and they are obtainable for making measurements up to 1000° C.

2–10. TEMPERATURE VERSUS TIME

Any device in which energy is dissipated experiences an increase in temperature. Therefore, if the device is an electric device containing resistance, the resistance changes, as discussed in Art. 2–8. However, in many applications the manner in which the temperature changes with time and the ultimate temperature are of more interest than the change in resistance.

When energy is consumed in a device, two separate processes are involved in the accounting for the total energy: namely, storage and dissipation. Storage of heat within the device causes the temperature to rise: dissipation, by means of conduction, convection, and radiation, in any combination or singly, eventually limits the temperature rise.

At any time the summation of the incremental energy stored and the incremental energy dissipated must be equal to the incremental energy supplied. Therefore, because heat-energy dissipation is a function of the temperature difference between the device being heated and its surroundings, all the energy is stored and none dissipated when electric power is first supplied to a resistor that is at the temperature of its surroundings. As the temperature rises and the energy dissipation increases, the amount of energy stored decreases and the rate of temperature rise, which is a function of the energy stored, decreases. Thus, when the energy dissipated is equal to the energy supplied, the temperature can no longer increase and equilibrium will have been reached.

The preceding discussion of energy balance is well represented by the differential equation

$$P \, dt = C \, dx + Gx \, dt \qquad (2\text{--}10)$$

where P is the power input, C is a factor representing storage and having a unit of energy per unit temperature rise, G is a factor representing dissipation and having a unit of power per unit temperature rise, and x is the temperature rise at any time. Thus, the left side of the equation represents incremental energy input and the right side represents the summation of storage and dissipation.

TEMPERATURE VERSUS TIME

If P, C, and G of Eq. 2–10 are assumed to be constant, the initial rate of temperature rise and the ultimate temperature can be easily determined. Thus, at the instant of applying power no energy is dissipated, and Eq. 2–10 can be modified to give

$$\frac{dx}{dt} = \frac{P}{C} \tag{2-11}$$

or the initial rate of temperature rise is determined by the ratio of the power input to the storage factor. Similarly, when the energy dissipated is equal to the energy input, no energy can be stored and Eq. 2–10 can be modified to give

$$x = \frac{P}{G} \tag{2-12}$$

or the ultimate temperature rise is determined by the ratio of the power input to the dissipation factor. Therefore, the limits of the temperature-rise–time curve are as shown in Fig. 2–3a.

The solution of the differential equation Eq. 2–10 for the temperature rise in terms of time results in the relation

$$x = \frac{P}{G}(1 - \epsilon^{-Gt/C}) \tag{2-13}$$

which can be written as

$$x = X(1 - \epsilon^{-Gt/C}) \tag{2-14}$$

where X is the ultimate temperature rise, and P, G, and C are assumed to be constants. The plot of Eq. 2–14 is shown in Fig. 2–3b.

As noted in Fig. 2–3b, if the temperature rise had been linear at the

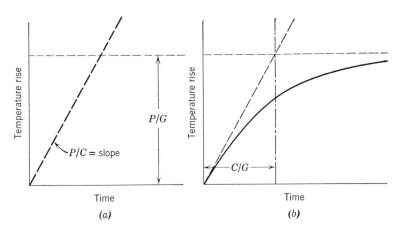

Fig. 2–3. Temperature rise versus time.

initial rate, the time to reach the ultimate value would be given by the ratio of the storage factor to the dissipation factor, determined by using the slope and ultimate value shown in Fig. 2–3a. However, because of the exponential relation, the time to reach the ultimate value is much longer. (The time to reach the ultimate value is mathematically infinite but this is not true in practice because a slight variation of the factors assumed to be constant changes the "ideal" exponential relation. The relation given, however, is a good approximation for the practical situation.)

The time given by the ratio of C to G is called the **time constant** and is useful in comparing devices when the temperature–time characteristics are of interest. Substituting this value of time in Eq. 2–14 and solving for the temperature rise gives

$$x = 0.632X \qquad (2\text{–}15)$$

or the temperature rise reaches 63.2 per cent of the ultimate in a time given by one time constant.

2–11. RHEOSTATS

In many circuits, the current in the load element is limited or controlled by the insertion of additional resistance. Resistors for this purpose may be of two types: fixed or variable. A resistor provided with means for readily varying its resistance is known as a **rheostat**. Rheostats are so constructed that the resistance in the circuit may be varied continuously or in steps.

One form of a continuously variable rheostat consists of an insulated metal or porcelain tube upon which resistance wire is wound. Such a rheostat has two fixed terminals, to which the ends of the resistance wire are connected, and one variable contact. The variable contact is

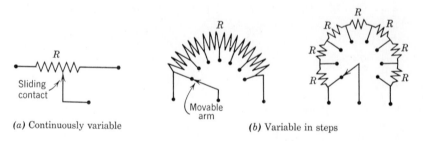

(a) Continuously variable (b) Variable in steps

Fig. 2–4. Schematic diagrams of rheostats.

so mounted that it can be slid over the resistance wire and thus make contact with the wire at small intervals over the length of the wire. Figure 2–4a shows a schematic diagram of such a rheostat, which may be identified by any of several names: tubular, sliding-contact, slide-wire, or potentiometer.

Figure 2–4b shows schematic diagrams for rheostats that are variable in steps. The movable terminal makes contact with fixed terminals connected at specific points on a continuous resistor or between individual resistors. The width of the movable terminal and the distance between the fixed terminals are such that the circuit is not opened while the resistance is being changed. Fixed terminals are also provided at the ends of this rheostat, which is often called a plate-type rheostat.

2–12. RATING OF RESISTORS

Fixed resistors are usually rated in terms of resistance and power; rheostats, in terms of total resistance and current. The power and current ratings are values that the resistor can handle without exceeding the rated temperature rise. This temperature rise is governed by the allowable ultimate temperature of the device and the standard ambient (or surrounding) temperature for which the resistor is designed. The rated temperature rise of commercial resistors is of the order of 300° C to 375° C, based upon a standard ambient temperature of 40° C.

If the temperature of the surroundings in which the resistor is located is lower than the standard ambient, the resistor can carry a current greater than rated without exceeding the permissible ultimate temperature. Conversely, if the ambient temperature is greater than the standard ambient, the resistor must be operated at less than rated current in order not to exceed the permissible ultimate temperature.

The factors that enter into the determination of the power or current rating of a resistor are the ambient temperature, the physical size, the proposed location, and the materials of which it is constructed.

The ratings of resistors are usually continuous ratings; that is, the resistor may be connected in a circuit continuously at the power or current values specified, without exceeding the rated temperature rise.

2–13. STRAIN GAGES

The strain in a structure under variable loading may be determined from the change in resistance of a particular design of resistor attached to the structure. The circuit of which such a resistor is a part is called a **strain gage.**

If the resistor of the strain gage is placed in tension or compression, the length of the resistor changes and therefore the resistance of the resistor changes. An additional change in resistance occurs because the stress affects the resistivity of the metal; that is, the change in resistance cannot be predicted solely from the change in dimensions expected for a given stress. Calibration curves showing the change in resistance for a given change in length are supplied with all commercial strain gages. The magnitude of the change in resistance to be expected is of the order of 2 to 4 per cent for a 1 per cent change in length.

The resistors used in strain gages are made of very fine wires of the order of 1 mil in diameter and are usually cemented to the part in which strains are to be measured. Such strain gages have wide application in the measurement of dynamic as well as static strains.

PROBLEMS

2-1. A bar of copper 1 ft long and 1 in. square has a resistance along its length of 0.00000814 ohm.
(a) Express ρ in square inch-ohms per inch. *Ans.* 0.00000068.
(b) Express ρ in circular mil-ohms per foot. *Ans.* 10.35.
(c) What is the resistance of the copper bar along its width?
Ans. 0.000000057 ohm.

2-2. A definite resistance is needed in an electric connection. Aluminum and copper are being considered for this connection. How would the copper conductor compare with the aluminum conductor in weight and diameter?
Ans. $W_{Cu} = 2.01 W_{Al}$; $D_{Cu} = 0.78 D_{Al}$.

2-3. The resistance of a 1-in. cube of a certain material is found to be 0.00055 ohm. What would be the resistance of a cable consisting of 7 strands of this material, each strand being 0.1 in. in diameter and 3 in. long?
Ans. 0.03 ohm.

2-4. The cube of material in Prob. 2-3 is drawn down to 36 gage without loss of material. What is the resistance of the resulting wire?
Ans. 1.43 megohms.

2-5. An electric connection 10 in. long is to be made, using a conductor having the form of a truncated pyramid (a four-sided pyramid with the top cut off by a plane parallel to the base). One end of the conductor has a cross section 0.2 in. by 0.2 in.; the other end, 0.4 in. by 0.4 in.
(a) If the resistivity of the material is designated as ρ ohm-in., what is the resistance between the ends of the conductor? *Ans.* 125 ρ ohms.
(b) What percentage error would be introduced in calculating the resist-

ance by assuming a constant cross-sectional area that is one-half the sum of the end areas? *Ans.* 20%.

2–6. The cross section of an insulated cable is shown in Fig. 2–5. The space between the outer conducting sheath and the conductor is filled with an insulating material having a resistivity of 4×10^{11} ohm-sq in. per in. If the diameters of the conductor and the sheath are $a = 0.1$ in. and $b = 0.272$ in., respectively, what is the insulation resistance between conductor and sheath per foot of cable? What is the insulation resistance per 1000 ft?
Ans. 6.36×10^{10} ohms; 6.36×10^7 ohms.

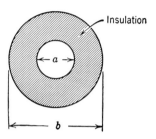

Fig. 2–5.

2–7. (a) A line is to supply a motor 200 ft distant from a panel at which the voltage is maintained constant at 120 volts. If the motor requires a current of 60 amperes and the voltage drop in the line is limited to 5% of the supply voltage, what is the standard size of the smallest copper wire that can be used? *Ans.* No. 4.

(b) If voltage drop were not a factor in (a), what would be the size of the smallest conductor that could be used? *Ans.* No. 8.

2–8. The copper armature winding of a motor is subjected to an operating temperature of 65° C. The room temperature is 25° C. What is the percentage change in resistance of the armature winding from cold starting condition to normal running? *Ans.* 15.3%.

2–9. The resistance of the field winding of a generator at a temperature of 70° F is 97 ohms. After the generator has been operating for several hours, the resistance is found to be 112 ohms. What are the rise in temperature and the final temperature of the winding? *Ans.* 71° F; 141° F.

2–10. (a) The resistance of a 60-watt, 120-volt tungsten-filament lamp is measured by connecting it to a 1.5-volt battery of negligible internal resistance. The current measured is 0.15 ampere. Explain why the value of lamp resistance so measured does not agree with the 60-watt 120-volt rating of the lamp.

(b) What would probably be the test results if the lamp had a carbon filament? Explain.

2–11. Given the following steady-state test data for a resistor:

I (amperes)	5.0	5.35	5.62
E (volts)	30	37.5	45

(a) Account for the nonlinearity of a plot of voltage versus current for this resistor material.

(b) Approximately what current does this resistor take immediately upon being connected to a 120-volt supply?

2–12. A resistor made of steel wire is connected to a constant voltage d-c supply. Sketch and briefly explain the curve of power versus time for the resistor.

2-13. A load that has a resistance of 10 ohms is to be connected to a supply that has a constant voltage of 120 volts. If it is desired that the current to the load be varied from 3 to 5 amperes, what are the resistance and the current rating of the series rheostat that permit this variation?

Ans. 30 ohms; 5 amperes.

2-14. (*a*) Draw a connection diagram showing a 50-ohm rheostat connected so as to vary the voltage on a 10-ohm resistor from zero to 120 volts when the supply voltage is 120 volts.

(*b*) What should be the current rating of the rheostat?

Ans. 14.4 amperes.

2-15. Series-connected three-terminal rheostats are often connected as shown in Fig. 2–6a instead of as shown in Fig. 2–6b. Is there any advantage to connection *a* over connection *b*?

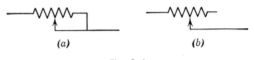

Fig. 2–6.

2-16. Assume that you are the manufacturer of solid resistors having a cylindrical shape, but that you produce only one physical size. If you decide to double the length and the diameter of the resistors, what change, if any, could you make in the resistor rating?

2-17. A certain 2-watt resistor is known to have a thermal time constant of 7 min. Could this resistor be used in a circuit in which it must consume 4 watts? Explain.

3 Electric Circuit Laws and D-C Circuits

3-1. INTRODUCTION

The discussion of the electric circuit has thus far been confined to the simple d-c circuit consisting of a battery and a combination of resistors. Although the circuits encountered in practice are often more complex than those already considered and may contain either a-c or d-c sources, or both, the laws governing the simple circuit apply to any circuit. It is the purpose of this chapter to consider these laws in detail as they are applied to both d-c and a-c circuits and, in addition, to consider some aspects applying particularly to the d-c circuit.

3-2. FUNDAMENTAL CIRCUIT LAWS

The two fundamental laws for the electric circuit are known as Kirchhoff's laws. One of the laws states, in effect, that *the summation of the potential differences taken successfully around a closed traverse is equal to zero.* Thus,

$$\Sigma e = 0 \qquad (3\text{--}1)$$

Since the summation of the potential differences must equal zero, the statement of the law implies that definite algebraic signs must be applied to the potential differences under consideration. Thus, a voltage rise may be assumed to be algebraically negative; a voltage drop, algebraically positive. This convention is followed in this book although it would be just as correct to make the reverse assumption. The algebraic sign of a potential difference is therefore related to (a) the polarity of the potential difference, (b) the direction of traverse with respect to the potential difference, and (c) the original assumption as to the algebraic sign to be assigned to a voltage rise and a voltage drop.

The statement of the Kirchhoff voltage law includes the word traverse rather than circuit. The summation is made of all the potential differences encountered in any closed path; portions of this path need not be physical parts of the electric circuit. Thus, for the circuit of Fig. 3–1, the summation could be taken for a traverse including the voltage across the open switch, represented as e_{xy}, or the voltage between two conductors, represented as e_{bf}.

The law as stated indicates that the summation is of the potential differences "taken successively." Thus, each potential difference is considered as it is encountered in the traverse, instead of by random selection. This procedure seems only logical, but it is the deviation from this method that most often leads to error and confusion in circuit calculations. Such deviation is encouraged by statements of the voltage law which differ from that given; for example, "the sum of the voltage rises is equal to the sum of the voltage drops." Although this statement of the law is entirely correct, the procedure suggested by it becomes cumbersome for circuits that are only moderately complex, and error and confusion may result.

Some voltage summations for the circuit of Fig. 3–1 are as follows:

$$e_{ab} + e_{bc} + e_{cd} + e_{de} + e_{ef} + e_{fg} + e_{gh} + e_{ha} = 0$$

$$e_{ab} + e_{bf} + e_{fg} + e_{gh} + e_{ha} = 0$$

$$e_{cx} + e_{xy} + e_{yz} + e_{zw} + e_{we} + e_{ed} + e_{dc} = 0$$

$$e_{bf} + e_{fe} + e_{ed} + e_{dc} + e_{cb} = 0$$

The actual values substituted for the symbolic potential differences in the voltage equations must be consistent with the convention adopted. Thus, those potential differences that are voltage drops are positive; those that are voltage rises are negative.

The adoption of a convention that assigns a positive value to a voltage drop and the use of double subscripts to indicate the terminals of the potential difference under consideration can be so combined that, when the unknowns in any of the preceding summations are determined, the significance of the result can be instantly recognized. Thus, the potential difference given by e_{ab} is the voltage measured between points a and b and is numerically positive when a is positive with respect to b.

PROCEDURE IN D-C CIRCUIT SOLUTIONS

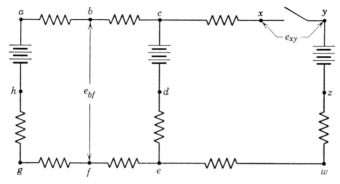

Fig. 3-1. An electric circuit.

The second law of Kirchhoff states, in effect, that *the summation of currents at a junction is equal to zero.* Thus,

$$\Sigma i = 0 \qquad (3\text{-}2)$$

To effect the algebraic summation indicated by the statement of the law, a conventional assumption as to algebraic sign is that current to a junction is algebraically positive and that current away from a junction is algebraically negative. This convention is followed in this book.

The current law may also be stated: *The current flow away from a junction is equal to the current flow toward the junction.* The procedure suggested by this statement of the law gives no more difficulty than the procedure indicated by the original statement, and either may be applied successfully.

3-3. PROCEDURE IN D-C CIRCUIT SOLUTIONS

Electric-circuit problems are similar to all other types of problems; all aspects of the problem must be carefully analyzed, and the problem must be correctly prepared for the mathematical operations that lead to a solution. The procedure suggested for preparing the d-c electric-circuit problem is set down here step by step, as follows:

(*a*) Draw a complete circuit diagram, and completely label all circuit components. This should be done even though the circuit may be so simple as to be easily visualized. The labeling should include all known magnitudes, polarities, and current directions.

(*b*) Assume current directions for those portions of the circuit for which the current directions are unknown. (If the actual current direction is opposite to the direction assumed, the solution will indicate a negative current.)

(c) Assign resistor-voltage polarities consistent with the assumed current directions.

(d) Write the necessary voltage and current equations according to the two laws of Kirchhoff. For some circuits it is possible to write many more equations than are actually needed in the solution. In general, it is necessary to write only (1) $n - 1$ current equations, where n is the number of current junctions, and (2) as many voltage equations as there are simple circuit loops, or loops that cannot be further subdivided.

For example, if the switch in the circuit of Fig. 3–1 is closed and only the voltages within the circuit are of interest, a solution can be obtained for the circuit by writing one current equation (since there are only two current junctions) and two voltage equations. The voltage equations are the summations for the two small loops including the middle branch of the circuit. A voltage equation could be written for the large outside loop, but this equation is the sum or difference of the two voltage equations already written and is unnecessary for a solution.

(e) Write any other equations that may be pertinent to a solution (such as Ohm's law, or power equations).

(f) Solve for the desired quantities.

The procedure outlined is applicable to the solution of any d-c circuit problem. However, there are other methods (such as superposition, Maxwell's mesh, Thévenin's theorem) that may provide a solution for some circuit problems with more ease and in a shorter time than the procedure suggested. Nevertheless, the method outlined provides a fundamental approach to the problem and, accordingly, is the only method considered at this time.

The mathematical procedure in solving a circuit problem is dependent upon the nature of the problem. The problem may be so simple as to require only ordinary arithmetic; it may be so complex as to require complicated mathematical methods.

3–4. EQUIVALENT CIRCUITS

The solution of complex circuits is often simplified if components of the circuit are combined, wherever possible, to reduce the number of terms in the resulting equations. However, the combinations must be such that the resultant circuit is an equivalent circuit. For a circuit or a portion of a circuit to be the equivalent of another, it must be possible

EQUIVALENT CIRCUITS

to interchange the two without affecting the remainder of the circuit.

Circuits containing resistors in series or parallel offer the best possibility for combination. Application of Kirchhoff's law of voltage to a series circuit of n resistors being supplied by a source having an emf of E volts gives

$$E - IR_1 - IR_2 - IR_3 - \cdots - IR_n = 0 \qquad (3\text{-}3)$$

Dividing Eq. 3–3 by the circuit current I gives

$$\frac{E}{I} - R_1 - R_2 - R_3 - \cdots - R_n = 0 \qquad (3\text{-}4)$$

Since the equivalent, or total, resistance of the circuit, R_t, is equal to the circuit voltage divided by the circuit current, or E/I, then, from Eq. 3–4,

$$R_t = R_1 + R_2 + R_3 + \cdots + R_n \qquad (3\text{-}5)$$

or the total resistance of a series circuit is equal to the sum of the component resistances.

Similarly, applying Kirchhoff's law of current to a circuit of n resistors in parallel, supplied by an emf of E volts, gives

$$I_t - \frac{E}{R_1} - \frac{E}{R_2} - \frac{E}{R_3} - \cdots - \frac{E}{R_n} = 0 \qquad (3\text{-}6)$$

Since the total current to the parallel circuit I_t is equal to the circuit voltage divided by the equivalent resistance, Eq. 3–6 may be written

$$\frac{E}{R_t} - \frac{E}{R_1} - \frac{E}{R_2} - \frac{E}{R_3} - \cdots - \frac{E}{R_n} = 0 \qquad (3\text{-}7)$$

Eliminating E from Eq. 3–7 and transposing gives

$$\frac{1}{R_t} = \frac{1}{R_1} + \frac{1}{R_2} + \frac{1}{R_3} + \cdots + \frac{1}{R_n} \qquad (3\text{-}8)$$

Thus, the reciprocal of the equivalent resistance of a parallel circuit of resistors only is equal to the sum of the reciprocals of the component resistances. (Note that the parallel circuit must consist only of resistors if Eq. 3–8 is to be applicable; the relation cannot be applied to parallel branches containing emf's.)

A group of resistors in series or parallel can be replaced by a single resistor having the value determined by the proper application of Eqs. 3–5 or 3–8. The single resistor is the equivalent of the original group of resistors, because the current and power taken by the single resistor are the same as the current and power taken by the original group.

If the resistance values of the original resistors are not functions of

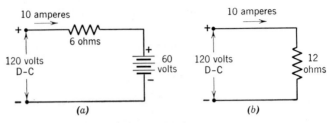

Fig. 3-2. Equivalent circuits.

the current, the single resistor of the preceding discussion is the equivalent of the original group under all circuit conditions; that is, changes may be made in the source voltage or in the remainder of the circuit without affecting the equivalence established for the portion of the circuit under consideration. This is not true for all equivalent circuits, however; some are equivalent only under certain conditions. Thus, although the circuit of Fig. 3–2b is equivalent to that of Fig. 3–2a, a change in the supply voltage from 120 volts to 90 volts requires that the 12-ohm resistor of Fig. 3–2b be replaced by one of 18 ohms in order that the two circuits remain equivalent. Circuits containing voltage sources in addition to the supply voltage are not reducible to equivalent resistor circuits that maintain their equivalence under all circuit conditions.

3–5. THÉVENIN'S THEOREM

A useful theorem known as **Thévenin's theorem** states that any network of constant sources and constant resistances can be replaced by an equivalent series circuit containing one source and one resistor, each constant. Thus, the solution of a circuit problem may be simplified if the values desired are those for only one element or those for a simple circuit connected between two terminals, and the remainder of the network is made up of constant sources and resistors.

The source voltage E_0 of the Thévenin equivalent circuit can be determined by opening the circuit between the two terminals under consideration and calculating the voltage that appears between these two terminals under these conditions.

The series resistor R_0 of the Thévenin equivalent circuit can be determined by two methods. The first method is to place a short circuit between the terminals under consideration and to calculate the current through the short circuit. The resistor R_0 then has a value given by the voltage E_0 previously determined, divided by the short-circuit current,

THÉVENIN'S THEOREM

as it would have if the original network had actually been a series circuit having one source and one resistor.

The second method of determining the value of the resistor R_0 is (1) to open the circuit between the two terminals under consideration, (2) to replace all sources by their internal resistances (if the internal resistance is negligible, the source is replaced by a short circuit), and (3) to calculate the net resistance now connected between the two terminals. The value of the resistance thus determined is the resistance value of R_0.

The circuit shown as Fig. 3–3a is a useful one to demonstrate the application of Thévenin's theorem and to compare the solution by that method with the solution by the method previously considered. It is assumed for purposes of this problem that various values of resistance are to be connected between the terminals a and b and that the values of the current through and the voltage across these resistors are to be determined. The first value of resistance to be considered is 10 ohms.

Solution by Thévenin's Theorem. 1. DETERMINATION OF E_0. With the 10-ohm resistor removed, and the 100-volt battery assumed to have negligible internal resistance, the current in the remainder of the circuit is given by

$$I = \frac{E}{R_1 + R_2} = \frac{100}{20 + 80} = 1 \text{ ampere}$$

The voltage between terminals a and b is therefore the voltage across the 80-ohm resistor or

$$E_{ab} = IR_2 = 1 \times 80 = 80 \text{ volts}$$

and this is the value of E_0.

2. DETERMINATION OF R_0. *Method A.* With the 10-ohm resistor replaced by a short circuit, the current through the short-circuit is given by

$$I_{ab} = \frac{E}{R_1} = \frac{100}{20} = 5 \text{ amperes}$$

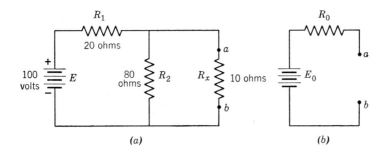

Fig. 3–3. Circuit for illustrative problem.

because no current flows through the 80-ohm resistor under these conditions. Therefore, the resistance of R_0 is given by the relation

$$R_0 = \frac{E_0}{I_{sc}} = \frac{80}{5} = 16 \text{ ohms}$$

Method B. With the 10-ohm resistor removed and the battery replaced by a short circuit, the resistance between terminals a and b consists of the 20-ohm resistor in parallel with the 80-ohm resistor, and the equivalent resistance is given by the relation

$$\frac{1}{R_0} = \frac{1}{R_1} + \frac{1}{R_2} = \frac{1}{20} + \frac{1}{80}$$

or

$$R_0 = \frac{20 \times 80}{20 + 80} = 16 \text{ ohms}$$

Thus, method B gives the same result as method A.

3. SOLUTION OF PROBLEM. With the Thévenin equivalent circuit of Fig. 3–3b having the values calculated, and with the 10-ohm resistor connected between the terminals a and b, the current in the circuit is given by

$$I_{ab} = \frac{E_0}{R_0 + R_x} = \frac{80}{16 + 10} = 3.08 \text{ amperes}$$

and the voltage across the 10-ohm resistor by

$$E_{ab} = I_{ab}R_x = 3.08 \times 10 = 30.8 \text{ volts}$$

Solution by Direct Application of Kirchhoff's Laws. One method of solving the given circuit without the use of the Thévenin equivalent circuit is to write the voltage and current equations that are applicable and to solve the resulting simultaneous equations. Thus, if we consider I_2 to be the current flowing downward in R_2, and I_{ab} to be the current flowing downward in R_x, the loop equations necessary are (traversing in a clockwise direction):

$$-E + (I_2 + I_{ab})R_1 + I_2R_2 = 0 \qquad (1)$$

$$-I_2R_2 + I_{ab}R_x = 0 \qquad (2)$$

Substituting values, Eqs. 1 and 2 become

$$-100 + (I_2 + I_{ab})20 + 80I_2 = 0 \qquad (1)$$

$$-80I_2 + 10I_{ab} = 0 \qquad (2)$$

Rearranging and combining terms gives

$$-100 + 100I_2 + 20I_{ab} = 0 \qquad (1)$$

$$-80I_2 + 10I_{ab} = 0 \qquad (2)$$

Solving the simultaneous equations for I_{ab} gives

$$I_{ab} = 3.08 \text{ amperes}$$

and the voltage across the 10-ohm resistor is

$$E_{ab} = 3.08 \times 10 = 30.8 \text{ volts}$$

Although the second solution may appear to be as direct and perhaps simpler than that by the Thévenin equivalent, the problem chosen for illustrative purposes is not a difficult one and perhaps is not a good example of the usefulness of the Thévenin approach. However, the original problem stated that various resistors were to be inserted between a and b. Therefore, the current for any resistor at that point would be given by

$$I_{ab} = \frac{80}{16 + R_x}$$

and the voltage by

$$E_{ab} = I_{ab} R_x$$

Thus, the current and voltage could be determined quickly for any value of R_x. It is interesting to note that the same equations result if the simultaneous equations are solved for the current with the value for R_x left as a variable.

3-6. NONLINEAR D-C CIRCUITS

In many circuit problems it is assumed that the resistance of the resistors in the circuit remains constant, regardless of any changes that may occur in the current because of other circuit alterations. Such an assumption neglects the fact that a change in temperature accompanies a change in current and, therefore, that the resistance may change. For many resistors, however, a plot of resistor voltage versus resistor current is approximately linear over the operating range; therefore, the assumption of constant resistance introduces negligible error into the circuit calculations. The solution of circuits consisting entirely of such linear components is as previously indicated.

The resistance of some resistors is a function of the current through the resistor, and a plot of resistor voltage versus resistor current is not linear over the operating range. The assumption of constant resistance for circuit problems involving such resistors might therefore introduce large errors into the circuit calculations. The solution of circuits containing such nonlinear components must be at least partially graphical unless the resistance can be expressed as a function of current. If the

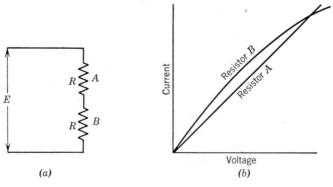

Fig. 3-4. A nonlinear circuit.

resistance-current equation is known, the solution is identical with that for linear elements.

A simple example of a nonlinear circuit is shown as Fig. 3–4a. In this circuit, resistor A has a resistance value that is assumed to remain constant, as shown by the plot of current versus voltage in Fig. 3–4b. Figure 3–4b also shows the voltage–current characteristic of resistor B, a nonlinear element having a resistance value that increases as the current through the element increases.

The procedure for solving the circuit of Fig. 3–4a depends upon the nature of the problem. If the value of applied voltage necessary to produce a given current is desired, the procedure is to find the voltage across each resistor on the respective current–voltage characteristics at the given current, and then to add these voltages to obtain the desired voltage.

To determine the current of the circuit when the voltage across the circuit is known, the procedure is somewhat different and may be more laborious. One procedure is first to assume a value of circuit current and then to determine, from the respective current–voltage characteristics, the voltage across each resistor for the assumed value of current. The sum of the voltages thus determined must be equal to the known applied voltage. If the first assumption does not satisfy this requirement, another current assumption must be made, and the process repeated. This "cut-and-try" procedure is continued until the problem is solved. Unless one is experienced in its use, the cut-and-try process may require many "tries" before the problem is finally solved. A more direct method of arriving at a solution is provided by a mathematical analysis of the circuit. The analysis proceeds as follows:

(a) For the circuit of Fig. 3–4, the current may be expressed mathe-

NONLINEAR D-C CIRCUITS

matically in two ways. Thus,

$$I = \frac{E_a}{R_a} \tag{3-9}$$

and

$$I = f(E_b) \tag{3-10}$$

where E_a and R_a are the voltage and resistance, respectively, of resistor A, and $f(E_b)$ indicates the nonlinear relationship between the current through resistor B and the voltage across resistor B, shown in Fig. 3–4b.

(b) Since the circuit is a series circuit, then

$$E = E_a + E_b \tag{3-11}$$

where E is the applied voltage and is known.

(c) Equation 3–9 can then be rewritten as

$$I = \frac{E - E_b}{R_a} \tag{3-12}$$

or

$$I = -\frac{E_b}{R_a} + \frac{E}{R_a} \tag{3-13}$$

Equation 3–13 is of the form

$$y = mx + b$$

which is the equation for a straight line having a slope m and an intercept b on the y axis. For this problem, the slope is $-1/R_a$, and the y-axis intercept is E/R_a. Although the slope and the y-axis intercept provide sufficient information for the construction of the straight line, the simplest procedure for problems of this type is to also determine the x-axis intercept. For this problem, the x-axis intercept is E, the applied voltage. The straight line is then drawn as shown in Fig. 3–5.

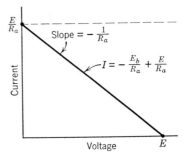

Fig. 3–5. Load-line construction.

(d) The current in the circuit is given by both the equation for a straight line and the equation for a nonlinear characteristic (Eqs. 3–10 and 3–13). Since there can be but one value of current in the circuit for the given conditions, the actual value of current is determined from the intersection of the straight line and the nonlinear characteristic, as shown in Fig. 3–6.

The graphical method of solution derived by the mathematical analysis

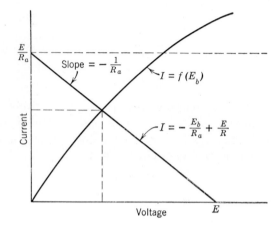

Fig. 3–6. Solution of a nonlinear circuit by load-line construction.

of the nonlinear circuit is known as the load-line method and finds wide application.

PROBLEMS

3–1. In order to provide the tensile strength necessary for certain electric connections, conductors are sometimes made with a round steel core and a copper coating. A conductor of this type, 0.1 in. in diameter, has a steel core having a diameter of 0.0808 in. The conductivity of the steel is 8%, and the resistivity of the copper is 10.37 circular mil-ohms per foot at 20° C.

(a) What is the resistance of 1000 ft of this type of conductor at 20° C?
Ans. 2.57 ohms.

(b) What is the conductivity of this type of conductor? *Ans.* 40%.

3–2. A conductor consisting of six strands of aluminum wire of 0.1 in. diameter covering a steel core of one strand of 0.1 in. diameter is known to have a conductivity of 53%.

(a) What voltage drop and power loss will occur if this conductor is required to carry a current of 50 amperes to a load that is 200 ft from the source? *Ans.* 5.6 volts; 280 watts.

(b) If the aluminum is assumed to have a conductivity of 60% what is the percentage conductivity of the steel? *Ans.* 11%.

3–3. Given a battery having a constant emf of E_b volts and assumed to have a constant internal resistance of R_b ohms. This battery is connected to a variable resistance load.

(a) What is the maximum current that this battery could deliver?
Ans. E_b/R_b amperes.

(b) What is the maximum power that this battery could deliver to a load connected to its terminals?
Ans. $E_b^2/4R_b$ watts.

3–4. (a) What voltage must be impressed on the circuit of Fig. 3–7 to have 10 amperes in the 4-ohm resistor?
Ans. 128.88 volts.

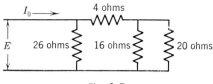

Fig. 3–7.

(b) What is the total current I_0?
Ans. 14.95 amperes.

(c) What is the equivalent resistance of the network?
Ans. 8.62 ohms.

3–5. (a) In a general circuit of the type shown as Fig. 3–8, there are three different circuit components possible for the element identified as X; a switch, a source, or a load. Give complete information on the possible circuit components when the magnitudes are as shown in Fig. 3–8.
Ans. Resistor load of 4.82 ohms; or battery load of 96.5 volts.

(b) Repeat (a) for a current through the 9.1-ohm resistor of 12 amperes instead of 11 amperes.

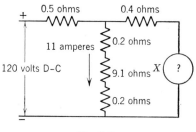

Fig. 3–8.

Ans. Open switch; or battery of 114 volts.

3–6. Draw an equivalent circuit consisting of one source and one resistor for the circuit shown as Fig. 3–9.

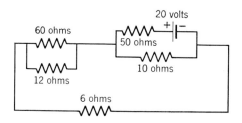

Fig. 3–9.

3–7. (a) In the circuit shown as Fig. 3–10, what should be the power rating of the 1-ohm load resistor?
Ans. 3600 watts.

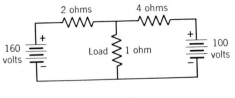

Fig. 3–10.

(b) What value of load resistance (replacing the 1-ohm resistor) causes the current in the 100-volt battery to be zero?   *Ans.* 3.33 ohms.

3–8. (a) For the circuit of Fig. 3–11, what is the minimum value of resistance that can be used at P for the given load?   *Ans.* 0.133 ohm.

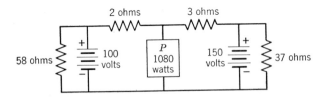

Fig. 3–11.

(b) What is the maximum current that can flow at P (under any condition of resistance at that point)?   *Ans.* 100 amperes.

(c) What is the maximum power that can be delivered at P, and what is the value of resistance for which this maximum exists?
   Ans. 3000 watts; 1.2 ohms.

(d) What is the significance of the two values of resistance in the solution of (a)?

3–9. Apply Thévenin's theorem to determine the current flowing in R_L of Fig. 3–12. Draw the Thévenin equivalent circuit, and show clearly how all results are obtained.   *Ans.* 4 amperes.

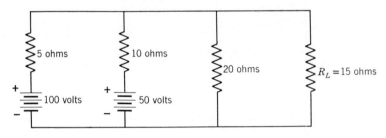

Fig. 3–12.

PROBLEMS

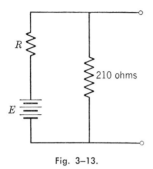

Fig. 3–13.

3–10. From electrical measurements at the two terminals of a box, it is concluded that the box contains a 100-volt d-c source and a 10-ohm resistor connected in series between the terminals. However, when the box is opened, it is discovered that, instead of the series combination expected, the box contains the circuit shown in Fig. 3–13. What are the values of E and R? *Ans.* 105 volts; 10.5 ohms.

3–11. (*a*) For the circuit of Fig. 3–14, what is the effect on the load voltage when two rheostat sliders (which are mechanically connected) are shifted to points B?
(*b*) Repeat (*a*) for the sliders halfway between points A and B.

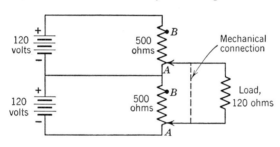

Fig. 3–14.

3–12. The circuit shown as Fig. 3–15 is known as a **three-wire circuit**. In this type of circuit, two supplies are connected in series as shown. It is thus possible to have a two-voltage distribution system using only three wires. The wire connected between the two sources is called the **neutral**. The same type of circuit, with a-c supplies replacing the batteries of Fig. 3–15, is used by power companies for residential power distribution.

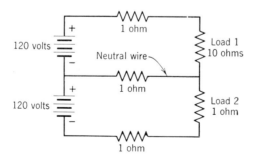

Fig. 3–15.

(a) When the magnitudes are as shown in Fig. 3–15, what is the voltage of each load? *Ans.* Load 1, 137 volts; load 2, 44.4 volts.

(b) Repeat (a) for the condition when the 1-ohm line resistors are replaced by line resistors of 0.05 ohm.

Ans. Load 1, 124 volts; load 2, 109.9 volts.

(c) Determine the voltage ratings of the loads for the line conditions of (b) if the 1-ohm load resistor is changed to one of 10 ohms.

Ans. Load 1, 119.4 volts; load 2, 119.4 volts.

(d) Repeat (b) for the condition of the middle line being broken.

Ans. Load 1, 216 volts; load 2, 21.6 volts.

(e) Repeat (c) for the condition of the middle line being broken.

Ans. Load 1, 119.4 volts; load 2, 119.4 volts.

(f) List the conclusions reached regarding a three-wire distribution system with regard to (1) line resistance, (2) load balance, (3) whether or not a fuse should be placed in the neutral wire.

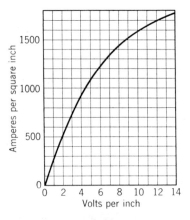

Fig. 3–16.

3–13. A certain resistance material has the current–voltage characteristic shown as Fig. 3–16. The characteristic is plotted in terms of current density (amperes per square inch) and voltage gradient (volts per inch of length).

(a) A 1-ohm resistor is connected in series with a 3-in.-by-0.02-sq.-in. resistor made of this material. If it is desired that the current through the 1-ohm resistor be 25 amperes, what voltage should be supplied to the combination? *Ans.* 43 volts.

(b) What is the percentage change in current to this combination if the voltage is increased to 56 volts (a 31% increase in voltage)?

Ans. 20% increase.

(c) What is the percentage change in current to this combination if the voltage is decreased to 30 volts (a 31% decrease in voltage)? *Ans.* 26% decrease.

3–14. Figure 3–17a is the schematic diagram of a proposed temperature-indicating device. Resistor R_3 is a thermistor, a nonlinear resistor having the resistance-temperature characteristic shown in Fig. 3–17b. Resistors R_1 and R_2 have zero resistance–temperature coefficients. If the temperature to which R_3 is subjected varies between 0° C and 50° C, what value should R_1 have so that the maximum current through R_3 is 1 ampere?

Ans. 1.43 ohms.

3–15. Figure 3–18 is the schematic diagram of a proposed temperature-indicating device. Resistor R_3 is a thermistor, a nonlinear resistor having the resistance-temperature characteristic shown in Fig. 3–17b. Resistors R_1 and R_2 have zero resistance–temperature coefficients. If the current through R_3 is 0.25 ampere, what is the temperature of R_3? *Ans.* Approximately 18° C.

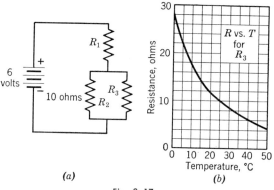

Fig. 3-17.

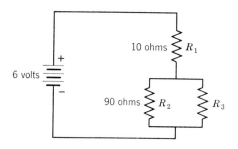

Fig. 3-18.

3-16. What should be the resistance and power rating of resistor R_3 in Fig. 3-19a in order that the currents through resistor R_2 and the nonlinear resistor be equal? The current–voltage characteristic for the nonlinear resistor is shown in Fig. 3-19b. *Ans.* 72 ohms; 18 watts.

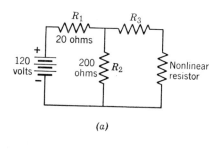
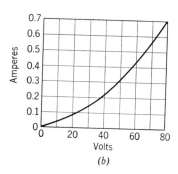

Fig. 3-19.

3-17. (*a*) In the circuit of Fig. 3–20*a*, resistor R_1 is a nonlinear resistor having the voltage–current characteristic shown in Fig. 3–20*b*. What value

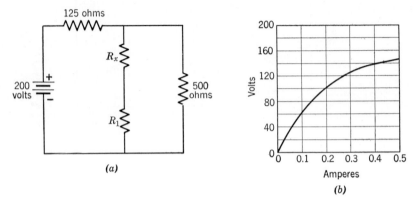

Fig. 3–20.

of R_x is necessary to cause a current of 0.1 ampere to be drawn by the nonlinear resistor. *Ans.* 900 ohms.

(*b*) What would be the effect on the value of R_x found in (*a*) of reducing the value of the 500-ohm resistor?

3-18. The circuit shown in Fig. 3–21 is known as a bridge circuit; it is used widely in measurement and control. In the example shown, the bridge is applied to measure an unknown resistance. The element R_m represents the resistance of the ammeter that indicates the state of balance of the bridge.

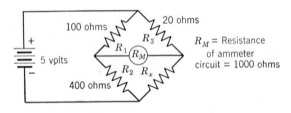

Fig. 3–21.

The three known resistors and the unknown resistor form the arms of the bridge. Bridge circuits operate in both the balanced (no current through ammeter) condition and the unbalanced (current in ammeter) condition.

(*a*) What is the value of the unknown resistor R_x when the bridge is balanced? *Ans.* 80 ohms.

(*b*) Derive a relation expressing R_x in terms of R_1, R_2, and R_3 for the condition of a balanced bridge. *Ans.* $R_x = (R_2/R_1)R_3$.

(c) If the unknown resistor is replaced by a battery having negligible internal resistance, what should be the battery emf and connection to cause the bridge to be balanced?

3–19. Discuss the possible application of a bridge circuit in the measurement of temperature.

3–20. In a factory making resistors, it is desired to select from production the 100-ohm size within a tolerance of ±10%. A bridge circuit for making the measurements is set up as shown in Fig. 3–22. The measuring instru-

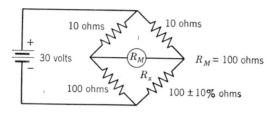

Fig. 3–22.

ment is a zero-center ammeter that has the current values representing the +10% limit and the −10% limit marked on the scale. When the current exceeds either of the limits, the resistor being measured is discarded by the operator. This method of operation results in efficient selection of the desired resistors.

Write the equations necessary for the solution of the value of current that should be marked on the ammeter scale as the +10% tolerance limit.

3–21. Is it possible to complete the circuit between points x and y of Fig. 3–23 in such a manner that no current flows through the 1-ohm resistor? If so, explain completely; if not, explain completely.

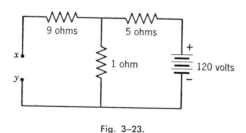

Fig. 3–23.

3–22. The steady-state current–voltage characteristic for a certain lamp is as shown in Fig. 3–24. This lamp and a 120-ohm resistor are to be connected in series to a 120-volt d-c source.

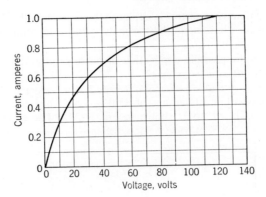

Fig. 3-24.

(a) Approximately what current flows in the circuit immediately after the connection is made? *Ans.* 0.78 ampere.

(b) What current flows after equilibrium is reached? *Ans.* 0.68 ampere.

3-23. Determine the voltage between points a and b of Fig. 3-25a when the voltage applied to the circuit is 120 volts. As indicated, each branch of the circuit consists of a 600-ohm resistor connected in series with a nonlinear resistor having the voltage-current characteristic shown in Fig. 3-25b.
Ans. 14 volts.

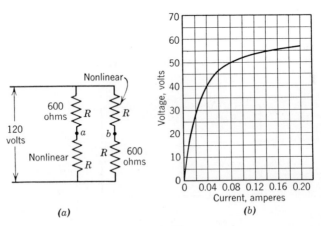

Fig. 3-25.

3-24. Between what resistance limits should rheostat R_x of Fig. 3-26 be operated so that the potential difference between points a and b does not exceed 10 volts? *Ans.* 150 and 550 ohms.

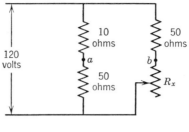

Fig. 3-26.

3-25. For the circuit of Fig. 3-27, what power is dissipated in the 600-ohm resistor when the slider of the 1000-ohm rheostat is at the midpoint of the rheostat? *Ans.* 3 watts.

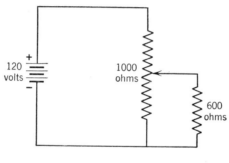

Fig. 3-27.

4 Magnetics and Magnetic Circuits

4-1. MAGNETICS

To a great extent, an understanding of the operation of electric equipment is dependent upon a knowledge of **magnetics,** "that branch of science which deals with the laws of magnetic phenomena." * Certain definitions and observations of the properties of magnets provide a satisfactory basis for this understanding.

A **magnet** is defined technically as "a body that produces a magnetic field external to itself." A magnet may also be defined as a body that has the ability to attract pieces of iron. The attraction of a magnet for pieces of iron is greater at certain places on the magnet than at others. These places are called the **poles** of the magnet, and a simple bar magnet, as shown in Fig. 4–1, has two such poles; one is called the north pole, the other the south pole.

The identity of the poles of a bar magnet may be established by suspending the magnet so that it is free to turn. A magnet suspended in this manner tends to be aligned with its long axis in a north–south direction; the north pole of the magnet is the one that tends to point toward the north. (For this reason, the north pole of a magnet is often referred to as the north-seeking pole.) A magnetic compass is a suspended magnet.

Magnets have characteristics similar to electric charges in that like magnetic poles repel each other and unlike magnetic poles attract each

---

* American Standard Definitions of Electrical Terms.

other. This force of repulsion or attraction is inversely proportional to the distance between the poles.

4–2. LINES OF FORCE

The region in the vicinity of a magnet in which a force of repulsion or attraction is noticeable when another magnet is brought near is called the **magnetic field.** If a compass is brought into the magnetic field and is moved from one pole toward the other in the direction in which the north pole of the compass points, a tracing of the compass path results in a smooth line, as shown in Fig. 4–1. The selection of another starting point and a repetition of the operation result in another smooth line. (Many such lines could be traced out in the area around the magnet, the number depending on how thin one can make the lines.) These lines, called **lines of force,** form a pattern similar to that which results when iron filings are sprinkled on a sheet of paper held over a magnet.

By definition, the direction of the lines of force is taken as the direction in which the north pole of the compass points when the compass is placed in the magnetic field. Thus, the direction of the lines of force outside the magnet is from the north pole to the south pole. It is assumed that lines of force form closed paths; therefore, inside the magnet, the direction of the lines of force is from the south pole to the north.

It is often convenient to think of lines of force as having substance. (As defined, they indicate only the direction of the force on the poles of a compass placed in a magnetic field.) Thus, lines of force are sometimes thought of as rubber bands; that is, lines of force tend to

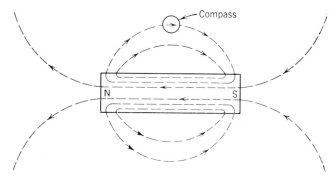

Fig. 4–1. Magnetic field associated with a bar magnet.

be as short as possible. Another characteristic implying physical reality is the tendency of lines of force to take the easiest magnetic path; this characteristic is comparable to an electric current taking the path of least resistance. Thus, if a piece of iron is placed in the magnetic field of a bar magnet, the lines of force converge on the piece of iron.

Another important characteristic of lines of force is that they never cross. A crossing of lines of force would indicate that there are two directions of force at a point, which is impossible.

4-3. FLUX AND FLUX DENSITY

The lines of force indicate the direction of the magnetic field, but they may also be applied to indicate its strength if they are spaced in proportion to the strength of the magnetic effect of the field. Thus, in those places where the magnetic effect is great, the lines are closely spaced; where there is little magnetic effect, the lines are spread out. The density of the lines therefore indicates the intensity of the magnetic effect, and the total number of lines is a measure of the total strength of the magnet. The term **flux**, used to indicate the number of lines of force, is represented by the symbol ϕ.

The intensity of the magnetic field, or **flux density**, is often of more importance in magnetics than the total flux. In a field such as that produced by the bar magnet of Fig. 4–1, the flux density varies throughout the path from one pole to the other. If the flux density is uniform over a given length of magnetic path, as it is in most of the magnetic applications to be considered in this book, it is given by the relation

$$B = \frac{\phi}{A} \tag{4-1}$$

where B is the flux density, ϕ is the flux in the path, and A is the area of the flux path perpendicular to the flux.

The unit of flux density depends upon the unit adopted for the area of the magnetic path. In this book, flux density is expressed in terms of lines per square inch.

4-4. PRODUCTION OF FLUX

There is always a magnetic field associated with an electric current. The direction and presence of this field can be established by holding a compass near a conductor that is carrying an appreciable current. The

PRODUCTION OF FLUX 55

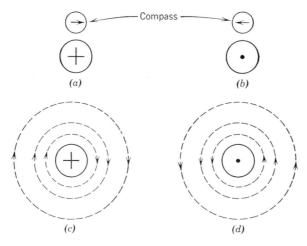

Fig. 4–2. Magnetic field associated with current in a conductor.

needle of the compass then takes the position shown in Fig. 4–2a. (The + in the conductor cross section indicates that the conductor is carrying current into the paper, away from the observer.) If the current in the conductor is reversed, the direction of the needle is reversed, as shown in Fig. 4–2b. (The dot in the conductor cross section indicates that the current is coming out of the paper, toward the observer.) The action of the compass needle therefore establishes the presence of a magnetic field; the direction of the field is indicated by the direction in which the north pole of the compass points, as previously defined and as shown in Figs. 4–2c and 4–2d.

The direction of the magnetic field produced by a current in a conductor can be determined by placing the right hand as if to grasp the conductor, with the thumb pointing in the direction of current flow. The fingers then point in the direction of the field.

If the current-carrying conductor is formed into a coil having closely wound turns, as shown in Fig. 4–3a, the resulting flux links the turns of the coil and, in effect, establishes magnetic poles at the open ends of the coil. The direction of the field through the coil depends upon the direction of the current in the conductor and can be determined by an analysis of the fields produced by the individual turns. Thus, if it is assumed that the coil is cut lengthwise through the center, a section appears as shown in Fig. 4–3b. The directions of the magnetic fields produced by two of the conductors at the top of the coil are such that the lines between the conductors tend to cancel, as shown in Fig. 4–3c; the resulting field pattern for the two conductors is then as shown in

Fig. 4–3d. If this method of analysis is applied for all the conductors, the field pattern for the coil is obtained and is as shown in Fig. 4–3e.

The direction of the field through the inside of a current-carrying coil can be determined in a manner similar to that used to determine the direction of the field around a current-carrying conductor. Thus, if the right hand is held in the same position as before but with the fingers pointing in the direction of the current in the coil, the thumb points in the direction of the magnetic field within the coil.

A coil such as that shown in Fig. 4–3a is called a **solenoid,** and it may have a core, or support, made of any type of material. (In this example, the core is air, and the coil is self-supported.) Although the fields in most magnetic applications are produced by solenoids, some magnetic fields are produced by magnets, called permanent magnets, made of ferrous materials that have the ability to retain magnetic properties after being subjected to the field produced by a solenoid.

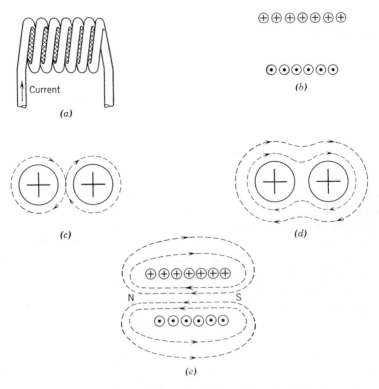

Fig. 4–3. Magnetic field produced by a current-carrying coil.

4-5. MAGNETOMOTIVE FORCE

An increase in the current of the solenoid of Fig. 4-3 results in an increase in the flux. A similar result occurs if the current is held constant and the coil is replaced by a coil having the same outside dimensions but more turns. Thus, the flux depends upon both the current and the number of turns, or upon the number of ampere turns. The number of ampere turns of the coil determines the **magnetomotive force** of the coil, that is, the capability of the coil to produce flux. Magnetomotive force (abbreviated **mmf**) in the magnetic circuit is analogous to electromotive force in the electric circuit.

The unit of magnetomotive force depends upon the system of units adopted for the magnetic circuit, that is, whether English (inch-pound-second), cgs (centimeter-gram-second), or mks (meter-kilogram-second), but for all systems the magnetomotive force is determined by the number of ampere turns in the circuit. In the English and the mks systems, the unit of magnetomotive force is the **ampere turn** (symbolized NI). Thus, the magnetomotive force is given by the relation

$$\mathcal{F} = NI \tag{4-2}$$

where $\mathcal{F}$ is the symbol for magnetomotive force.

4-6. RELUCTANCE

As previously indicated, a solenoid can have a core made of any material. The choice of material affects the amount of flux produced by a given number of ampere turns. For example, more flux is produced if the core is made of iron rather than of wood.

Materials are generally classified, magnetically, as being either magnetic or nonmagnetic. In general, magnetic materials are ferrous. Other materials used in electrical equipment, such as copper, aluminum, brass, wood, glass, rubber, etc., are considered to be the nonmagnetic materials. A material is said to be nonmagnetic if there is little or no change in the strength, or pattern, of a magnetic field when the material replaces air in that field.

The effect of a material upon the flux produced by a given mmf depends upon the **reluctance** of the material. Thus, iron has less reluctance to the production of flux than air or the other nonmagnetic materials; therefore, for a given mmf, more flux is produced in iron than in a nonmagnetic material. Reluctance, then, is analogous to the resistance of an electric circuit.

The similarity of magnetic reluctance and electrical resistance goes even further, in that they are both dependent upon the length, the area, and the material of the circuit path. The relation for magnetic reluctance, in terms of these quantities, is

$$\mathcal{R} = \nu \frac{l}{A} \qquad (4\text{--}3)$$

where l is the length of the magnetic path, A is the area perpendicular to the path, and ν (nu) is the reluctivity, a factor depending upon the material forming the magnetic path. The reciprocal of reluctance is called **permeance** (symbolized by $\mathcal{P}$) and is analogous to conductance in the electric circuit.

Instead of reluctivity, its reciprocal, **permeability,** is usually encountered in magnetic discussions; therefore, Eq. 4–3 is modified to be

$$\mathcal{R} = \frac{l}{\mu A} \qquad (4\text{--}4)$$

where μ (mu) is the symbol for permeability.

Units for permeance and reluctance have not been generally agreed upon, and therefore the units of permeability and reluctivity, which would normally be expressed in terms of the unit for permeance or reluctance, have not been established. However, units for all these quantities may be expressed in terms of other magnetic quantities from relations developed later in this chapter.

The permeability of nonmagnetic materials is considered a constant and has a value dependent upon the system of units. The permeability of magnetic materials is usually much greater than that for nonmagnetic materials and is not a constant but is dependent upon the type and the magnetic condition of the material. The permeability of magnetic materials is considered in more detail later in this chapter.

4–7. ELECTRIC-CIRCUIT ANALOGIES

It was pointed out earlier that one characteristic of flux lines is that they form closed paths. This characteristic gives rise to the concept of a magnetic circuit in which the flux lines correspond to the current in an electric circuit. The magnetic circuit and the electric circuit are analogous in many ways, and these analogies are helpful in the

ELECTRIC-CIRCUIT ANALOGIES 59

analysis and solution of magnetic problems. A major difference between the two types of circuit, however, is the fact that the flux lines do not represent the flow of anything, whereas current is a flow of electrons. Analogous electric and magnetic circuits are shown in Fig. 4–4.

For the electric circuit, Fig. 4–4a, it is assumed that the leads connecting the battery and load resistor have a much lower resistance than the load resistor. Therefore, since the current in the circuit is dependent upon the emf of the battery and the total resistance of the circuit, the resulting voltage distribution is such that most of the voltage drop occurs across the load resistor.

In the analogous magnetic circuit of Fig. 4–4b, the coil, having mmf, corresponds to the battery of Fig. 4–4a. The form upon which the coil is wound is made of iron and has an air gap. Since the air gap has a greater magnetic reluctance than the iron portion of the flux path, the iron portion corresponds to the connecting leads in the electric circuit and the air gap corresponds to the load resistor. The flux in the magnetic circuit is analogous to the current in the electric circuit, and, since the circuit is a series circuit, the same flux exists in all parts of the circuit.

Just as an electric current is dependent upon emf and resistance, so flux is dependent upon mmf and reluctance. The analogy is such, in fact, that an expression similar to that for Ohm's law can be applied to the magnetic circuit. Thus,

$$\phi = \frac{\text{mmf}}{\mathcal{R}} \qquad (4\text{–}5)$$

When Eq. 4–5 is applied to a magnetic circuit such as that in Fig. 4–4b, the value of reluctance should be the total reluctance of the circuit when the mmf equals that of the coil. However, Eq. 4–5 is applicable for portions of the magnetic circuit if the proper values of mmf and reluctance are inserted. Thus,

$$\phi = \frac{\text{mmf}_{(\text{iron})}}{\mathcal{R}_{(\text{iron})}} = \frac{\text{mmf}_{(\text{air})}}{\mathcal{R}_{(\text{air})}} \qquad (4\text{–}6)$$

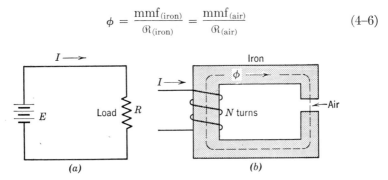

Fig. 4–4. Analogous electric and magnetic circuits.

where $\text{mmf}_{(iron)}$ and $\text{mmf}_{(air)}$ represent the fractions of the total mmf used to produce the flux in the iron and air, respectively. Thus, $\text{mmf}_{(iron)}$ and $\text{mmf}_{(air)}$ may be considered to be mmf drops. Since the air gap has a greater reluctance than the iron, the mmf distribution is such that most of the mmf drop occurs across the air gap.

4–8. MAGNETIC RELATIONS

For a magnetic path having a constant area and having the same material throughout, Eq. 4–5 can be written

$$\phi = \frac{NI}{l/\mu A} = \frac{NI\mu A}{l} \quad (4\text{–}7)$$

where NI is the mmf across the path, μ is the permeability, and l and A are the length and area, respectively. Solving Eq. 4–7 for flux density gives

$$B = \frac{\phi}{A} = \left(\frac{NI}{l}\right)\mu \quad (4\text{–}8)$$

The ratio of the mmf to the length in Eq. 4–8 is known as the **magnetizing force**. The electrical analogy for magnetizing force is called potential gradient and is expressed in terms of volts per unit length. Calculation of the potential gradients for the circuit of Fig. 4–4a would show a large potential gradient for the load resistor, whereas that for the leads would be small. Similarly, the magnetizing forces for the iron and air in the magnetic example are different; the magnetizing force for the air path is the larger.

Equation 4–8 may be rewritten

$$B = \mu H \quad (4\text{–}9)$$

where H is the magnetizing force. In the English system, the unit of magnetizing force is the ampere turn per inch.

4–9. THE MAGNETIZATION CURVE

Magnetic problems are usually such that the flux density of the circuit or a portion of the circuit is known, and it is required to find the number of ampere turns to produce this flux density, or vice versa. If the permeability and length of the circuit under consideration are known,

THE MAGNETIZATION CURVE

and if the material is nonmagnetic, Eq. 4–8 may be applied for the solution. Thus, for nonmagnetic materials, Eq. 4–8 becomes

$$B = 3.19 \frac{NI}{l} \qquad (4\text{--}10)$$

where 3.19 is the value of μ for nonmagnetic materials in the English system.

Transposing the terms of Eq. 4–10 gives

$$NI = 0.3132Bl \qquad (4\text{--}11)$$

where NI represents the number of ampere turns required to produce a flux density of B lines per square inch in a magnetic circuit composed of a nonmagnetic material having a length of l inches.

When the problem involves magnetic materials, however, the permeability is not a constant and is usually not known; therefore, Eq. 4–8 is not directly applicable. Problems of this type necessitate a **magnetization curve**, a plot of flux density versus magnetizing force, for the particular material involved. Representative magnetization curves (commonly called **B–H curves**) are shown in Fig. 4–5.

For the curves of Fig. 4–5, the flux density increases almost linearly with an increase in magnetizing force up to the "knee" of the curve. Beyond the knee, a continued increase in the magnetizing force results in a relatively slight increase in the flux density. When a magnetic material experiences only a slight increase in flux density for a relatively large increase in magnetizing force, the material is said to be saturated.

The characteristic of saturation appears only in magnetic materials and may be "explained" on the basis of a theory advanced for the behavior of magnetic materials. According to this theory, a magnetic material is composed of many very small magnets that take such random positions when no magnetizing force is applied that the material exhibits little external magnetic effect. Application of a magnetizing force, however, causes some of the magnets to align in the direction of the resultant magnetic field. For the lower portion of the magnetization curve, the alignment of the small magnets is approximately proportional to the increase in magnetizing force, and there are many magnets to be aligned. After the knee of the curve is reached, however, few magnets remain to be "regimented"; therefore, large increases in magnetizing force result in only small increases in flux density. When no magnets remain to be aligned, the increase in flux density for a given increase in magnetizing force is the same as that for nonmagnetic materials.

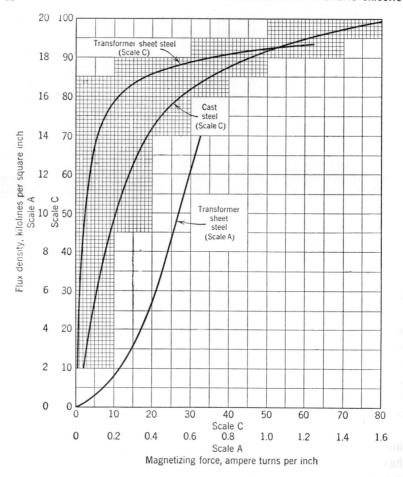

Fig. 4–5. Magnetization curves.

Expansion of the scale of the magnetization curves for most magnetic materials discloses that the curves are not linear near the origin. Such an expansion for the curve of sheet steel is shown in Fig. 4–5.

4–10. PERMEABILITY OF MAGNETIC MATERIALS

Transposition of the terms of Eq. 4–9 gives

$$\mu = \frac{B}{H} \qquad (4\text{–}12)$$

MAGNETIC CIRCUITS

and indicates one method for determining the permeability of a magnetic material. Thus, the slope of the line from the origin to a point on the *B–H* curve is the permeability of the material for the given condition of magnetization. This method of determining permeability is satisfactory for purposes of this book, although several other methods might be applied. Application of this method to the curves of Fig. 4–5 shows that the permeability of magnetic materials is not a constant but that it increases until the knee of the curve is reached and then starts to decrease.

4-11. MAGNETIC CIRCUITS

Laws similar to Kirchhoff's laws for current and voltage in the electric circuit can also be applied to the magnetic circuit. Thus, the magnetic equivalent of Kirchhoff's law of voltage can be stated: *The summation of the magnetic potential differences taken successively around a closed magnetic traverse is equal to zero.* Expressed mathematically,

$$\Sigma AT = 0 \qquad (4\text{-}13)$$

where *AT* designates ampere turns. Thus, just as the volt is the unit for emf and potential differences in the electric circuit, so the ampere turn is the unit for mmf and magnetic potential difference in the magnetic circuit. (The symbol *AT* is adopted rather than *NI* or $\mathfrak{F}$, since *NI* and $\mathfrak{F}$ are more frequently associated with a coil carrying a current.) The concept of magnetic potential difference has not been discussed previously in this book, but the establishment of such a concept leads to simplification of the analysis and solution of magnetic-circuit problems.

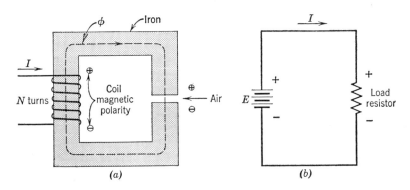

Fig. 4-6. Analogous magnetic and electric circuits.

A current-carrying coil may be considered to be a magnetic source, because the coil is capable of producing a magnetic field. Although the coil does not necessarily make contact with the magnetic circuit, it is, nevertheless, a fundamental part of the circuit. Such a coil and circuit are shown in Fig. 4–6a; the analogous electric circuit is shown as Fig. 4–6b.

The coil of Fig. 4–6a has magnetic polarity, since a change in the direction of the current through the coil changes the direction of the magnetic field. In addition, the coil has magnetic potential difference, since a change in the magnitude of the current or in the number of turns changes the strength of the magnetic field. Therefore, the symbols ⊕ and ⊖ are applied to the coil of Fig. 4–6a to indicate the relative magnetic polarity of the ends of the coil and the fact that there is a magnetic potential difference between the ends. The convention thus assumed for magnetic polarity is that *the direction of flux is from ⊖ to ⊕ through a magnetic source.* As previously indicated, the magnetic potential difference of a magnetic source depends upon the number of turns in the coil and the current through these turns.

The iron and air portions of the magnetic circuit are analogous to the leads and the load resistor, respectively, of the electric circuit of Fig. 4–6b and may be considered to be magnetic receivers. The magnetic polarity and potential difference of a magnetic receiver depend upon the direction of flux and the amount of flux, respectively, in the portion of the magnetic circuit under consideration. The convention assumed for the magnetic source necessitates that *the direction of flux is from ⊕ to ⊖ through a magnetic receiver.* The polarity of the air portion of the magnetic circuit of Fig. 4–6a is so indicated.

The electric potential difference across the load resistor or the connecting leads in the electric circuit of Fig. 4–6b is equal to the product of the current in the circuit and the resistance of the element under consideration. Similarly, the magnetic potential difference of the air or iron portion of the magnetic circuit is given by the relation

$$AT = \phi \mathcal{R} \qquad (4\text{–}14)$$

where AT represents the number of ampere turns required to produce ϕ lines in the portion of the magnetic circuit having a reluctance $\mathcal{R}$.

For the air path of Fig. 4–6a, the permeability is known and, therefore, Eq. 4–14 reduces to

$$AT = 0.3132 Bl \qquad (4\text{–}15)$$

The length l is the distance between the iron portions on each side of the air gap; the flux density B is determined from the total flux and the area of the iron path adjacent to the air gap, if there is no fringing of flux. If the length of the air gap is very large, the flux does not stay within the bounds set by the adjacent area but spreads out, as shown in Fig. 4–7. This spreading is called fringing and has the effect of reducing the flux density in the air gap. The assumption that there is no fringing is therefore an important one for the application of Eq. 4–15 in magnetic calculations. In the majority of the magnetic problems in this book, the assumption of "no fringing" may be made without seriously affecting the accuracy of the solution.

As previously indicated, the permeability of magnetic materials is usually not known; therefore, Eq. 4–14 is not directly applicable. However, the magnetization curve for the material provides a relation between the flux density and the magnetizing force, and Eq. 4–14 can be modified to be

$$AT = Hl \qquad (4\text{--}16)$$

where AT represents the number of ampere turns required to develop a magnetizing force H in a uniform path having a length l. If the magnetization curve for the given material is available, Eq. 4–16 can be applied to the magnetic portion of the circuit either to find the magnetic potential difference required to produce a given flux or, conversely, to find the flux produced by a given magnetic potential difference. Equation 4–16 applies only to those portions of the magnetic path having uniform cross section, since a change in cross section results in a change in flux density and magnetizing force.

For the circuit of Fig. 4–6a, the length l in Eq. 4–16 is the mean length of the iron path and includes the portion of the path under the coil. This portion of the path may be considered to be analogous to

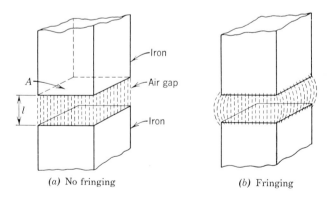

Fig. 4–7. Flux paths in an air gap.

the internal resistance of the battery which, although not shown in the diagram of Fig. 4–6b, would have to be considered in an exact solution of the circuit. In many magnetic calculations, particularly those for circuits without air gaps, serious error might result if the portion of the path surrounded by the coil is omitted from the calculation.

Since the law applied to magnetic potential differences calls for the summation of these potential differences to be equal to zero, algebraic signs must be applied to the potential differences encountered in a traverse of the magnetic circuit. Therefore, to complete the analogy of the magnetic circuit to the electric circuit, traverse from ⊖ to ⊕ is assumed to constitute a magnetic potential rise and is designated as being algebraically negative in the circuit relations written for the traverse. Similarly, traverse from ⊕ to ⊖ is assumed to constitute a magnetic potential drop and is designated as being algebraically positive. Thus, a summation of the magnetic potential differences for a clockwise traverse of the circuit of Fig. 4–6a results in the relation

$$-NI + H_i l_i + H_a l_a = 0 \qquad (4\text{–}17)$$

where NI represents the magnetic potential difference produced by the coil of N turns carrying a current of I amperes, and $H_i l_i$ and $H_a l_a$ are the potential differences for the iron and air paths, respectively.

In a magnetic circuit, *the summation of the flux at a junction is equal to zero.* Thus,

$$\Sigma \phi = 0 \qquad (4\text{–}18)$$

In order to effect the algebraic summation implied, as in the application of Kirchhoff's law of current to the electric circuit, flux in a direction toward a junction in a magnetic circuit is assumed to be algebraically

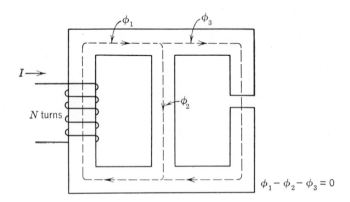

Fig. 4–8. Series–parallel magnetic circuit.

MAGNETIC CIRCUITS

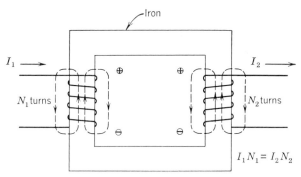

Fig. 4-9. Leakage flux in a magnetic circuit.

positive; flux in a direction away from a junction is assumed to be algebraically negative. The magnetic circuit shown as Fig. 4-8 is an example of a circuit having a junction in its magnetic path.

A significant failure of the analogy between electric and magnetic circuits is the fact that there are no open magnetic circuits; there are no magnetic insulators. Magnetic circuits may be such, however, that a portion of the circuit has no flux even though an mmf is applied in other parts of the circuit; an example is shown as Fig. 4-9. In the circuit of Fig. 4-9, the two coils tend to produce flux in opposite directions, but, since the coils have the same mmf, no flux is produced in the iron path linking the two coils. However, flux is produced in the air path adjacent to the coils, since the mmf of the coils acts on the air path as well as on the iron path. The flux in the air path is called **leakage flux,** if the flux in the common iron path is considered to be the useful flux of the magnetic application. If one of the coils is not energized, the amount of flux in the iron path due to the energized coil would probably be so large in comparison to the leakage flux that the leakage flux could be neglected. Whether leakage flux can be considered as negligible or not depends upon the magnetic application in which it occurs.

If one of the coils of the circuit of Fig. 4-9 has a greater mmf than the other, the direction of flux in the circuit is determined by the coil having the greater mmf. Figure 4-10 shows the circuit conditions when coil 1 has a greater mmf than coil 2. The ampere-turn relation that applies under these conditions is

$$-NI_1 + H_i l_i + NI_2 = 0 \qquad (4\text{-}19)$$

In the circuit of Fig. 4-10, coil 2 acts as a magnetic receiver, according to the convention previously established, because the flux direction

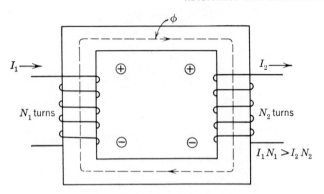

Fig. 4–10. Magnetic circuit with opposing mmf's.

is from ⊕ to ⊖ through the coil. This condition is analogous to the charging of a battery in an electric circuit. That a coil is considered to be a magnetic receiver, however, has no significance other than the role it plays in simplifying the magnetic analysis.

4–12. THE SOLUTION OF MAGNETIC-CIRCUIT PROBLEMS

Because of the similarity of electric and magnetic circuits, the procedure for the solution of electric-circuit problems can also be applied, with modifications, to the solution of magnetic-circuit problems. The analysis and solution of a magnetic circuit are simplified if an analysis is first made of its analogous electric circuit. To assist in setting up the analogy, some analogous elements and relations for magnetic and electric circuits are tabulated as follows:

Magnetic	Electric
mmf, $\mathcal{F}$	emf
flux, ϕ	current
flux density, B	current density
magnetizing force, H	potential gradient
Hl	IR
$\Sigma AT = 0$	$\Sigma e = 0$
$\Sigma \phi = 0$	$\Sigma i = 0$

The procedure recommended for the solution of magnetic-circuit problems is outlined as follows:

1. Draw a complete circuit diagram, completely labeled.
2. Assume flux directions for those portions of the circuit for which the flux directions are unknown.

THE SOLUTION OF MAGNETIC-CIRCUIT PROBLEMS

3. Assign magnetic polarities consistent with the assumed flux directions.
4. Write the necessary flux and ampere-turn relations.
5. Prepare a tabular form which incorporates space for both the known and unknown quantities to be required in the solution. The columns for such a tabular form might be headed as follows:

Part   Material   Length   Area B ϕ H Hl $\mathcal{F}$

The column headings should include the units, as in most tabular forms.
6. Solve for the desired quantities.

The actual solution of the problem (step 6 of the procedure) depends upon the nature of the problem. Many problems are possible for a given magnetic circuit, even for a simple series magnetic circuit such as that shown in Fig. 4–6a. Two possible problems for this circuit are as follows:

(a) Given the dimensions of the magnetic circuit, determine the mmf required to produce a given flux density in the air gap.

(b) Given the dimensions of the magnetic circuit, determine the flux density produced by a given mmf.

The steps in the solution of problem (a) are as follows:

(1) Calculation of the total flux in the circuit.
(2) Calculation of the flux density for each part of the magnetic circuit.
(3) Determination of the magnetizing forces necessary to produce the flux densities found in step 2.
(4) Calculation of the magnetic potential differences required to produce the magnetizing forces found in step 3.
(5) Solution for the mmf required.

The solution of problem (b) is not so direct as that of problem (a), since there is no direct way of determining how the given mmf is distributed over the magnetic circuit. However, the solution may be obtained by a cut-and-try method. In the cut-and-try method, a flux density is assumed for the air gap, and the procedure for the solution of problem (a) is followed. The mmf thus determined is compared to the given mmf; if there is a difference between the calculated mmf and the given mmf, the assumed value of flux density is adjusted in accord with the difference, and the procedure is repeated. This process is continued until the calculated mmf is equal to the given mmf; the value of assumed flux density is then the value sought.

The number of "trys" in the cut-and-try process can often be reduced to two by plotting values of the assumed flux density against the resultant calculated mmf. Thus, after two trys, if both fail to give a solution, a straight line is drawn between the two plotted points. The approximate value of flux density is then found on this straight line at the value of the given mmf. Unless the first two trys were too far apart, the plotting procedure usually results in a reasonably accurate solution of the problem. However, the value of flux density found on the plot should be used in another solution of the problem to make certain that it is a reasonably accurate value.

4-13. HYSTERESIS

A magnetic material that exhibits no external magnetic effect when the externally applied mmf is zero may be said to be demagnetized. The magnetization curves shown in Fig. 4-5 are plotted for materials that were originally demagnetized. An increase in the magnetizing force on a demagnetized material produces a magnetization curve similar to those of Fig. 4-5 and as shown in Fig. 4-11a. If, after reaching a given value of magnetizing force, the magnetizing force is then decreased to zero, the B–H curve is not retraced, but the curve takes a path as indicated in Fig. 4-11b. As shown in Fig. 4-11b, higher values of flux density are obtained for a given value of magnetizing force after the magnetizing force has been increased to a maximum and then decreased. Thus, the flux density to be obtained with a given magnetizing force depends upon the magnetic "history" of the material. The property of a material by virtue of which the flux density for a given magnetizing force depends upon the previous conditions of magnetization is called **hysteresis.**

The value of flux density remaining after the magnetizing force has

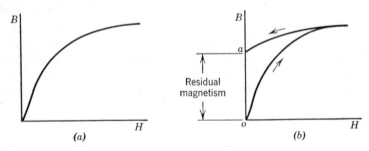

Fig. 4-11. Hysteresis effect.

HYSTERESIS

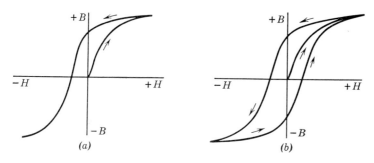

Fig. 4–12. Development of a hysteresis loop.

been decreased to zero (indicated by *oa* in Fig. 4–11*b*) is called **residual magnetism**. The amount of residual magnetism in a magnetic material is a function of the type of material as well as the original condition of magnetization. Materials employed for permanent magnets must exhibit large values of residual magnetism after the magnetizing force is reduced to zero.

If the current through the coil that produced the magnetizing forces for the curves of Fig. 4–11 is reversed and then increased to the same maximum value as before, the *B–H* curve extends as shown in Fig. 4–12*a*. The ultimate value of flux density is the same as that obtained when the magnetizing force was a positive maximum.

Decreasing the current from the negative maximum to zero and then increasing it to a positive maximum again causes the curve to take the path shown in Fig. 4–12*b*.

The complete loop formed when the magnetizing force is changed from a positive maximum to a negative maximum and then back to a positive maximum again is called a **hysteresis loop.**

In addition to the effects described, each complete reversal of the magnetizing force causes an energy loss, which appears as heat in the magnetic material. It has been found that the magnitude of the energy loss is a function of the area of the hysteresis loop. The energy loss is therefore called hysteresis loss.

One explanation for hysteresis loss is based upon the theory of the small magnets. The small magnets may be considered to be changing position as the magnetizing force is changed, and thus friction develops between the magnets. The energy loss is thus attributed to magnetic friction within the material.

As previously indicated, hysteresis loss is a function of the area of the hysteresis loop. The relationship is such that the loss becomes larger as the loop increases in area. The loss is also a function of the frequency of the reversals of magnetizing force. For a given magnetic circuit, the power loss due to hysteresis can be expressed by the relation

$$P_h = KB^n f \qquad (4\text{--}20)$$

where P_h = hysteresis power loss.
K = constant dependent upon the material, the volume of material, and the system of units.
B = maximum flux density.
n = an exponent dependent upon the material and the degree of magnetization.
f = frequency of the reversals of magnetizing force.

For the description of hysteresis loss, the example consists of a changing magnetizing force and a fixed material in a magnetic circuit. The loss also occurs if the material of the magnetic circuit is rotated and the magnetizing force is fixed. Thus, if a cylinder of magnetic material is rotated in a magnetic field perpendicular to the axis of rotation, hysteresis loss occurs in the cylinder.

4–14. ELECTROMAGNETIC FORCES

An important application of magnetic fields in electric equipment is the production of mechanical force. The electromagnetic forces fall into two general classifications: (*a*) the pull of a magnet, and (*b*) the force on a conductor.

Pull of a Magnet. When each of the coils of the magnetic circuits shown in Fig. 4–13 is energized, an electromagnetic force is produced. In each case, the force acts to shorten the air path of the magnetic circuit. The action of the force is due to the characteristic tendency of flux lines to be as short as possible. A better explanation is that the north and south poles produced on each side of the air gap experience a force of attraction which acts to shorten the gap. For each of the applications shown in Fig. 4–13, it is assumed that the portion of the magnetic circuit incorporating the coil remains fixed. Thus, in Fig. 4–13*a*, the force is exerted on the pivoted member to pull it toward the main member; in Fig. 4–13*b*, the force is applied to lift the unattached member; in Fig. 4–13*c*, the iron bar is drawn into the cylindrical coil.

The force on the pivoted member of Fig. 4–13*a* is a function of the flux density at the pole face and the area of the pole face and is given by the relation.

$$F = KB^2 A \qquad (4\text{--}21)$$

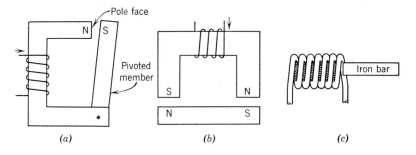

Fig. 4–13. Applications of force due to pull of a magnet.

For force in pounds, flux density in lines per square inch, and area in square inches,

$$K = \frac{1}{72{,}130{,}000} \qquad (4\text{--}22)$$

In the example shown in Fig. 4–13b, there are two poles and, therefore, two pole faces. Therefore, if the flux density is the same at each pole face and the pole-face areas are the same, the total force exerted on the unattached member is $2F$.

Force on a Conductor. A current-carrying conductor in a transverse magnetic field experiences a force that is a function of (a) the strength of the field, (b) the current in the conductor, and (c) the length of the conductor in, and normal to, the field. Thus, if a conductor, carrying current i, is placed in a magnetic field of flux density B, the force exerted on the conductor is given by the relation

$$F = K'Bil \qquad (4\text{--}23)$$

where l is the length of the conductor normal to the field.

For force in pounds, flux density in lines per square inch, current in amperes, and length of conductor normal to the field in inches,

$$K' = \frac{1}{11{,}300{,}000} \qquad (4\text{--}24)$$

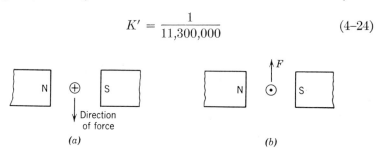

Fig. 4–14. Directions of force on a current-carrying conductor.

The direction of the force exerted on a current-carrying conductor in a magnetic field is perpendicular to the field and to the current and depends upon the relative directions of the field and the current. Figures 4–14a and 4–14b show the direction of the force on the conductor for different relative directions of field and current: Fig. 4–14a, for current into the conductor; Fig. 4–14b, for current out of the conductor.

One method for determining the direction of the force on a current-carrying conductor requires an analysis of the resultant field produced by the combination of the main field and the field produced by the current in the conductor. Figure 4–15a shows the uniform field pattern of the main field alone; Fig. 4–15b shows the field pattern produced by the current-carrying conductor alone. When the two fields are combined, the resultant field pattern is as shown in Fig. 4–15c. Above the conductor the lines are distorted and crowded because the lines of the two fields are in the same direction and are additive; below the conductor the field is weak because the lines of the two fields are in opposite directions and are subtractive. If the lines are considered to act as stretched rubber bands, the resultant force on the conductor is downward, as indicated in Fig. 4–15c. Thus, for this method, the direction of the force on a current-carrying conductor in a magnetic field is determined by the "stretched rubber bands" of the resultant field.

Figure 4–16a shows the cross section of a one-turn current-carrying coil mounted on a cylinder that can be rotated; the combination of cylinder and coil is located in a magnetic field. By application of the rubber-band method, the forces on the individual conductors are found to be in a direction to rotate the cylinder counterclockwise. When the coil reaches the position where its plane is perpendicular to the main field, as shown in Fig. 4–16b, the forces on the conductor no longer cause rotation. If the inertia of the cylinder causes the coil to rotate beyond the position shown in Fig. 4–16b, the forces on the conductors are such that the coil is rotated in a clockwise direction. Thus, the coil oscillates about the position shown in Fig. 4–16b and finally comes to rest in this position.

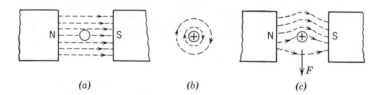

Fig. 4–15. Electromagnetic force on a conductor.

ELECTROMAGNETIC FORCES

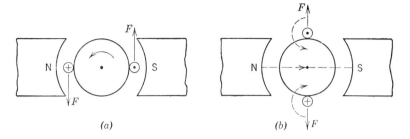

Fig. 4–16. Electromagnetic force on a current-carrying coil.

When the coil comes to rest, the flux produced by the current in the coil is in the same direction through the coil as the flux of the main field. This fact is the basis for another method of determining the direction of the force on a current-carrying coil in a magnetic field. Thus, the force on a current-carrying coil in a transverse magnetic field acts to cause the flux of the main field and the flux produced by the coil to align in the same direction through the coil. A more general statement is: "The force on a conductor carrying current in a transverse magnetic field is such as to maximize the flux linking the circuit of which the conductor is a part."

PROBLEMS

4–1. (*a*) Sketch the approximate curve of permeability versus magnetizing force for brass, and explain its shape.
(*b*) Sketch the approximate curve of permeability versus magnetizing force for sheet steel, and explain its shape.
(*c*) What final value of permeability does the curve of (*b*) approach?

4–2. Two cast-iron rings are linked by a common coil of 200 turns which is carrying 10 amperes. The cross-sectional area of ring 1 is 4 sq in.; that of ring 2 is 2 sq in. The mean diameters of the rings are the same. How do the flux densities compare? *Ans.* $B_1 = B_2$.

4–3. The exciting coil for a certain relay has a resistance of 120 ohms and is connected directly to a 120-volt d-c supply. The magnetic circuit consists mainly of sheet steel. If one-half the turns of the coil are short-circuited, how will the new flux density compare with the old? Explain.

4–4. Repeat Prob. 4–3 for the same relay connected through a 60-ohm resistor to the 120-volt supply.

4–5. (*a*) In Fig. 4–17, a permanent magnet is so pivoted that it is free to turn around its center. Four pole pieces with identical exciting coils are placed as shown. Indicate the position of the permanent magnet for the following conditions of current in the coils (positive means current as shown; negative means current reversed).

(1) $I_1 = I_3 = +5$
    $I_2 = I_4 = 0$

(2) $I_1 = I_3 = +2.5$
    $I_2 = I_4 = +2.5$

(3) $I_1 = I_3;\ I_2 = I_4$
    $I_1 < I_2$

(4) $I_1 = I_3 = 0$
    $I_2 = I_4 = +5$

(b) What might be a practical application of such a device as that shown in Fig. 4–17?

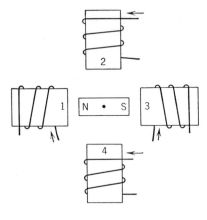

Fig. 4–17.

4–6. (a) Is it possible to determine, without a magnetization curve, the magnetic potential difference between points a and b of Fig. 4–18a? If so, what is it? If not, why not?

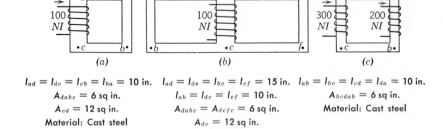

$l_{ad} = l_{dc} = l_{cb} = l_{ba} = 10$ in.
$A_{dabc} = 6$ sq in.
$A_{cd} = 12$ sq in.
Material: Cast steel

$l_{ad} = l_{de} = l_{bc} = l_{cf} = 15$ in.
$l_{ab} = l_{dc} = l_{ef} = 10$ in.
$A_{dabc} = A_{defc} = 6$ sq in.
$A_{dc} = 12$ sq in.
Material: Cast steel

$l_{ab} = l_{bc} = l_{cd} = l_{da} = 10$ in.
$A_{bcdab} = 6$ sq in.
Material: Cast steel

Fig. 4–18.

(b) Repeat (a) for Fig. 4–18b.
(c) Repeat (a) for Fig. 4–18c.

4-7. Find the number of ampere turns necessary to produce a flux density of 10 kilolines per sq in. in the air gap of Fig. 4-19. The dimensions are given in tabular form; area in square inches, lengths in inches.

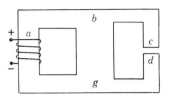

Fig. 4-19.

Part	Material	Area	Length
bag	Cast steel	6.25	15.0
gd	Cast steel	4	6.0
dc	Air	4	0.02
cb	Cast steel	4	6.0
bg	Cast steel	6	4.30

Ans. 424 NI.

4-8. Figure 4-20 shows a magnetic circuit with two magnetizing coils. Coil 1 has 440 turns and carries 2.3 amperes; coil 2 has 280 turns. What current should coil 2 carry, and what should be the polarity of terminal x,

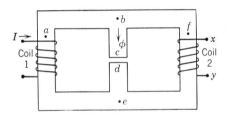

Fig. 4-20.

to have a flux in the air gap of 45,000 lines with the direction as shown? The specifications for the magnetic circuit are tabulated below, with areas in square inches and lengths in inches.

Part	Material	Area	Length
bae	Cast steel	1.5	17
bc	Cast steel	1.5	3
de	Cast steel	1.5	3
bfe	Cast steel	1.5	17
cd	Air	1.5	0.05

Ans. 1.12 amperes, —.

4-9. Repeat Prob. 4-8 for a flux of 85,000 lines in the air gap.
Ans. 4.05 amperes, —.

4-10. Repeat Prob. 4-8 for a flux of 75,000 lines in the air gap.
Ans. 3 amperes, —.

4–11. In the magnetic circuit shown as Fig. 4–21, what length of air gap should be cut into path *bge* to make the flux in that path equal to the flux in

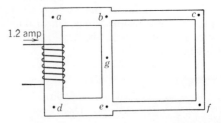

Path	Length	Area
bade	12	10
bge	6	5
bcfe	18	5

Material: Cast steel

Fig. 4–21.

path *bcfe* when the current in the 1000-turn coil is 1.2 amperes? (The dimensions of the magnetic circuit are tabulated: lengths in inches, areas in square inches.) *Ans.* 0.017 in.

4–12. Determine the number of turns and the connections (*x* to *y* or *x* to *z*?) of coil 1 in Fig. 4–22 in order that there be no flux in the central leg, path *be*. The dimensions are tabulated: lengths in inches, areas in square inches. *Ans.* 400 turns; *x* to *y*.

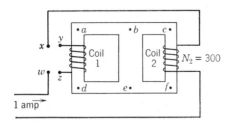

Path	Length	Area
bade	20	6
bcfe	15	6
be	5	12

Material: Cast steel

Fig. 4–22.

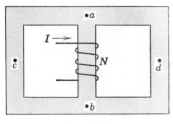

Fig. 4–23.

4–13. For the cast-steel magnetic circuit of Fig. 4–23, briefly explain the effect, if any, on the flux and flux density in path *acb* of independently

(*a*) Doubling the ampere turns of the coil.

(*b*) Cutting an air gap in path *adb*.

(*c*) Doubling all dimensions of the magnetic circuit, including the ampere turns of the coil.

(*d*) Adding a coil having one-half the ampere turns of coil 1 around path *adb* such that the added mmf acts in the direction from *b* to *a*.

4-14. Except for the coil arrangements the magnetic circuits shown in Fig. 4-24 are identical. Compare the magnitude of I_1 and I_2 for a given flux density in the air gap. *Ans.* $I_1 = 2I_2$.

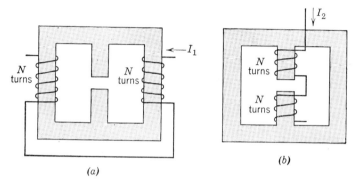

Fig. 4-24.

4-15. In the magnetic circuit shown in Fig. 4-25, the cast-steel path has a length of 12 in. and a cross-sectional area of ½ sq in. Determine the length to which the air gap must be adjusted so that a weight W of 4 lb will just be lifted when a current of 5.7 amperes flows through the exciting coil of 1000 turns. *Ans.* 0.75 in.

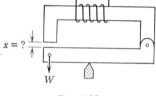

Fig. 4-25.

4-16. (*a*) Figure 4-26 shows a magnetic relay. The relay is so constructed that the hammer can be moved either to the right or to the left, depending upon the net force exerted on it. The hammer arm is built so that it saturates at very low values of coil mmf. If there is no flux leakage, in what direction is the hammer pulled when the flux distribution is as shown in Fig. 4-26?

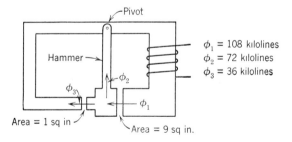

Fig. 4-26.

(b) In what direction would the hammer move if the total flux were decreased 10%? Explain.

4–17. Figure 4–27a shows the schematic sketch of a cast-steel lifting magnet that just holds a 2500-lb load of cast steel when the current in the 100-turn coil is 8 amperes. Is it possible to determine, without a magnet-

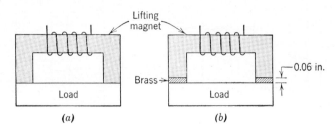

Fig. 4–27.

ization curve, the coil current required to just hold the load if a 0.06-in. brass plate is inserted between the magnet and the load, as shown in Fig. 4–27b? If so, what is it? If not, why not? (The area of the magnetic path may be considered to be uniform at 36 sq in., and leakage and fringing may be considered to be negligible.)

4–18. If the length of the magnetic path in Fig. 4–27a is known to be 90 inches and the magnetization curve of Fig. 4–5 is applicable, what load will the magnet of Prob. 4–17 support with the brass plate inserted and the current held constant at 8 amperes? *Ans.* Approx. 200 lb.

4–19. Figure 4–28 shows the cross section and side view of a magnetic track brake similar to those used on modern electric railways. When not energized, the brake is held above the track by springs (not shown) so that there is about ¼-in. separation between the brake and the track. When energized, the brake is pulled against the track, thus producing braking action. (The magnetization curve for cast steel can be extended with the data supplied below.)

Curve Extension Data

B (kilolines per sq in.)	101.3	102.8	104	105	106
H (ampere turns per in.)	90	100	110	120	130

(a) Assuming the dimensions given, what number of ampere turns is required to produce an initial downward pull of 95 lb on the brake? (Assume a flux leakage across the gap between poles of the brake of 66% of the flux in the air gap between the brake and track.) *Ans.* 1911 NI.

(b) What force is exerted when the brake is in contact with the track, if the ampere turns are the same as in (a) and if there is no leakage flux?
Ans. Approximately 7500 lb.

(c) What must be the characteristic of any material that fills the space between the brake poles?

PROBLEMS

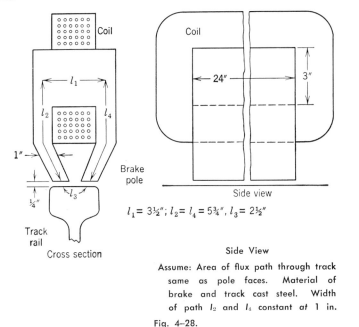

$l_1 = 3\frac{1}{2}''$; $l_2 = l_4 = 5\frac{3}{4}''$, $l_3 = 2\frac{1}{2}''$

Assume: Area of flux path through track same as pole faces. Material of brake and track cast steel. Width of path l_2 and l_4 constant at 1 in.

Fig. 4–28.

4–20. Figure 4–29a shows the schematic diagram of a magnetic blowout applied in a motor starter. The purpose of the blowout is to extinguish the arc formed when contact B moves away from the fixed contact A. When the motor is operating, the contacts are together, completing the circuit as shown in Fig. 4–29b. Before the contacts are opened, the motor draws 50 amperes, and the blowout coil produces a flux density of 60,000 lines per sq in. at the

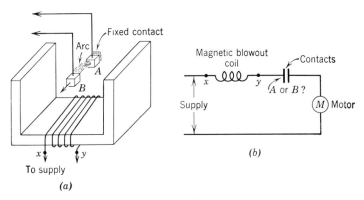

Fig. 4–29.

arc. (Assume that the width—not length, which is variable—of the arc is $\frac{1}{16}$ in.)

(a) What force per unit area is exerted on the arc to extinguish it?

Ans. 4.25 lb per sq in.

(b) Show, by means of a simple sketch, the principle by which the blowout force is produced.

(c) Determine whether A or B should be connected to terminal y in order that the arc be forced upward.

4-21. It is desired that a valve inside a fuel-oil supply pipe be opened only when a coil outside, and around, the pipe is energized. In addition, it is desired that the amount of fuel oil that flows through the opened valve be a function of the current in the exciting coil. By means of sketches and explanations, show how the requirements might be satisfied.

■ 5 Electromagnetic Induction

5–1. ELECTROMAGNETIC INDUCTION

Electric energy in large quantity and at low cost is obtained primarily from devices in which mechanical energy is converted to electric energy and in which the electromotive force is produced by **electromagnetic induction.** Electromagnetic induction is the process by which an electromotive force is induced, or generated, in an electric circuit when there is a change in the magnetic flux linking the circuit. The magnitude of the voltage induced is proportional to the time rate of change of the flux linkage.

5–2. FLUX LINKAGES

As the name implies, a flux linkage exists when the flux and the electric circuit are connected like two links of a chain. The circuit may consist of a single conductor or a number of conductors; the linking flux may be produced by current within the circuit or by an external means.

Several examples of flux linkages are shown in Fig. 5–1. These examples are representative of the practical application of electromagnetic induction. In Fig. 5–1a, a current-carrying coil (shown in cross section) is linked by the flux produced by the current in the coil. In Fig. 5–1b, which shows a portion of a magnetic circuit having two coils, the flux produced by the energized coil links each of the coils. The linkage of the coil shown in Fig. 5–1c is similar to that of the unenergized coil in Fig. 5–1b in that the linking flux is produced externally; in Fig. 5–1c, however, the linked coil is mounted on a cylinder that can be rotated. In the example shown as Fig. 5–1d, a fixed coil is linked by the flux produced by a U-shaped permanent magnet; the bar of magnetic material above the magnet forms a part of the magnetic circuit and can be moved upward or downward with respect to the magnet.

The magnitude of a flux linkage is determined by the number of times a circuit is linked by flux. Thus, in the example of Fig. 5–1a, each flux line links the circuit once; therefore, the total flux linkage is equal to the total flux. In Fig. 5–1b, however, each flux line links the circuit of the unenergized coil as many times as there are turns in the unenergized coil; thus, the total flux linkage is $N\phi$, where N represents the number of turns in the unenergized coil and ϕ is the flux linking the unenergized coil. The examples of Fig. 5–1c and Fig. 5–1d are similar to that of Fig. 5–1b, except that the linked coil of Fig. 5–1c has only one turn.

The examples of Fig. 5–1 also illustrate the various methods of producing a change in flux linkage, a requirement for electromagnetic induction. Thus, in Fig. 5–1a, a change in the current in the coil causes the flux, and therefore the flux linkage, to change. Similarly, the flux linkage of each coil in Fig. 5–1b is changed when the current in the energized coil is changed.

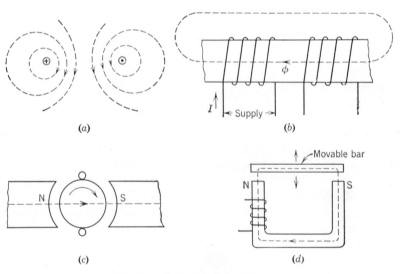

Fig. 5–1. Examples of flux linkages.

MAGNITUDE OF INDUCED VOLTAGE

The flux linkage in Fig. 5–1c may be changed by rotation of the cylinder upon which the coil is mounted. Thus, if the cylinder is rotated clockwise from the position shown in Fig. 5–1c, the amount of flux linking the coil, and therefore the flux linkage, decreases until the plane of the coil is coincident with the axis of the poles; when the coil is in this position, no flux links the coil. Continued rotation of the coil causes the flux linkage to increase until the plane of the coil is again perpendicular to the axis of the poles, the position shown in Fig. 5–1c. A similar change in flux linkage occurs if the cylinder remains fixed and the magnetic field poles are revolved around the cylinder.

If the position of the movable bar in Fig. 5–1d is varied, the reluctance of the magnetic circuit is changed; therefore, the flux and flux linkage are changed.

The methods of changing a flux linkage are summarized as follows:

(a) Change of mmf (Figs. 5–1a and 5–1b).
(b) Relative motion of conductor and field (Fig. 5–1c).
(c) Change of reluctance (Fig. 5–1d).

All three methods have practical applications which are considered later in the book.

5–3. MAGNITUDE OF INDUCED VOLTAGE

The flux linkages considered in the previous article are such that all the flux in the magnetic circuit links all the turns. Thus, the magnitude of the flux linkage is given by the product of the number of turns and the linking flux, or $N\phi$. In many applications, however, some turns are linked by more flux than others so that the actual number of flux linkages must be determined by a summation of the incremental linkages. This degree of refinement is not necessary for the discussions in this book, and, for simplification, all turns are considered to be linked by the same flux. For these conditions, the magnitude of the voltage induced by electromagnetic induction is expressed by the relation

$$e = KN \frac{d\phi}{dt} \tag{5-1}$$

where e is the induced voltage, N is the number of turns, ϕ is the flux linking the turns, and K is a constant dependent upon the system of units.

If the flux linking a one-turn coil changes at the rate of 100 million lines per second, 1 volt is induced in the coil. Therefore, Eq. 5–1 may be written

$$e = N \frac{d\phi}{dt} \times 10^{-8} \qquad (5\text{-}2)$$

when e is the induced voltage in volts, N is the number of turns, ϕ is the linking flux in lines, and t is the time in seconds.

If the time rate of change of flux linkage is not constant, the application of Eq. 5–2 results in the induced voltage for the particular instant of time chosen. Revised to give the average induced voltage for a given time interval, Eq. 5–2 is expressed as

$$E_{\text{avg}} = \frac{N\phi}{T} \times 10^{-8} \qquad (5\text{-}3)$$

where $N\phi$ is the total change in flux linkage for the total time T. The units for Eq. 5–3 are the same as for Eq. 5–2.

5–4. POLARITY OF INDUCED VOLTAGE

The polarity of a voltage induced because of a change in flux linkage may be correctly determined in several ways. One method is based upon the fact that, when there is a change in the flux linking a circuit, the voltage induced *tends* to oppose the change. (The change in flux linkage may be either an increase or a decrease.)

To oppose a change in the flux linking a circuit, current must flow in the circuit in such a direction that it produces an opposing mmf. However, current is not mentioned either in the definition of electromagnetic induction or in the statement concerning the polarity of the voltage so induced; in fact, current is not necessary to the existence of

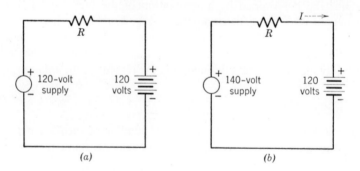

Fig. 5–2. Circuits with emf's in opposition.

POLARITY OF INDUCED VOLTAGE

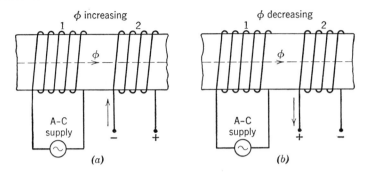

Fig. 5-3. Polarity of induced voltage.

any emf, as shown in the circuit of Fig. 5–2a. In the circuit of Fig. 5–2a, the voltage of the supply and the voltage of the battery are equal; therefore, no current flows in the circuit. In Fig. 5–2b, the voltage of the supply is greater than that of the battery; therefore, current flows as indicated. In each of the examples, the voltage of the battery *tends* to oppose the flow of current. Similarly, the voltage induced because of a change in flux linkage tends to oppose the change, and current may or may not flow in the circuit in which the change in linkage takes place.

After a consideration of the examples of Fig. 5–2, the expression "... voltage ... tends to oppose the change" may be interpreted to mean "If the circuit were complete and if there were no other emf's in the circuit, the voltage induced would cause current to flow so as to produce an mmf opposing the change in flux linkages." The method of polarity determination to be described is based upon this interpretation. In Fig. 5–3a, when the flux linking coil 2 is increasing, a voltage is induced in coil 2 which tends to cause current to flow as indicated, thus creating an mmf to oppose the increase in flux linkage. Since current flows from − to + through a source, the polarity of the induced voltage is as indicated in Fig. 5–3a. The flow of current in coil 2 is not necessary to the existence of the induced voltage; the current flow is indicated to simplify the polarity determination.

Similarly, when the flux that links coil 2 is decreasing, as indicated in Fig. 5–3b, a voltage is induced that tends to cause current to flow to create an mmf opposing the decrease in flux linkage. The resultant polarity of the induced voltage is indicated in Fig. 5–3b.

The change in flux linkage of coil 2 occurs at the same time as a similar change in the flux linkage of coil 1. The analysis of the induced voltage resulting in coil 1 is considered in Art. 5–6.

5-5. FLUX CUTTING

With regard to the voltage induced in a circuit because of a change in the flux linking the circuit, the American Standard Definitions of Electrical Terms states: "When the change in flux linkage is caused by the motion, relative to a magnetic field, of a conductor forming part of an electric circuit, the electromotive force induced in the circuit is proportional to the rate at which the conductor cuts the flux." This statement introduces the concept of flux cutting, which, in some instances, is more readily applied than the concept of a changing linkage.

A change in the flux linkage of a coil mounted on a cylinder that can be rotated, as shown in Fig. 5–4a, is an example of a changing linkage caused by the movement of a conductor relative to a magnetic field. As the cylinder rotates from the position shown in Fig. 5–4a to that shown in Fig. 5–4c, the flux linkage of the coil is decreasing; therefore, the polarity of the voltage induced in the coil during this period of rotation is such as to tend to produce the current flow shown in Fig. 5–4b. Figure 5–4b represents an intermediate position of the coil and serves to illustrate the polarity of the voltage induced during the entire transition from the position shown in Fig. 5–4a to that shown in Fig. 5–4c.

No flux links the coil when it is in the position shown in Fig. 5–4c; therefore, continued rotation from this position to that shown in Fig. 5–4e causes an increase in flux linkage in the opposite direction. The polarity of the induced voltage during this entire transition period is such as to tend to cause current flow as indicated for an intermediate

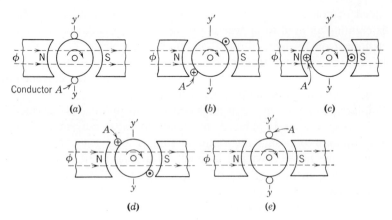

Fig. 5–4. Change in flux linkage by relative motion.

FLUX CUTTING

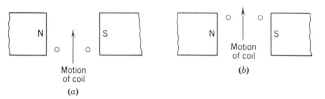

Fig. 5–5. An example of movement of a coil through a magnetic field with no change in flux linkage.

position in Fig. 5–4d. As shown, the polarity is such as to oppose the increase in flux linkage and is the same as when the coil was rotated from the position shown in Fig. 5–4a to that shown in Fig. 5–4c. Thus, while conductor A of the coil moves from position y to position y', the polarity of the coil remains the same.

In moving from y to y', conductor A may be considered to "cut" the field flux and, by so doing, to have a voltage induced in it. This is the concept of flux cutting. However, while the conductor cuts flux, there is also a change in the flux linkage of the coil; a change in the flux linkage of a circuit is necessary if a voltage is to be induced in a circuit, even though a conductor of the circuit cuts flux. Thus, in the example shown in Fig. 5–5, if the whole coil is moved laterally with respect to the field, from the position shown in Fig. 5–5a to that shown in Fig. 5–5b, there is no change in the flux linkage of the coil; therefore, no voltage is induced in the coil, even though each conductor cuts flux.

In the flux-cutting concept, the polarity of the voltage induced depends upon the direction of motion of the conductor with respect to the direction of the field. Thus, since conductor A of Fig. 5–4 always cuts the flux in the same relative direction while moving from y to y', the polarity of the conductor should remain unchanged. The same conclusion is reached by the application of the concept of a changing flux linkage.

The two conductors of Fig. 5–5 cut flux in the same direction and therefore may be considered to have induced voltages of the same polarity. Since the two conductors are connected to form a coil, the two induced voltages are in opposition, and the net induced voltage in

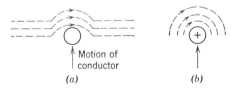

Fig. 5–6. Polarity of induced voltage.

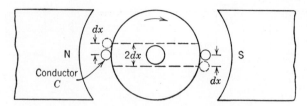

Fig. 5-7. Rotating coil in a magnetic field.

the coil is zero. This conclusion is also reached by the application of the concept of a changing flux linkage.

One method for determining the polarity of the voltage induced in a conductor that is cutting flux involves the rubber-band concept applied to flux. Thus, as the conductor is moved through a magnetic field, the lines are assumed to take the distorted form shown in Fig. 5–6a and to produce a force opposing the motion of the conductors. The voltage induced in the conductor tends to cause current to flow in such a direction as to produce a field pattern similar to the distorted field caused by the motion of the conductor; or, the voltage induced tends to cause current to flow so as to produce a force opposing the motion of the conductor, as shown in Fig. 5–6b.

The magnitude of the voltage induced in a conductor that cuts flux can be determined by an analysis of the rotating coil shown in Fig. 5–4 and redrawn as Fig. 5–7.

For the position shown in Fig. 5–7, the conductor C is moving perpendicular to the field; therefore, the change in flux linkage of the coil when the conductor moves a distance dx can be expressed as

$$B \, dA \tag{5-4}$$

or

$$B(2l \, dx) \tag{5-5}$$

where B is the flux density (assumed uniform), dA is the change in area of the flux path through the coil, and l is the length of the conductor normal to the field. (While conductor C moves a distance dx, the other conductor of the coil moves the same distance, but in the opposite direction, thus accounting for the "2" in Eq. 5–5. Because the coil has only one turn, N is not included in either expression.) The voltage induced in the coil is therefore given by the relation

$$e = \frac{2Bl \, dx}{dt} \times 10^{-8} \tag{5-6}$$

where dt is the time required for the conductor to move the distance dx.

EDDY CURRENTS

Since the coil has two conductors, the voltage per conductor is given by

$$e = \frac{Bl\,dx}{dt} \times 10^{-8} \qquad (5\text{–}7)$$

which can be converted to

$$e = Blv \times 10^{-8} \qquad (5\text{–}8)$$

where v is the conductor velocity (dx/dt) perpendicular to the field. In Eq. 5–8, e is the induced voltage in volts, B is the uniform flux density in lines per square inch, l is the conductor length normal to the field in inches, and v is the conductor velocity normal to the field in inches per second.

5–6. EDDY CURRENTS

If the rotating cylinder considered in Art. 5–5 is made of a conducting material, a voltage is induced in the cylinder as well as in the coil. The

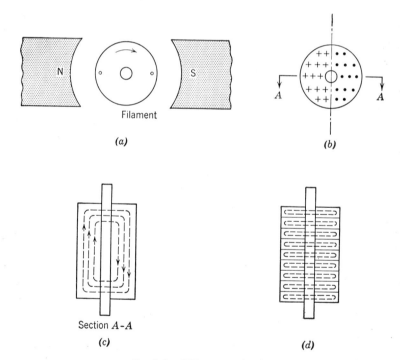

Fig. 5–8. Eddy-current details.

resulting current that flows in the cylinder is called an **eddy current.**

The manner in which the voltage is induced in the cylinder and the path of the resulting current may be determined by considering the cylinder to be made up of filaments stretching along its length, each filament having its counterpart in a similar position on the cylinder but diametrically opposite so as to form a coil. Two such filaments are shown in Fig. 5–8a.

As the cylinder rotates, the coil formed by the two filaments in Fig. 5–8a has a change in flux linkage; therefore, a voltage is induced, and, because the circuit is complete, a current flows. Since the cylinder may be considered to be composed of many such filamentary coils, the resulting current pattern is similar to that shown in Fig. 5–8b. The current crosses from one side of the cylinder to the other by means of the end connections of the filaments, as shown in Fig. 5–8c, which shows a top view of the cylinder.

One result of eddy currents in the cylinder is a loss in the form of heat, but whether the presence of eddy currents is desirable or undesirable depends on the application in which they occur. For the example shown, which represents an elementary motor or generator and in which the cylinder would be made of a magnetic material, the presence of eddy currents is undesirable because the resulting loss causes a decrease in efficiency as well as an increase in the temperature of the machine. In such applications, the eddy-current loss is usually made as small as practicable by laminating the cylinder instead of using a solid cylinder. The eddy-current loss is thus reduced because of the increased resistance of the eddy-current paths, as shown in Fig. 5–8d. The currents are restricted to the paths shown in the figure by the varnish layer that separates the laminations. Similar laminating is applied in a-c equipment where the magnetic material may be subjected to an alternating magnetic field, which would produce the same results as in the example discussed, where the field is fixed and the magnetic material is rotated.

Some applications in which eddy currents are desirable are those that rely on the braking action produced when the current due to an induced voltage reacts with the field that causes the induced voltage. Examples are eddy-current brakes, often made of a conductor material such as copper or brass, and the familiar aluminum disk in a watt-hour meter. The purpose of the disk, as discussed in more detail in Art. 10–12, is to provide a retarding force that is a function of rotational velocity.

Another application in which eddy currents are desirable is in the process known as induction heating, where eddy currents are produced

SELF-INDUCED VOLTAGES

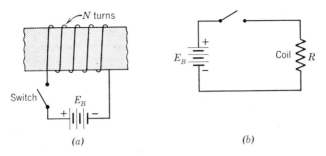

Fig. 5-9. Coil circuit and an electric equivalent.

in a conducting material by a rapidly alternating magnetic field for purposes of heating the material.

5-7. SELF-INDUCED VOLTAGES

Upon the opening or closing of the switch shown in the circuit of Fig. 5-9a, a voltage is induced in the coil because of the change in flux linkage produced by the change in current within the coil. This voltage is said to be self-induced. (A similar voltage is induced in coil 1 of Fig. 5-3 because of the changing flux linkage produced by the alternating current flowing in the coil.)

Before the switch in Fig. 5-9a is closed, the electric equivalent of the coil circuit is as shown in Fig. 5-9b. At this time, the circuit consists of the battery, the switch, and a resistor which represents the resistance of the coil.

At the instant the switch is closed, the current in the circuit is zero. The change in current from zero produces a change in flux linkage and causes a voltage to be induced in the circuit. The induced voltage tends to oppose the change in flux linkage, and the coil circuit may then be

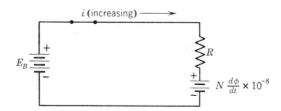

Fig. 5-10. Coil-circuit equivalent when current is increasing.

represented by the circuit shown as Fig. 5–10. As shown, the induced voltage is represented by a battery having an emf of $N\, d\phi/dt \times 10^{-8}$ volt and having a polarity to oppose the flow of current in the circuit. The voltage equation applicable to the circuit after the switch is closed may then be written

$$-E_B + iR + \left(N \frac{d\phi}{dt} \times 10^{-8}\right) = 0 \qquad (5\text{–}9)$$

where i is the instantaneous value of current.

At the instant the switch is closed, the induced voltage is exactly equal to the battery voltage, since the current and the resistance drop are zero. As the current increases in magnitude, the resistance drop increases, and, therefore, the induced voltage decreases. A decrease in induced voltage means that the rate of change of current is decreasing, since the rate of change of flux that determines the induced voltage depends upon the rate of change of current. The rate of change of current, and therefore the induced voltage, becomes zero when the current reaches the value determined by the voltage of the battery and the resistance of the circuit; that is, when the resistance drop is equal to the battery voltage.

The manner in which current varies with time after the switch is closed is dependent upon the magnetic circuit as well as the electric circuit, as indicated by the factors in Eq. 5–9. Of particular importance is the relationship between flux and current; this relationship, and its effect upon the increase in current, are discussed in detail in Chapter 6.

When the switch in the circuit of Fig. 5–10 is opened, the current and the flux linkage decrease to zero. As the current decreases, a voltage is induced in the circuit, such as to oppose the decrease in current.

The voltage induced immediately as the switch is opened is not easily determined and could be of any value. Thus, if the current is decreased to zero instantaneously, the induced voltage is infinite; if the switch is opened slowly so that an arc is formed, the rate of change of current, and therefore the induced voltage, is much less.

Since the induced voltage opposes the change in flux linkage, the voltage induced immediately as the switch is opened attempts to keep the current at the same value as before the switch is opened. Thus, if the coil is paralleled by a resistor, as shown in Fig. 5–11, the magnitude of the induced voltage immediately as the switch is opened is given by the relation

$$e = I(R + R_x) \qquad (5\text{–}10)$$

where I is the value of current in the coil circuit before the switch is opened.

SELF-INDUCED VOLTAGES

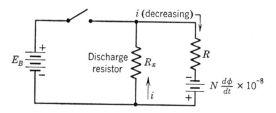

Fig. 5–11. Connection of a discharge resistor.

A resistor such as R_x, which is connected in a circuit to reduce the value of voltage induced when the circuit is opened, is called a **discharge resistor**.

After the switch is opened, the final value of current in the circuit is zero; therefore, the final value of induced voltage is zero. The manner in which the current decreases to zero is considered in Chapter 6.

PROBLEMS

5–1. Figure 5–12 shows the cross-sectional view of a coil mounted so that it can be rotated about a central axis. The drum on which the coil is

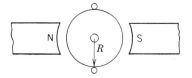

Fig. 5–12.

mounted is made of a magnetic material. The flux density is uniform in the air gap under the pole faces, and the field may be considered to be radial in the gap. The data applying to this device are as follows:

> Total flux per pole: 650,000 lines.
> Area of pole face: 16 sq in.
> Length of pole face along axis of rotation: 4 in.
> Speed of rotation: 1800 rpm.
> Distance R: 6 in.

(a) What is the maximum emf induced in each conductor during one complete revolution of each conductor? *Ans.* 1.84 volts.

(b) What is the average emf induced in a conductor during one-half a revolution of the conductor? *Ans.* 0.39 volt.

(c) Plot the approximate curve of emf versus time for one conductor for one revolution.

5–2. A coil of 100 turns is linked by a flux that varies as shown in Fig. 5–13. Plot the induced emf versus time for the coil, indicating magnitudes.

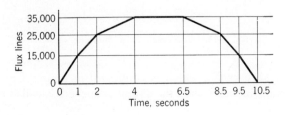

Fig. 5–13.

5–3. A coil of 100 turns is linked by a flux which varies according to the relation $\phi = 80{,}000 \sin 2\pi t$. Plot the induced emf versus time for the coil, showing the time relation of the flux and induced emf, and indicating magnitudes.

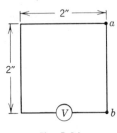

Fig. 5–14.

5–4. Assume that there is a uniform magnetic field perpendicular to, and with direction into, this page and that the field covers the entire page. The field has a density of 10,000 lines per sq in. In the field is a rigid one-turn coil, having sides 2 in. long and 2 in. apart. A voltmeter is connected to the ends of the coil, as shown in Fig. 5–14.

(a) Explain, completely, the reaction of the voltmeter if the coil as shown in Fig. 5–14 is moved 2 in. to the left in 0.01 sec.

(b) Explain, completely, the reaction of the voltmeter if the coil is turned 90° out of the page in 0.01 sec, with a–b as the axis of rotation.

5–5. The 100-turn coil in the circuit of Fig. 5–15 has a resistance of 10 ohms. The switch S is so designed that it makes contact at a when it is closed and immediately makes contact at b when it is opened.

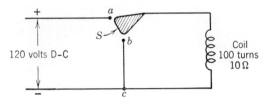

Fig. 5–15.

(a) What is the time rate of change of flux at the instant of closing the switch? *Ans.* 1.2×10^8 lines per sec.

(b) After the switch has been closed for a certain length of time, the current has a value of 5 amperes. What is the value of the induced voltage at this time? *Ans.* 70 volts.

(c) What is the value of the induced voltage immediately upon opening the switch? *Ans.* 120 volts.

(d) What is the value of the induced voltage when the current has decreased to 5 amperes after opening the switch? *Ans.* 50 volts.

(e) If the wire lead from b to c were replaced by a 10-ohm resistor, what would be the value of induced voltage immediately upon opening the switch? *Ans.* 240 volts.

(f) Repeat (e) for a 1000-ohm resistor. *Ans.* 12,120 volts.

(g) Repeat (f) for the value of the voltage at the coil terminals. *Ans.* 12,000 volts.

5–6. If placing your fingers between points a and b of the relay circuit shown in Fig. 5–16 is equivalent to the connection of a 10,000-ohm resistor between these points, calculate the approximate voltage to which your fingers are subjected when the relay contact opens. List all assumptions made in making the calculation. *Ans.* 600 volts.

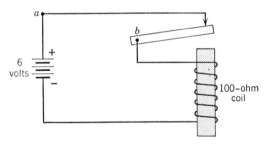

Fig. 5–16.

5–7. When the switch in the circuit of Fig. 5–17 is opened after being closed for a long time, the maximum value of voltage between terminals

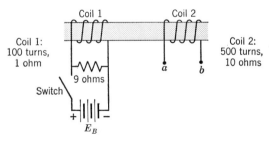

Fig. 5–17.

a and b is found to be 1000 volts. What is the battery voltage E_b? (Leakage flux may be considered to be negligible.) *Ans.* 20 volts.

5-8. Figure 5-18 shows a schematic diagram of a proposed ignition system. For purposes of this problem it is assumed that

(1) The magnetic circuit is complete; that is, there is no leakage flux.

(2) G is the gap of a spark plug which does not break down; that is, there is no arc across the gap.

(3) S is a switch that opens without arcing.

Is it possible to calculate the maximum voltage appearing across the gap G after the switch S is opened? If not, why not? If so, what is it?

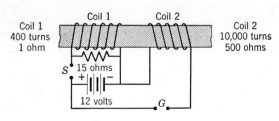

Fig. 5-18.

5-9. Show that an alternating magnetic field assumed to be uniformly distributed over and with direction perpendicular to the flat surface of a thin conducting disk will produce the largest eddy current at the periphery of the disk.

5-10. In the circuit of Fig. 5-19 switch S is closed after the current in the rest of the circuit has become steady.

(a) What is the voltage between points a and b immediately after closing S? *Ans.* 16.67 volts.

(b) What is the steady-state value of E_{ab} with S closed? *Ans.* 17.14 volts.

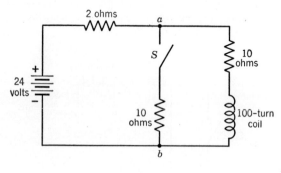

Fig. 5-19.

5-11. A W-lb steel cylinder is rolling down a $\theta°$ incline on two steel rails separated by L ft, as shown in Fig. 5-20. A magnetic field vertically downward between the rails has a uniform density of B lines per sq in. A resistor

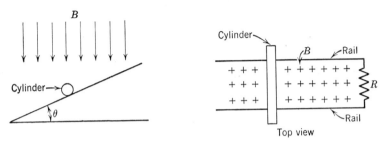

Fig. 5-20.

of R ohms is connected between the ends of the rails. Neglecting friction and the resistance of the cylinder and rails, calculate the steady-state velocity of the cylinder in inches per second.

$$\text{Ans.} \quad V = \frac{7.85WR \sin \theta}{B^2 L^2} \times 10^{12}.$$

■ 6　Inductance and Capacitance

6-1. INTRODUCTION

Electric circuits may have, singly or in combination, the properties of resistance, inductance, and capacitance. These properties are called the basic parameters of all circuits. The property of resistance has already been considered; the remaining parameters are considered in this chapter.

6-2. COEFFICIENT OF SELF-INDUCTANCE

In circuits having a self-induced voltage, the change in flux linkage necessary to cause the voltage to be induced is produced by a change in the current within the circuit. Thus, a proportionality exists between the self-induced voltage and the time rate of change of current. The coefficient relating the induced voltage and the time rate of change of current is called the **coefficient of self-inductance** and is symbolized by the letter L. Thus, in mathematical form, the relationship is expressed

$$e = L \frac{di}{dt} \qquad (6-1)$$

The factor L is described as a coefficient and not as a constant, although under some circuit conditions it may be constant. In abbreviated form, the coefficient of self-inductance is often called **self-inductance**, or simply **inductance**. (A device that introduces inductance into an electric circuit is called an **inductor**.)

PHYSICAL ASPECTS OF INDUCTANCE

The basic unit of inductance is the **henry**. A circuit has an inductance of 1 henry if 1 volt is induced in the circuit when the current is changing at the rate of 1 ampere per second. Therefore, in Eq. 6–1, when L is in henrys, i is in amperes, and t is in seconds, the induced voltage is in volts.

6–3. PHYSICAL ASPECTS OF INDUCTANCE

A self-induced voltage may be mathematically expressed either by

$$e = N \frac{d\phi}{dt} \times 10^{-8} \tag{6-2}$$

or by

$$e = L \frac{di}{dt} \tag{6-3}$$

Equating expressions 6–2 and 6–3 and solving for the inductance gives

$$L = N \frac{d\phi}{di} \times 10^{-8} \tag{6-4}$$

Equation 6–4 indicates that the inductance depends upon the ratio of the change in flux to the change in current; therefore, the inductance is constant only when the flux varies uniformly with current. This condition is met when the flux paths are composed entirely of non-magnetic materials; the condition is approximately satisfied when magnetic materials are involved if the magnetization is such as to be on the linear portion of the magnetization curve.

Equation 6–4 leads to an expression for inductance in terms of the physical characteristics of the circuit if a substitution is made for $d\phi$. Thus, since

$$\phi = \frac{\text{mmf}}{\mathcal{R}} = \frac{Ni\mu A}{l} \tag{6-5}$$

then

$$d\phi = \frac{N\mu A}{l} di + \frac{NiA}{l} d\mu \tag{6-6}$$

and

$$L = N^2 \left(\frac{\mu A}{l} + \frac{iA}{l} \frac{d\mu}{di} \right) \times 10^{-8} \tag{6-7}$$

Equation 6–7 verifies that the inductance of an electric circuit is constant only when the permeability of the magnetic circuit is constant. If the permeability is constant, Eq. 6–7 reduces to

$$L = \frac{N^2 \mu A}{l} \times 10^{-8} \tag{6-8}$$

6–4. EFFECTS OF SELF-INDUCTANCE

Since the property of self-inductance causes a voltage to be induced in a circuit only when the current is changing, the effects of inductance depend upon the nature of the circuit. In a-c circuits, the inductance is always effective because the current is continuously changing in magnitude and direction. (The effects of self-inductance in a-c circuits are considered in detail in Chapter 7.) In a d-c circuit, inductance is effective only when the circuit is being energized or de-energized, or when a circuit change results in a change in current. Thus, when the switch is closed in the d-c circuit shown as Fig. 6–1, the current in the circuit increases, and a voltage is induced that tends to oppose the increase in current.

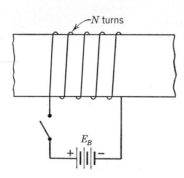

Fig. 6–1. Coil circuit.

The voltage relation applying to the circuit when the switch is closed may be written

$$-E_B + iR + L\frac{di}{dt} = 0 \tag{6-9}$$

A transposition of the terms in Eq. 6–9 gives the relation

$$\frac{di}{dt} = \frac{E_B - iR}{L} \tag{6-10}$$

According to Eq. 6–10, the inductance of a d-c circuit affects the time rate at which the current increases when the circuit is energized. Thus, if a circuit has negligible inductance, the current reaches its final, or Ohm's law, value almost immediately upon closing the switch; if the inductance is not negligible, some time elapses before the current reaches its final value. Although the time to reach the final value of current may be small, the delay may have considerable effect upon the operation of the circuit being energized.

The manner in which the current in the circuit of Fig. 6–1 builds up to its final value can be determined by solving Eq. 6–10 for the current

EFFECTS OF SELF-INDUCTANCE

in terms of time. For a circuit having a constant battery voltage, constant resistance, and constant inductance, the solution proceeds as follows:

(a) Rearranging Eq. 6–10 to separate the variables,

$$\frac{di}{(E_B/R) - i} = \frac{R\, dt}{L} \tag{6-11}$$

(b) Integrating Eq. 6–11,

$$-\ln\left(\frac{E_B}{R} - i\right) = \frac{Rt}{L} + C \tag{6-12}$$

where C is the constant of integration.

(c) At $t = 0$, $i = 0$; therefore,

$$C = -\ln\frac{E_B}{R} \tag{6-13}$$

and Eq. 6–12 becomes

$$-\ln\left(\frac{E_B}{R} - i\right) = \frac{Rt}{L} - \ln\frac{E_B}{R}$$

or

$$\ln\left(\frac{E_B}{R} - i\right) - \ln\frac{E_B}{R} = -\frac{Rt}{L} \tag{6-14}$$

which can be simplified to

$$\frac{(E_B/R) - i}{E_B/R} = \epsilon^{-Rt/L} \tag{6-15}$$

(d) Solving Eq. 6–15 for current,

$$i = \frac{E_B}{R}(1 - \epsilon^{-Rt/L}) \tag{6-16}$$

or, since the factor E_B/R is the final value of current in the circuit,

$$i = I(1 - \epsilon^{-Rt/L}) \tag{6-17}$$

where I represents the final value of current. A plot of this equation is shown as Fig. 6–2.

The curve plotted as Fig. 6–2 shows that the rate of change of current is maximum at the instant the switch is closed, and that it decreases as the current increases, as previously indicated. The maximum value of the rate of change of current can be determined by differentiating Eq. 6–17 with respect to time and evaluating the resulting expression for $t = 0$. Thus,

$$\frac{di}{dt} = \frac{IR}{L}\epsilon^{-Rt/L} \tag{6-18}$$

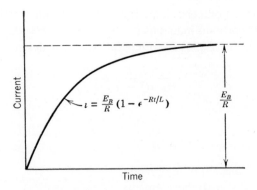

Fig. 6–2. Growth of current with time in an inductive circuit.

and, at $t = 0$,

$$\frac{di}{dt} = \frac{IR}{L} = \frac{E_B}{L} \tag{6-19}$$

Equation 6–19 indicates that the rate at which current starts to increase in the circuit depends only upon the inductance of the circuit and the applied voltage; the resistance of the circuit is not a factor. However, the resistance is a factor in determining the **time constant** of the circuit.

For an electric circuit containing constant resistance and inductance (in addition to the source), the time constant can be defined as the time required for the current to reach its final value if it continues to change at its initial rate. Thus, if a straight line having a slope corresponding to the initial rate of current increase is drawn through the origin of the current–time curve, the straight line intersects the final value of current at a time corresponding to the time constant, as shown in Fig. 6–3.

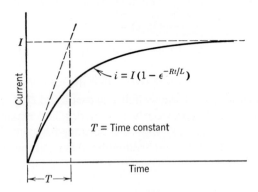

Fig. 6–3. Graphical determination of time constant.

EFFECTS OF SELF-INDUCTANCE

In terms of current and time, the initial rate of current increase is given by the relation

$$\frac{di}{dt} = \frac{I}{T} \tag{6-20}$$

where T is the time constant. Equation 6–19 shows the same initial rate to be

$$\frac{di}{dt} = \frac{IR}{L} \tag{6-21}$$

Combining Eqs. 6–20 and 6–21 and solving for T yields

$$T = \frac{L}{R} \tag{6-22}$$

Thus, the time constant depends only upon the inductance and the resistance of the circuit.

The actual value of current in the circuit after an elapsed time equal to the time constant can be found by substituting the value of the time constant for t in Eq. 6–17. Thus,

$$i = I(1 - \epsilon^{-Rt/L}) = I(1 - \epsilon^{-(R/L)(L/R)})$$
$$= I(1 - \epsilon^{-1}) = I\left(1 - \frac{1}{2.718}\right)$$
$$= 0.632I \tag{6-23}$$

The fact that the current is 63.2 per cent of its final value after an elapsed time of one time constant is the basis for another definition of time constant. Thus, the time constant of a circuit containing constant resistance and inductance is the time required for the current to change

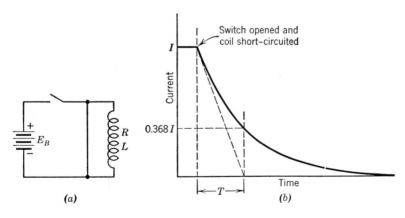

Fig. 6–4. Decay of current in an inductive circuit.

by an amount equal to 63.2 per cent of the total change that will occur.

Although in the preceding paragraphs the value of the time constant is determined by a consideration of the circuit when it is energized, the value and the definitions of the time constant are equally applicable to the circuit when it is de-energized. If the coil is short-circuited at the instant the switch is opened, as shown in Fig. 6–4a, the voltage relation for the circuit is

$$iR + L\frac{di}{dt} = 0 \tag{6-24}$$

The equation for current is found by solving Eq. 6–24; thus,

$$i = I\epsilon^{-Rt/L} \tag{6-25}$$

which is plotted as Fig. 6–4b. From Eq. 6–25, the value of current is 36.8 per cent of the initial value after an elapsed time equal to the time constant (thus, the change in current is 63.2 per cent of the total change that will occur).

According to Eqs. 6–17 and 6–25, the time to reach the final value of current in a circuit is mathematically infinite. For all practical purposes, however, the current approximates its final value in a very short time. For example, the current changes approximately 98.2 per cent of its total change in a time equal to four time constants. Thus, for a circuit having a time constant of 1 second, the time for the current to change 98.2 per cent of its total change is 4 seconds.

The preceding discussion of the time constant of a circuit is based upon a circuit of constant inductance. If the magnetic circuit associated with the electric circuit consists, in part, of a magnetic material, the inductance may or may not be constant. If the inductance is not constant, the current does not have the same time variation as the circuit having constant inductance; therefore, the time constant herein defined

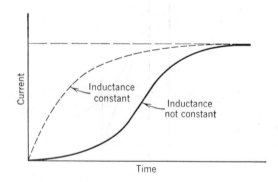

Fig. 6–5. Growth of current with time in inductive circuits.

does not apply. There is nothing comparable in meaning to time constant for such circuits. An example of a current–time curve for a circuit having nonconstant inductance is shown as Fig. 6–5; for comparative purposes, the curve previously developed for a constant-inductance circuit is also shown.

6–5. A MECHANICAL ANALOGY

The fact that inductance delays the build-up of current in a d-c circuit is the basis for an analogy comparing inductance in an electric circuit to mass in a mechanical system.

The force applied to a mechanical system may be considered to have two components; one accelerates the system, the other overcomes the friction force developed because of the velocity of the system. (Resistance in the electric circuit is analogous to friction in the mechanical system.) When the force is initially applied to the system, the friction component of the force is zero, and the entire force accelerates the system. As the system velocity increases, however, the friction force increases, and less force is available for acceleration. When the friction force equals the applied force, the acceleration ceases, and the system reaches a constant velocity. (The action described is analogous to the electric circuit: the applied force corresponds to the applied voltage; the friction force, to the resistance drop; the accelerating force, to the induced voltage; and the velocity, to the current.) If the mass of a mechanical system is large, the application of a force results in a low initial acceleration, and a relatively long time elapses before the system reaches a constant velocity. Similarly, a large value of inductance in an electric circuit causes a low rate of change of current, and a relatively long time elapses before the current becomes constant.

The magnitude of the voltage induced when the electric circuit is opened depends upon the inductance of the circuit and the rate at which the current is decreased. If the inductance is large and the current is decreased rapidly, the induced voltage may be large enough to break down the insulation of the circuit. In comparing the electric and mechanical systems, this induced voltage is analogous to the force developed in a mechanical system when a large moving mass is stopped almost immediately.

The analogy between mechanical systems in motion and electric systems can be extended to apply the principle of conservation of momentum in mechanical systems to systems. Thus, conservation of Mv in mechanical systems has as an analog the conservation of LI in electric systems. Conservation of LI can, in turn, be expressed as conservation of $N\phi$, or conservation of flux linkages. The fact that flux linkages must be instantaneously conserved is contained in some of the previous discussions of induced voltage but is not specifically mentioned in this terminology. In many problems, a complete analysis can only be made by application of this principle.

In Art. 5-4 the discussion of the polarity of the voltage induced places great stress on the tendency of the induced voltage to oppose a change in flux linkage. Thus, because the turns of the coil in which the voltage is induced are fixed, the tendency is to maintain the flux at the value it had before the change started to take place. Therefore, the initial reaction is to maintain the flux linkage or to conserve flux linkage. Similar reactions occur in circuits in which self-induced voltages are produced, as discussed in Art. 5-6 and Art. 6-4.

6-6. ENERGY RELATIONS

The large force developed in suddenly stopping a moving mass is due to the energy stored in the mass while accelerating the mass to its final constant velocity. Similarly, the voltage induced upon de-energizing an electric circuit is due to the energy stored in the magnetic field while the current is building up to its final value.

The power relation that applies to the circuit of Fig. 6-1 after the switch is closed may be written

$$- E_B i + i^2 R + Li \frac{di}{dt} = 0 \tag{6-26}$$

and is obtained by multiplying the voltage equation by the instantaneous current. In the power relation, $E_B i$ is the power delivered by the battery, $i^2 R$ is the power dissipated as heat in the resistance, and $Li(di/dt)$ is the power stored in the magnetic field. The incremental energy storage in the magnetic field is given by the relation

$$dW = Li \left(\frac{di}{dt} \right) dt \tag{6-27}$$

An integration of the incremental energies gives the total energy stored in the magnetic field. Thus, for a circuit having constant inductance,

$$W = \tfrac{1}{2} LI^2 \tag{6-28}$$

where I is the value of current in the circuit.

The energy stored in the magnetic field of a d-c circuit when the circuit is energized, or when the current in the circuit increases, is re-

MUTUAL INDUCTANCE   109

leased to the circuit only when the circuit is de-energized, or when the current in the circuit decreases. In an a-c circuit, since the current is continuously changing in magnitude and direction, energy is alternately stored in, and returned by, the magnetic field.

6–7. MUTUAL INDUCTANCE

In Chapter 5, a circuit similar to that shown in Fig. 6–6 is considered with regard to the polarity of the voltage induced in coil 2 by the change in flux created by coil 1. The magnitude of the voltage induced in coil 2 is given by the relation

$$e_2 = N_2 \frac{d\phi_{12}}{dt} \times 10^{-8} \tag{6-29}$$

where e_2 is the voltage induced in coil 2, N_2 is the number of turns of coil 2, and ϕ_{12} is the flux created by coil 1 that links coil 2. The flux produced by coil 1 that links coil 2 is called the **mutual flux.**

Since the change in mutual flux is due to a change in the current in coil 1, a proportionality exists between the voltage induced in coil 2 and the time rate of change in current in coil 1. The voltage induced in coil 2 may therefore be expressed by the relation

$$e_2 = L_{M21} \frac{di_1}{dt} \tag{6-30}$$

where L_{M21} is the **mutual inductance** of coil 2 with respect to coil 1, and i_1 is the current in coil 1.

Combining Eqs. 6–29 and 6–30 and solving for L_{M21} gives

$$L_{M21} = N_2 \frac{d\phi_{12}}{di_1} \times 10^{-8} \tag{6-31}$$

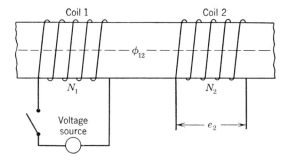

Fig. 6–6. Coils having mutual inductance.

Similarly, the mutual inductance of coil 1 with respect to coil 2 is found to be

$$L_{M12} = N_1 \frac{d\phi_{21}}{di_2} \times 10^{-8} \qquad (6\text{–}32)$$

where ϕ_{21} is that portion of the flux created by coil 2 that links coil 1, and i_2 is the current in coil 2.

6–8. EFFECTS OF MUTUAL INDUCTANCE

Figure 6–7 shows two coils connected in series to a voltage source. The coils are so connected that the mmf's are additive; therefore, the mutual flux is a function of the arithmetic sum of the mmf's.

When the current in the circuit of Fig. 6–7 changes, the voltage induced in a coil depends not only upon the change in flux linkage caused by the change in current within the coil but also upon the change in flux linkage caused by the change in current in the other coil. Thus, since the changes in flux linkage are in the same direction, the voltage induced in coil 1 is given by the relation

$$e_1 = L_1 \frac{di}{dt} + L_{M12} \frac{di}{dt} \qquad (6\text{–}33)$$

where L_1 is the self-inductance of coil 1, and L_{M12} is the mutual inductance of coil 1 with respect to coil 2. The current is common to both coils. Similarly, the voltage induced in coil 2 is given by the relation

$$e_2 = L_2 \frac{di}{dt} + L_{M21} \frac{di}{dt} \qquad (6\text{–}34)$$

The total induced voltage in the circuit is the sum of the voltages

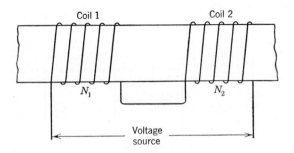

Fig. 6–7. Coils having additive mmf's.

EFFECTS OF MUTUAL INDUCTANCE

induced in the individual coils and is given by the expression

$$e_T = e_1 + e_2 = L_1 \frac{di}{dt} + L_{M12}\frac{di}{dt} + L_2 \frac{di}{dt} + L_{M21}\frac{di}{dt} \quad (6\text{-}35)$$

If the total induced voltage is also expressed by

$$e_T = L_T \frac{di}{dt} \quad (6\text{-}36)$$

then the total inductance of the two coils connected series-additive is

$$L_T = L_1 + L_2 + L_{M12} + L_{M21} \quad (6\text{-}37)$$

It can be shown that L_{M12} is equal to L_{M21} if the permeability of the magnetic circuit is constant. Under these conditions, Eq. 6–37 reduces to

$$L_T = L_1 + L_2 + 2L_M \quad (6\text{-}38)$$

where L_M is the mutual inductance and is the same for both coils.

If the two coils of Fig. 6–7 are connected series-subtractive, so that the resultant mutual flux is a function of the difference in coil mmf's, the induced voltage in coil 1 is expressed by the relation

$$e_1 = L_1 \frac{di}{dt} - L_{M12}\frac{di}{dt} \quad (6\text{-}39)$$

since the change in flux linkage due to the change in current within coil 1 is opposite to the change in flux linkage caused by the change in current in coil 2. Similarly, the voltage induced in coil 2 is

$$e_2 = L_2 \frac{di}{dt} - L_{M21}\frac{di}{dt} \quad (6\text{-}40)$$

The total induced voltage is therefore

$$e_T = L_1 \frac{di}{dt} - L_{M12}\frac{di}{dt} + L_2 \frac{di}{dt} - L_{M21}\frac{di}{dt} \quad (6\text{-}41)$$

and the total inductance is

$$L_T = L_1 + L_2 - L_{M12} - L_{M21} \quad (6\text{-}42)$$

For a magnetic circuit of constant permeability, Eq. 6–42 reduces to

$$L_T = L_1 + L_2 - 2L_M \quad (6\text{-}43)$$

Equations 6–37 and 6–42 indicate that the total inductance of two coils connected in series depends not only upon the self- and mutual inductances of the individual coils but also upon the manner in which the coils are connected. Similar conclusions apply to the parallel connection of the coils; however, the mathematical expressions for the parallel connection are different from those developed for the series connection.

Like self-inductance, mutual inductance is of greater importance in a-c circuits than in d-c circuits. The effects of mutual inductance in a-c circuits are considered in more detail later.

6–9. CAPACITANCE

Capacitance is defined as "that property of a system of conductors and dielectrics which permits the storage of electricity when potential differences exist between the conductors. Its value is expressed as the ratio of a quantity of electricity to a potential difference." *

A **dielectric** is an insulating material that has the ability to store electric energy when a potential difference exists across it. Examples of dielectrics employed in **capacitors** (devices that introduce capacitance into an electric circuit) are air (and vacuum), glass, mica, mineral oil, paper, paraffin, polystyrene, and porcelain. The construction of a capacitor is shown schematically in Fig. 6–8. As indicated, the dielectric is inserted between two plates, which serve as the terminals of the capacitor.

According to the last statement in the definition, capacitance is the ratio of a quantity of electricity to a potential difference. Thus,

$$C = \frac{Q}{E} \tag{6-44}$$

where C represents the capacitance and Q the quantity of electricity.

The practical unit of quantity is the **coulomb,** which represents the quantity of electricity that passes any section of an electric circuit in 1 second when the current is 1 ampere. Thus, in a circuit in which the current remains constant, the quantity of electricity that flows into the

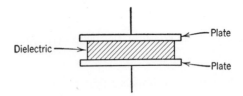

Fig. 6–8. Capacitor construction.

* American Standard Definitions of Electrical Terms.

EFFECTS OF CAPACITANCE

circuit in a given time is expressed by

$$Q = It \qquad (6\text{--}45)$$

where I is the constant value of current, and t is the time. When the current is in amperes and the time is in seconds, the quantity is in coulombs.

In a circuit having a varying current, the total quantity of electricity flowing into the circuit is the summation of the incremental quantities. Thus,

$$Q = \int_0^t i\,dt \qquad (6\text{--}46)$$

Equation 6–44 may then be written

$$C = \frac{\int_0^t i\,dt}{e} \qquad (6\text{--}47)$$

where e is the voltage between the terminals of the capacitor at time t.

When the quantity is in coulombs and the potential difference is in volts, the unit of capacitance is the **farad**. However, the farad is not a practical unit because of the physical size of the capacitor it represents and, therefore, the microfarad, one millionth of a farad, is more commonly used.

6–10. EFFECTS OF CAPACITANCE

Differentiating Eq. 6–47 and rearranging the terms gives

$$i = C\frac{de}{dt} \qquad (6\text{--}48)$$

As indicated by Eq. 6–48, current flows in a circuit involving capacitance only when there is a change in the voltage of the circuit. Thus, current always flows in a-c circuits that contain capacitance; current flow in d-c circuits involving capacitance, however, depends upon a transient in the circuit (such as the energizing or de-energizing of the circuit or a change in the voltage distribution in the circuit).

Actually, no current flows through a capacitor at any time if it is a perfect capacitor. (A perfect capacitor is one that is a perfect insulator.) The current flow in the remainder of the circuit is due to the transfer of electrons in the circuit as one plate of the capacitor

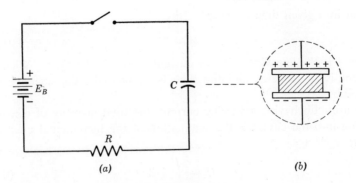

Fig. 6–9. Capacitor circuit.

becomes positively charged and the other becomes negatively charged. Thus, when the switch is closed in the d-c circuit shown in Fig. 6–9a, the electrons on the plate connected to the positive terminal of the battery are attracted to this positive terminal; the removal of electrons from the plate causes it to become positively charged. Similarly, electrons are "piled up" on the plate connected to the negative terminal of the battery. This plate then becomes negatively charged, and a potential difference exists between the plates, as indicated by the schematic diagram shown as Fig. 6–9.

The movement of electrons from and to the plates constitutes a current, which continues until the difference in potential between the plates becomes equal to the supply voltage. Thus, in a d-c circuit containing capacitance, current flows until the capacitor is fully charged, that is, until the quantity of electricity stored is sufficient to cause the potential between the plates of the capacitor to equal the voltage applied.

The manner in which the capacitor in Fig. 6–9a becomes charged can be determined by an analysis of the voltage relation applying to the circuit after the switch is closed. Thus,

$$E_B = e_R + e_C$$

$$= iR + \frac{q}{C} = iR + \frac{\int i\, dt}{C} \qquad (6\text{–}49)$$

where q represents the quantity of electricity at any time t. The solution of Eq. 6–49 gives

$$i = \frac{E_B}{R} \epsilon^{-t/RC} \qquad (6\text{–}50)$$

Therefore, the voltage between the terminals of the capacitor at any

EFFECTS OF CAPACITANCE

time t is

$$e_C = E_B - e_R = E_B - iR$$
$$= E_B - \left(\frac{E_B}{R}\epsilon^{-t/RC}\right)(R)$$
$$= E_B(1 - \epsilon^{-t/RC}) \qquad (6\text{-}51)$$

Equations 6–50 and 6–51 are plotted as Figs. 6–10a and 6–10b, respectively.

Figure 6–10a shows that the current in the circuit is maximum at the instant the supply switch is closed (it is assumed that the inductance of the circuit is negligible) and that the current decreases as time elapses and the capacitor becomes fully charged. Meanwhile, the voltage across the capacitor increases until it reaches an ultimate value equal to the voltage of the battery, as shown in Fig. 6–10b.

If the switch in the circuit of Fig. 6–9a is opened after the capacitor is charged, the capacitor retains its charge until its terminals are connected to a circuit that will dissipate the energy stored. When the capacitor is so discharged, the current flows in a direction opposite to its original direction and continues to flow until the voltage across the capacitor is zero. The manner in which the voltage of a charged capacitor decreases upon discharge can be determined by an analysis of the voltage relation applying upon discharge. Thus, if the fully charged capacitor is connected across the resistor of Fig. 6–9a when the switch is opened,

$$iR + \frac{q}{C} = iR + \frac{\int i\,dt}{C} = 0 \qquad (6\text{-}52)$$

The solution of Eq. 6–52 is

$$i = -\frac{E_B}{R}\epsilon^{-t/RC} \qquad (6\text{-}53)$$

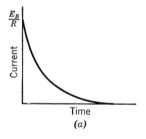
Time
(a)

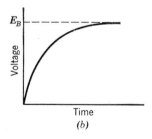
Time
(b)

Fig. 6–10. Decay of current and growth of capacitor voltage when capacitor is being charged.

Therefore, the voltage across the capacitor at any time t during discharge is

$$e_C = E_B \epsilon^{-t/RC} \qquad (6\text{-}54)$$

Equation 6–53 shows that the current upon discharge is in the opposite direction from the original current and that, except for the negative values, the form of the current–time curve is identical with that for the current–time curve during charging of the capacitor. The form of the voltage–time curve is indicated in Fig. 6–11.

Since the curves for the build-up and the decay of the capacitor voltage are identical in form with the curves of build-up and decay of current in the resistance–inductance circuit, the definitions of the time constant for current in the resistance–inductance circuit are applicable to capacitor voltage in the resistance–capacitance circuit. Thus, for a resistance–capacitance circuit, the time constant is the time required for the capacitor voltage to reach its final value if it continues to change at its initial rate; or the time constant is the time required for the capacitor voltage to change by an amount equal to 63.2 per cent of the total change that will occur. Application of these definitions to the resistance–capacitance equations shows that the time constant is given by the relation

$$T = RC \qquad (6\text{-}55)$$

Just as the energy stored by inductance may be considered to be stored in the magnetic field that is always associated with a current-carrying inductor, a capacitance has energy stored in an electric field that is associated with any charged capacitor. In charging a capacitor C to a voltage E, the power at any time is

$$p = ei = eC \frac{de}{dt} \qquad (6\text{-}56)$$

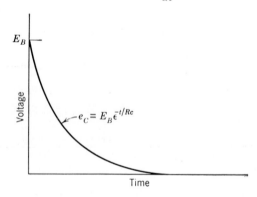

Fig. 6–11. Decay of capacitor voltage upon discharge.

CONNECTION OF CAPACITORS

and the total energy stored is the integral of power over time. Thus, from Eq. 6–56,

$$p \, dt = Ce \, de \tag{6-57}$$

and the total energy is

$$W = \int_0^T p \, dt = C \int_0^E e \, de = \tfrac{1}{2} C E^2 \tag{6-58}$$

This result is similar to Eq. 6–28, the corresponding expression for an inductance. In the case of the inductance, a magnetic field is produced by a current (proportional to the current for nonmagnetic materials), and the energy stored in this field is proportional to the square of that current; for the capacitor, an electric field is produced by the voltage between the terminals of a capacitor, and the energy stored in this field is proportional to the square of that voltage. A discussion of electric fields is found in Art. 20–10.

When connected in an a-c circuit, a capacitor is alternately charged and discharged as the voltage supply varies in magnitude and polarity; therefore, an alternating current flows in the remainder of the circuit. The effects of capacitance in a-c circuits are considered in more detail later.

6–11. CONNECTION OF CAPACITORS

The capacitance of a capacitor is a function of the cross-sectional area, thickness, and material of the dielectric. For a capacitor having parallel plates and a cross-sectional area that is large compared to the thickness of the dielectric, these factors may be related in an expression of the form

$$C = K'K \frac{A}{d} \tag{6-59}$$

where K is the dielectric constant (a factor depending on the dielectric material), A is the cross-sectional area of the dielectric, d is the thickness of the dielectric, and K' is a constant relating the units of capacitance, area, and thickness. The dimensions of the capacitor are shown schematically in Fig. 6–12.

The equivalent capacitance of n capacitors connected in parallel is given by the relation

$$C_T = C_1 + C_2 + C_3 + \cdots + C_n \tag{6-60}$$

The equivalent capacitance of n capacitors connected in series is given by

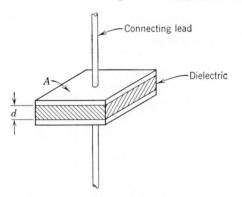

Fig. 6–12. Capacitor details.

the relation

$$\frac{1}{C_T} = \frac{1}{C_1} + \frac{1}{C_2} + \frac{1}{C_3} + \cdots + \frac{1}{C_n} \qquad (6\text{–}61)$$

Equations 6–60 and 6–61 may be explained on the basis of Eq. 6–59. Thus, if a number of capacitors are connected in parallel, the equivalent area of the dielectric is increased, compared to the connection of one of the capacitors; therefore, the equivalent capacitance is increased, as indicated by Eq. 6–60. Similarly, if a number of capacitors are connected in series, the equivalent thickness of the dielectric is increased, compared to the connection of one of the capacitors; therefore, the equivalent capacitance is decreased, as indicated by Eq. 6–61.

PROBLEMS

6–1. Calculate the rate of change of current for parts (*a*), (*b*), and (*d*) of Prob. 5–5, assuming that the coil has an inductance of 0.2 henry.
  Ans. (*a*) 600 amperes per sec. (*b*) 350 amperes per sec.
  (*d*) 250 amperes per sec.

6–2. Under what conditions can the coefficient of self-inductance be expressed by the relation $(N\phi/I) \times 10^{-8}$?

6–3. The flux linked with a certain circuit is proportional to the current and has a value in lines 1000 times the value of the current in amperes. The circuit has 100 turns.
  (*a*) What is the self-inductance of the circuit?   *Ans.* 0.001 henry.
  (*b*) Calculate, by two methods, the emf of self-induction when the current is changing at the rate of 50,000 amperes per sec.   *Ans.* 50 volts.

PROBLEMS

6-4. Figure 6-13 shows a schematic diagram of the ignition system of an automobile. For purposes of this problem, it is assumed that:

(1) The magnetic circuit is complete; that is, there is no leakage flux.

(2) G is the gap of a spark plug that doesn't break down; that is, there is no arc across the gap.

(3) S is a switch (corresponds to breaker points). If, upon the opening of switch S, the current falls to zero in 0.0005 sec, what is the average value of voltage appearing across the terminals of the gap G during that time?

Ans. 1194 volts.

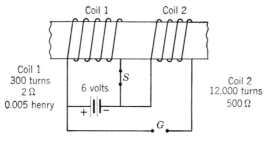

Fig. 6-13.

6-5. Assuming that the current in coil 2 of Fig. 6-14 can be interrupted instantaneously by the opening of the switch, plot current in coil 1 versus time for the period from before the switch is opened until steady state is reached, indicating values where possible. Briefly explain the plot.

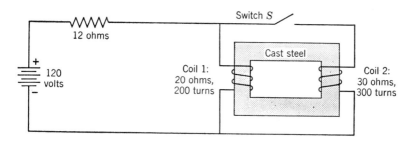

Fig. 6-14.

6-6. In the circuit of Fig. 6-15, coils 1 and 2 are wound on a common core assumed to have constant permeability. The operation of the circuit is such that (1) switch 1 is closed, and (2) after steady state is reached, switch 2 is opened. Plot current versus time from the initial closing of switch 1 until steady state is reached after the opening of switch 2. Indicate values where values are possible and explain the plot. (It may be assumed that leakage flux is negligible and that switch 2 opens without arcing.)

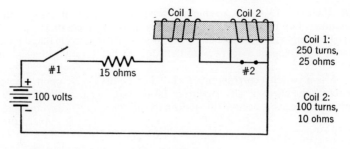

Fig. 6–15.

6–7. The plot of current versus time after the coil of the lifting device shown in Fig. 4–25 is energized is approximately as shown in Fig. 6–16. Explain the shape of the plot.

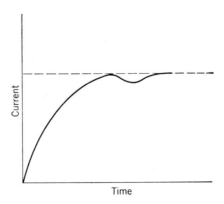

Fig. 6–16.

6–8. For the circuit of Fig. 6–17, what is the current in the circuit immediately upon closing the switch? What is the current when steady state is reached? *Ans.* 3.0 amperes; 2.4 amperes.

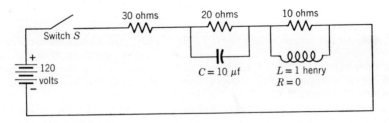

Fig. 6–17.

6-9. In the circuit shown in Fig. 6-18, all circuit elements are pure elements, and no current flows in any part of the circuit before the switch S is closed.

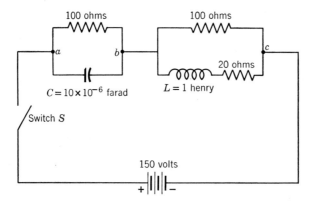

Fig. 6-18.

(a) What are the voltages e_{ab} and e_{bc} at the instant the switch is closed?
Ans. 0 volts; 150 volts.
(b) What are the voltages e_{ab} and e_{bc} after steady state is reached?
Ans. 128.5 volts; 21.5 volts.
(c) After steady state is reached, what are the voltages e_{ab} and e_{bc} immediately upon opening the switch? *Ans.* 128.5 volts; 107.5 volts.
(d) Which voltage found in (c) decays faster and why?

■ 7 Alternating Currents

7-1. INTRODUCTION

An alternating current is defined as a current that has regular changes in magnitude and direction and that has an average value of zero over a period. Figures 7-1a and 7-1b illustrate two types of alternating current.

The curves of Fig. 7-1 represent the time variation in magnitude and direction of a current. Thus, the vertical distance above or below the time axis represents the magnitude of the current. A positive (above-axis) value indicates that the current is in the direction originally assumed to be positive; a negative (below-axis) value indicates that the current is in the direction opposite to that originally assumed to be positive.

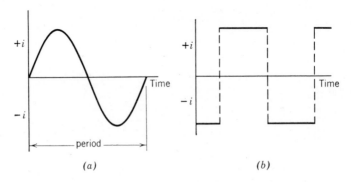

Fig. 7-1. Examples of alternating current.

DOUBLE-SUBSCRIPT NOTATION

The curves of Fig. 7–1 may represent the time variation in magnitude and polarity of an alternating voltage. The magnitude of the voltage is given by the vertical distance above or below the axis. A positive value of voltage indicates that the relative polarity of a given terminal with respect to another terminal is positive; a negative value, that the polarity of the given terminal is negative with respect to the other.

7–2. DOUBLE-SUBSCRIPT NOTATION

A system of double-subscript notation is helpful in specifying current directions and voltage polarities in a given a-c circuit. The application of double-subscript notation to currents, voltages, and powers is discussed below.

Voltage Notation. The plot of voltage versus time for the a-c circuit of Fig. 7–2a is shown as Fig. 7–2b.

The double-subscript notation applied to the voltage plot of Fig. 7–2b serves two purposes: first, the subscripts xy and ab indicate the terminals between which the voltage is measured; second, and equally important, the subscripts indicate the relative polarities of these terminals by the sequence in which the subscripts are written. Thus, in this book, the voltage e_{ab} is one measured between the terminals a and b with terminal a being positive with respect to b when the plot of voltage versus time is positive.

One curve of voltage versus time in Fig. 7–2b is labeled as both e_{xy} and e_{ab}. That this plot can have two designations is apparent when it is realized that there is only one voltage between any two points in a circuit, and that points x and y are, respectively, the same as points a and b, if the connecting leads have negligible resistance and induct-

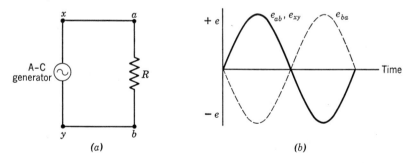

Fig. 7–2. Voltage relations in an a-c circuit.

ance. Therefore,
$$e_{ab} = e_{xy} \tag{7-1}$$

A plot of e_{ba} is also included in Fig. 7–2b. Inspection of the figure reveals that
$$e_{ab} = -e_{ba} \tag{7-2}$$
which may also be shown by applying double-subscript notation in the voltage equation for the circuit. Thus,
$$e_{xy} + e_{ba} = 0 \tag{7-3}$$
Since e_{xy} is equal to e_{ab} (Eq. 7–1), Eq. 7–3 may be written
$$e_{ab} + e_{ba} = 0 \tag{7-4}$$
or
$$e_{ab} = -e_{ba} \tag{7-5}$$

By the application of double-subscript notation, voltage equations may be as correctly and as simply written for a-c circuits as for d-c circuits, with the voltages arranged in order of their encounter around a closed traverse and with each pair of subscripts taken as they appear. In such voltage equations, there is no necessity for minus signs.

Double subscripts could have a different meaning in regard to polarity than that given above. As for other conventions in electrical terminology, different methods are used to indicate the same thing. These methods may be entirely correct, and the same conclusions may be reached, if their method of application is consistent.

Current Notation. In Fig. 7–2, the voltage e_{ab} represents the voltage applied to the given resistor. Current flows from a to b when a is positive with respect to b and is reversed when the polarity is reversed. The plot of current versus time is sinusoidal in shape, since the voltage is given as sinusoidal. Whether the combined plot of voltage and current versus time is shown as Fig. 7–3a or 7–3b depends upon the convention to be followed. In this book, the current i_{ab} is the current between

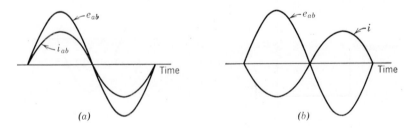

Fig. 7–3. Voltage and current versus time for a-c circuit of Fig. 7–2.

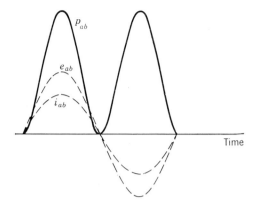

Fig. 7-4. Power versus time for resistor circuit.

points a and b and is in the direction from a to b when the plot of current versus time is positive. Thus, the plot shown as Fig. 7-3a agrees with the convention.

It is important to note that, although the current is flowing from a to b through the resistor, it flows from y to x through the generator. Therefore, the current plotted as i_{ab} in Fig. 7-3a could also be plotted as i_{yx}.

If the circuit had only two terminals identified, for example a and b, confusion could be created by labeling a current i_{ab} unless the current were further identified as to the element through which it was considered to flow. To safeguard against misinterpretation, the circuit should be labeled with as many subscripts as necessary to make the current designation clear.

Power Notation. The power at any instant in the resistor of Fig. 7-2a is the product of the voltage and current at that instant. The average power in the resistor is given by the expression

$$P_{\text{avg}} = \frac{1}{T} \int_0^T ei \, dt \qquad (7\text{-}6)$$

If, by applying the voltage and current curves shown in Fig. 7-3a, a plot is made of power versus time, the plot appears as shown in Fig. 7-4. (This is the plot, therefore, of $e_{ab}i_{ab}$.) The plot is all positive (above axis); therefore, the average power is positive.

Since energy is being dissipated in the resistor and since the average power is positive on the diagram, a convention may be established to identify the action of any circuit component under consideration: that is, whether the component is a source or a receiver of electric energy.

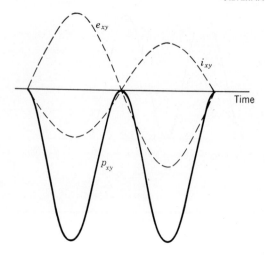

Fig. 7-5. Power versus time for an a-c generator supplying a resistor load.

Thus, in this book, P_{ab} is the power in the circuit between points a and b and is given by the product of e_{ab} and i_{ab}. If the average of this product over a complete period is positive, the component under consideration acts as a receiver of electric energy; if negative, the component acts as a source of electric energy.

The determination of the average power delivered by the a-c source of Fig. 7-2 serves as a check on the power convention just established. The instantaneous power of the source is given by

$$p_{xy} = e_{xy} i_{xy} \qquad (7\text{-}7)$$

It has already been shown that

$$e_{xy} = e_{ab} \qquad (7\text{-}8)$$

and that

$$i_{xy} = i_{ba} \qquad (7\text{-}9)$$

Therefore, since

$$i_{ba} = -i_{ab} \qquad (7\text{-}10)$$

the product of e_{xy} and i_{xy} is negative, as shown in Fig. 7-5. The average of this product is negative; thus, the known source has been identified as a source by the convention established for power notation.

7-3. DEFINITIONS

WAVE FORM. The wave form of an alternating quantity is the shape of the plot of the time variation of the quantity. The curve shown in

DEFINITIONS

Fig. 7–1a is called a sine wave; that shown in Fig. 7–1b is a square wave, or a rectangular wave. The wave forms of currents and voltages in a-c circuits are not limited to these two; these are merely included as examples. The wave forms most frequently encountered in practice more nearly approach the form of the sine wave, but they may be vastly different and still satisfy the definition of an alternating current.

Except where otherwise indicated, sine waves of voltage and current are assumed in all discussions of a-c circuits in this book. The definitions that follow, however, apply to all alternating currents or voltages. Figure 7–6 shows a sine wave labeled with some of the terms encountered in the discussion of alternating currents or voltages.

AMPLITUDE. The amplitude of an alternating quantity is the maximum value of the quantity.

CYCLE. The complete series of values of an alternating quantity is called a cycle.

ALTERNATION. One half of a cycle is called an alternation when it includes all positive or all negative values.

TIME PERIOD. The time period (symbolized by T) of an alternating quantity is the time necessary to complete one cycle.

FREQUENCY. The frequency (symbolized by f) of an alternating quantity is the number of cycles per unit time; thus, the frequency of an alternating current is the reciprocal of the time period, expressed in terms of cycles per second. The most common power frequency in the United States is 60 cycles per second, although frequencies other than this (usually 25 cycles per second) are sometimes encountered. When frequency is discussed, the "per second" is generally omitted. Thus, frequencies may be indicated as 60 cycles, 870 kilocycles (870,000 cycles), or 97.3 megacycles (97,300,000 cycles).

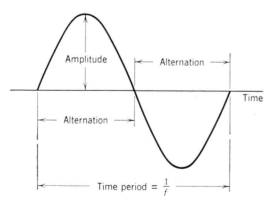

Fig. 7–6. One cycle of a sine wave.

EFFECTIVE VALUE. The effective value of an alternating current is defined as the continuous, or d-c, value that produces the same average i^2R loss. Thus, with I representing both the effective value of an alternating current and the value of direct current,

$$I^2R = \frac{1}{T}\int_0^T i^2R\, dt \tag{7-11}$$

or

$$I = \sqrt{\frac{1}{T}\int_0^T i^2\, dt} \tag{7-12}$$

As indicated in Eq. 7-12, the effective value is also the root-mean-square, or rms, value.

The effective value E of an alternating voltage may be found in a manner similar to that applied to find the effective value of current and is given by

$$E = \sqrt{\frac{1}{T}\int_0^T e^2\, dt} \tag{7-13}$$

AVERAGE VALUE. The average value of an alternating quantity is the average of the values taken throughout one period. As defined, a pure alternating current or voltage has an average value of zero over one period. However, other varying voltages and currents may have average values. For example, it is possible to reverse every other alternation of an alternating current or voltage, or to eliminate every other alternation entirely; the average value of these wave forms is not zero but is dependent upon the operation performed. (The deflections of some a-c instruments depend upon the average value of the pulsating current produced when alternate alternations are reversed.)

7-4. PRODUCTION OF SINE WAVES

One method of producing a sinusoidal wave form of voltage is shown in Fig. 7-7a. As indicated, a magnet is so mounted that it can be rotated inside a fixed coil. Rotation of the magnet changes the flux linkage of the coil and therefore causes a voltage to be induced in the coil. If the flux distribution in the air gap is sinusoidal, the induced voltage is sinusoidal, as shown in Fig. 7-7b.

Because the flux linkage is a maximum, the magnet position shown in Fig. 7-7a corresponds to zero induced voltage in the coil. As the magnet continues its rotation, the flux linkage decreases until the magnet reaches a position 90 degrees from that shown, at which time

EFFECTIVE AND AVERAGE VALUES OF SINE WAVES

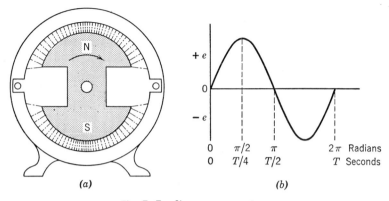

Fig. 7–7. Sine-wave generator.

the flux linkage is zero and the induced voltage is a maximum. Beyond the 90-degree position of the magnet, the flux linkage again increases until the magnet has moved 180 degrees from the original position, at which time the flux linkage is again a maximum and the induced voltage is again zero. A complete analysis of the voltage induced in the coil in one revolution of the magnet shows that the voltage varies sinusoidally, as indicated in Fig. 7–7b, and that one cycle of the alternating voltage is completed in the time required for the rotor to make one revolution. The abscissa of the curve shown as Fig. 7–7b is scaled in terms of both the time period and the angle that the rotor has moved from its original position.

The sinusoidal variation of the alternating voltage shown in Fig. 7–7b is expressed mathematically as

$$e = E_m \sin \omega t \tag{7-14}$$

or

$$e = E_m \sin 2\pi f t \tag{7-15}$$

where e = value of voltage at any time t.
E_m = maximum value of voltage.
ω = angular velocity of the rotor in radians per second.
f = frequency in cycles per second.
t = time in seconds.

7–5. EFFECTIVE AND AVERAGE VALUES OF SINE WAVES

The effective value of a sine wave of voltage or current is 0.707 times the maximum value.* Thus, for sine waves,

* Proof of this statement is left as a problem for the student.

$$I = 0.707I_m \qquad (7\text{–}16)$$
and
$$E = 0.707E_m \qquad (7\text{–}17)$$

The average value of one alternation of a sine wave of voltage or current is 0.637 times the maximum value.* Thus, for one alternation of a sine wave,
$$I_{avg} = 0.637I_m \qquad (7\text{–}18)$$
and
$$E_{avg} = 0.637E_m \qquad (7\text{–}19)$$

7–6. PHASE RELATIONS

The phase of a particular value of an alternating quantity is defined as "the fractional part of a period or cycle through which the quantity has advanced from a chosen origin." Thus, in Fig. 7–7b, the phase of the maximum value is $T/4$, since the time of observation started when the voltage was zero. If the time of observation of the wave had started when the voltage was a maximum, as shown in Fig. 7–8, the phase of the maximum value would be zero.

Phase angle is the angle equivalent of the phase in radians or degrees. For Fig. 7–7b, the phase angle of the maximum value is 90 degrees, or $\pi/2$ radians. For Fig. 7–8, the phase angle of the maximum value is zero. Similarly, the phase angle of the zero value of voltage that occurs before the value becomes positive is indicated in Fig. 7–8 as ϕ radians.

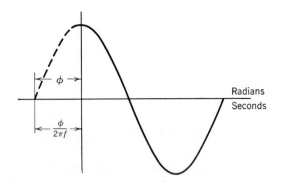

Fig. 7–8. Sine wave having phase angle ϕ.

* Proof of this statement is left as a problem for the student.

PHASE RELATIONS

Because the time of starting the observation has changed for the voltage plot shown as Fig. 7–8 (as compared to the time for Fig. 7–7), the time variation of the voltage is given as

$$e = E_m \sin(2\pi ft + \phi) \tag{7-20}$$

The angle ϕ represents the angle through which the magnet (or rotor) of the sine-wave generator of Fig. 7–7 moved before the observation of the voltage wave started.

The voltage–time curves for two sine-wave generators having equal frequencies but different rotor positions at the time of observation are shown as Fig. 7–9.

The angle ϕ in Fig. 7–9 is the angular phase difference. The phase difference between two sinusoidal quantities that have the same period is defined as the fractional part of a period (not greater than a half) through which the one wave would have to be advanced with respect to the other in order that, when plotted in percentage of amplitude, they coincide. Thus, for Fig. 7–9, one wave would have to be moved ϕ radians in order that the two waves reach their maximums at the same time.

Other terms involving phase are listed and defined as follows:

In phase—zero angular phase difference.

Phase quadrature—angular phase difference corresponding to $\pi/2$ radians (as shown in Fig. 7–9).

Phase opposition—angular phase difference corresponding to π radians (as shown for e_{ab} and i in Fig. 7–3b).

Phase sequence—the sequence with which two or more sinusoidal quantities which have the same period but are not in phase reach cor-

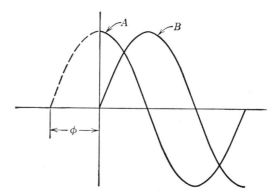

Fig. 7–9. Sine Waves having angular phase difference.

responding phases of their cycles. The phase sequence of the voltages in Fig. 7–9 is A–B.

Because the wave shown as B in Fig. 7–9 does not reach its maximum value until some time after A has reached its maximum, B is said to **lag** A. Using similar reasoning, A is said to **lead** B. The angle of lead or lag is the angular phase difference.

7–7. ADDITION OF SINE WAVES

Two sine-wave generators of the same frequency are connected in series, as shown in Fig. 7–10a. As indicated, the terminals of generator 1 are labeled a and b; those of generator 2, c and d. The voltage–time diagrams for the two generators are shown in Fig. 7–10b.

The total voltage of the generator combination of Fig. 7–10 is given by the expression

$$e_{ad} = e_{ab} + e_{cd} \qquad (7\text{–}21)$$

The addition indicated in Eq. 7–21 is an algebraic one, because the transition of the curves from positive to negative values is accompanied by a change in the polarity of the terminals. Thus, when the curve of e_{cd} is negative and that of e_{ab} is positive, the terminal c is negative with respect to d, and terminal a is positive with respect to b. The voltage e_{ad} at the same instant is therefore the arithmetic difference between the two voltages, and the polarity of a with respect to d is dependent upon the relative magnitudes of e_{ab} and e_{cd} at that instant.

The plot of e_{ad} shown in Fig. 7–10b is obtained by performing an algebraic addition at each instant of time. An inspection of the plot

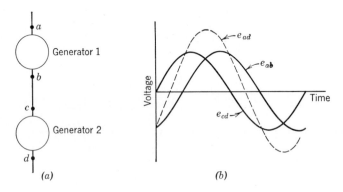

Fig. 7–10. Sine-wave generators in series.

VECTORS

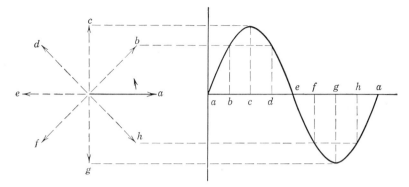

Fig. 7-11. Vector development of a sine wave.

reveals that the frequency of this voltage is the same as the frequency of the individual generators. The graphical addition therefore provides a proof for a statement that can also be proved mathematically; that is, the summation of two or more sine waves of the same frequency is a sine wave of that frequency. As shown in Fig. 7-10b, the sine waves need not be of the same phase.

The calculation of the amplitude and phase of the resultant sine wave produced when two sine waves are added can be made with trigonometric relationships or, more simply, by the application of vector representation.

7-8. VECTORS

A plot against time of the vertical projection of a vector rotating at constant speed is a sinusoidal wave form. The relationship of the rotating vector to the sinusoidal wave form is shown as Fig. 7-11.

In Fig. 7-11, when the vector, which is assumed to rotate counterclockwise, is at a, the vertical projection of the vector is zero, as indicated on the sine wave. At a time corresponding to a rotation of $\pi/4$ radians, indicated by point b, the vertical projection of the vector is 0.707 times the length of the vector. At a time corresponding to a rotation of $\pi/2$ radians, the vertical projection is equal to the length of the vector. The equation of the sine wave, then, is

$$v = V \sin \omega t \qquad (7\text{-}22)$$

where v is the vertical projection of the vector at time t, V is the length of the vector, and ω is the angular velocity of the rotating vector.

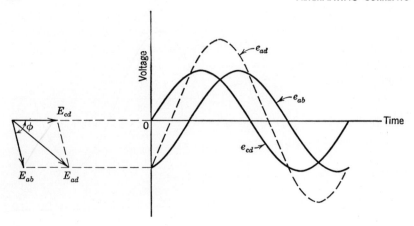

Fig. 7–12.

Equation 7–22 is of the same form as the expressions that represent an alternating current or voltage; therefore, a rotating vector may represent an alternating current or voltage. The vertical projection of the vector represents the instantaneous value of the quantity; the length of the vector represents the amplitude.

The voltages of the two generators connected in series in Fig. 7–10a are represented by vectors and sine waves in Fig. 7–12. The voltage E_{ab} lags the voltage E_{cd} by the angle ϕ, which is indicated on both diagrams. The position of the vectors in Fig. 7–12 corresponds to the time $t = 0$.

The significance of the double-subscript notation in connection with the vector is the same as previously indicated. When applied to a vector, the notation E_{ab} represents the voltage measured between terminals a and b, with a being positive with respect to b when the vertical projection of the vector is positive. (The capital letter is used for voltage because the length of the vector corresponds to the amplitude of the sine wave, not its instantaneous value.) Since the plot of voltage versus time is positive when the vertical projection of the vector is positive, the statement of the double-subscript convention applied to vectors is identical to that previously considered.

7–9. ADDITION OF VECTORS

The total voltage of the series-connected generators is also shown in Fig. 7–12. In the vector diagram, the voltage E_{ad} is found by adding

ADDITION OF VECTORS

the two vectors, E_{ab} and E_{cd}; on the sine-wave diagram, it is found by algebraically adding the two voltages at each instant of time. Both operations are in accord with the mathematical expression of Kirchhoff's law of voltage for this circuit. Thus, as previously indicated for voltages at any instant,

$$e_{ad} = e_{ab} + e_{cd} \tag{7-23}$$

Kirchhoff's law is similarly applied to vectors, with the result,

$$\overline{E_{ad}} = \overline{E_{ab}} + \overline{E_{cd}} \tag{7-24}$$

The bar over the letter designation indicates that the quantity is a vector quantity and that the addition is a vector addition.

The magnitude and phase of the resultant vector E_{ad} can be conveniently found by a method involving the horizontal and vertical components of the original vectors. With the horizontal axis as the reference line, each vector is resolved into horizontal and vertical components with respect to the reference line. The magnitude of the resultant vector is then given by the expression

$$E_{ad} = \sqrt{(\Sigma H)^2 + (\Sigma V)^2} \tag{7-25}$$

where H and V refer to the horizontal and vertical components. (The summations indicated are algebraic.) The graphical resolution of the voltage vectors into horizontal and vertical components is shown as Fig. 7–13.

Because the vertical component of the resultant vector is at all times equal to the algebraic sum of the vertical components of the original vectors, the resultant vector represents a sine wave that is the sum of the original sine waves.

When there are many vectors to be added, it is often advantageous to

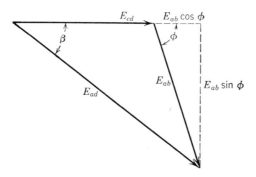

Fig. 7–13. Resolution of vectors into components.

tabulate the vertical and horizontal components. Thus, for the example given, the tabulation is as follows:

Vector	H	V
E_{ab}	$E_{ab} \cos \phi$	$E_{ab} \sin \phi$
E_{cd}	E_{cd}	0
E_{ad}	$E_{ab} \cos \phi + E_{cd}$	$E_{ab} \sin \phi$

The phase angle β of the resultant vector E_{ad} with respect to the reference axis may be found from any of the following expressions:

$$\tan \beta = \frac{V}{H} = \frac{E_{ab} \sin \phi}{E_{ab} \cos \phi + E_{cd}} \quad (7\text{-}26)$$

$$\cos \beta = \frac{H}{\sqrt{(\Sigma H)^2 + (\Sigma V)^2}} = \frac{E_{ab} \cos \phi + E_{cd}}{\sqrt{(E_{ab} \cos \phi + E_{cd})^2 + (E_{ab} \sin \phi)^2}} \quad (7\text{-}27)$$

$$\sin \beta = \frac{V}{\sqrt{(\Sigma H)^2 + (\Sigma V)^2}} = \frac{E_{ab} \sin \phi}{\sqrt{(E_{ab} \cos \phi + E_{cd})^2 + (E_{ab} \sin \phi)^2}} \quad (7\text{-}28)$$

PROBLEMS

7-1. What facts about a sinusoidal voltage (or current) are required to describe it completely?

7-2. Express $E_m \cos \omega t$ as a sine function.

7-3. Prove that the effective value of a sinusoidal voltage is $0.707 E_m$.

7-4. Prove that the average value of one alternation of a sinusoidal voltage is $0.637 E_m$.

7-5. The instantaneous current and voltage in a given a-c circuit are expressed as follows:

$$i = I_m \sin \omega t$$
$$e = E_m \sin (\omega t - \theta)$$

(a) Does the current lead or lag the voltage?
(b) Determine the average power of this circuit in terms of the effective values of voltage and current. *Ans. EI* $\cos \theta$.

7-6. A sinusoidal alternating voltage rises from zero to one-half maximum value ($E_m = 100$ volts) in 0.005 sec.
(a) What is the time period of the emf? *Ans.* 0.06 sec.
(b) What is the frequency of the emf? *Ans.* 16.67 cps.
(c) Write the equation for the emf in terms of the determined constants.
Ans. $e = 100 \sin 104.6 t$.

7-7. An alternating current having the wave form shown in Fig. 7-14 flows through a resistor.

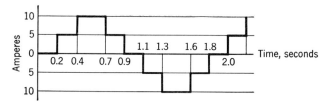

Fig. 7-14.

(a) Compute the effective value of the alternating current.
Ans. 6.66 amperes.

(b) Compute the average value of the current for one alternation.
Ans. 5.55 amperes.

7-8. Two sine-wave 60-cycle generators are to be connected in series. The maximum value of voltage of generator 1 is 100 volts, and that of generator 2 is 80 volts. If the phase relations of the voltages of the two generators can be changed, what are the limits (maximum values) of the output voltage of this combination? *Ans.* 180 volts; 20 volts.

7-9. The equation for an alternating voltage is given as 100 sin 377t. What is the frequency of the voltage? *Ans.* 60 cps.

7-10. Given two currents as follows in the circuit of Fig. 7-15:

$$i_{abc} = 30 \sin\left(377t - \frac{\pi}{6}\right)$$

$$i_{adc} = 40 \sin\left(377t + \frac{\pi}{3}\right)$$

Evaluate I_m, ω, and θ in

$$i_{cea} = I_m \sin(\omega t + \theta)$$

Ans. $I_m = 50$, $\omega = 377$, $\theta = +23°$.

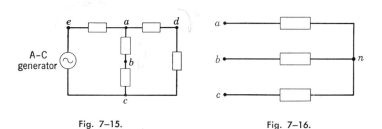

Fig. 7-15.   Fig. 7-16.

7-11. For the a-c circuit of Fig. 7-16, the vectors that correspond to the voltages (all of the same frequency) are tabulated as follows:

Voltage	Vector Magnitude	Phase Angle
an	100	0°
bn	100	+120°
cn	100	−120°

In the tabulation, the phase angle is with regard to the horizontal axis (a + angle is counterclockwise from the reference axis). Calculate the magnitude and phase angle of the vectors that correspond to the voltages *ab*, *bc*, and *ca*.

$$Ans. \quad ab = 173.2; \ -30°.$$
$$bc = 737.2; \ +90°.$$
$$ca = 173.2; \ -150°.$$

■ 8 Single-Phase A-C Circuits

8–1. INTRODUCTION

The study of a-c circuits involves no new concepts or circuit elements; the laws, rules, and circuit elements previously studied are merely applied to circuits where the source of electric energy has an alternating voltage instead of a continuous one. New terms are encountered, but the basic concepts remain the same.

An important factor to be considered at the outset is that a given alternating voltage (or current) has three values associated with it: namely, the maximum value, the effective value, and the average value. Each value has an appropriate place in the study of electric circuits and is identified by the following symbols:

$$E_m = \text{maximum value of voltage}$$
$$E = \text{effective value of voltage}$$
$$E_{avg} = \text{average value of voltage}$$

The effective value of alternating voltage (and current) is the one most frequently encountered because the current and voltage ratings of a-c equipment and circuits are in terms of effective values. Thus, a 120-volt a-c motor is designed to be operated from a supply that has an effective value of 120 volts. The supply itself is identified as a 120-volt supply, although the maximum value of voltage encountered is approximately 170 volts (assuming the wave form is sinusoidal).

The maximum and average values of voltage are important in the study of electronics and rectifiers. The maximum value of voltage is frequently referred to as the peak value, in order to avoid confusion in discussions or problems where the maximum effective value of voltage may be under consideration.

Unless otherwise specified, the values of alternating currents and voltages in the discussions and problems in this book are effective values.

8–2. PURE RESISTIVE CIRCUITS

If a resistor having negligible inductance and capacitance is connected to an alternating voltage having an instantaneous value given by

$$e_{ab} = E_m \sin \omega t \qquad (8\text{–}1)$$

the instantaneous current through the resistor is

$$i_{ab} = \frac{e_{ab}}{R} = \frac{E_m}{R} \sin \omega t \qquad (8\text{–}2)$$

or

$$i_{ab} = I_m \sin \omega t \qquad (8\text{–}3)$$

Thus, from Eqs. 8–2 and 8–3,

$$I_m = \frac{E_m}{R} \qquad (8\text{–}4)$$

In terms of effective values, therefore,

$$I = \frac{E}{R} \qquad (8\text{–}5)$$

According to Eqs. 8–1 and 8–3, the voltage and current for the pure resistance circuit are in phase. Thus, the vector diagram is as shown in Fig. 8–1a. The vector diagram shown as Fig. 8–1b is similar to that shown as Fig. 8–1a, except that the vectors in Fig. 8–1b are identified by the effective values instead of the peak values. The labeling of the vectors in this manner is justified because the effective values are the values applied in practice and because the effective value of a sinusoidal wave form bears a fixed numerical relation to the peak value. In the

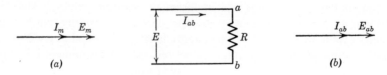

Fig. 8–1. Vector diagrams for a pure resistive circuit.

PURE INDUCTIVE CIRCUITS

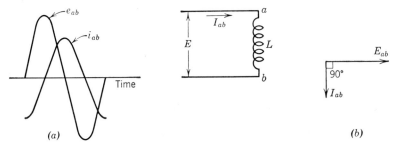

Fig. 8-2. Voltage and current relationships in a pure inductive circuit.

remainder of this book, all vector diagrams are therefore labeled in terms of effective values.

The instantaneous power in the resistive circuit is given by the relation

$$p_{ab} = e_{ab}i_{ab} = (E_m \sin \omega t)(I_m \sin \omega t) \tag{8-6}$$

which, when solved for the average power, gives

$$P = EI = I^2 R \tag{8-7}$$

8-3. PURE INDUCTIVE CIRCUITS

If the resistor of Fig. 8-1 is replaced by an inductor having constant inductance and negligible resistance and capacitance, as shown in Fig. 8-2, the induced voltage in the inductor must be equal to the supply voltage. Thus

$$e_{ab} = E_m \sin \omega t = L \frac{di_{ab}}{dt} \tag{8-8}$$

The expression for the instantaneous current is then found by integrating Eq. 8-8, as follows:

$$\frac{di_{ab}}{dt} = \frac{E_m}{L} \sin \omega t$$

$$i_{ab} = \frac{E_m}{\omega L} \int \sin \omega t \, d(\omega t)$$

$$i_{ab} = -\frac{E_m}{L\omega} \cos \omega t \tag{8-9}*$$

* Note that the constant of integration is omitted from Eq. 8-9. The value of the constant of integration depends upon the time in the voltage cycle that the inductor is energized. Thus, if the inductor is energized when the voltage

Therefore,

$$I_m = \frac{E_m}{\omega L} \qquad (8\text{--}10)$$

and, in terms of effective values,

$$I = \frac{E}{\omega L} = \frac{E}{2\pi f L} \qquad (8\text{--}11)$$

The plots of voltage and current versus time for the pure inductive circuit are shown as Fig. 8–2a, and the corresponding vector diagram is indicated in Fig. 8–2b. As shown in Fig. 8–2, the current in a pure inductive circuit lags the applied voltage by the time necessary for the voltage to complete one fourth of a cycle, or, in terms of angular position of the vectors, by 90 degrees.

For convenience, Eq. 8–11 is written

$$I = \frac{E}{X_L} \qquad (8\text{--}12)$$

where

$$X_L = 2\pi f L \qquad (8\text{--}13)$$

The quantity X_L is called the **inductive reactance** of the circuit, and, as indicated, it is proportional to the frequency and the inductance. When the frequency and the inductance are expressed in cycles and henrys, respectively, the reactance in Eq. 8–13 is in ohms.

As shown by Eq. 8–12, the effective voltage and current for an inductive circuit are related by an expression similar to the Ohm's law relation applied to resistive circuits. The similarity includes the fact that the current is in amperes when the potential difference is in volts and the resistance or reactance is in ohms. However, the current is in phase with the voltage in the resistive circuit, whereas the current lags the voltage by one fourth of a cycle in the pure inductive circuit.

---

is a maximum (for example, at $t = 1/4f$), the constant of integration is zero, because the current is zero at the instant the circuit is energized, and the value of cos ωt is zero. However, if the voltage is zero at the instant the inductor is energized (for example, at $t = 0$), then the constant of integration is $E_m/\omega L$, because the current is zero at that instant, and the value of cos ωt is 1. A constant of integration equal to $E_m/\omega L$ implies the existence of a d-c component; however, a d-c component of voltage is not present in the supply, and none can exist across the inductor. The practical significance of the constant of integration is that it determines the magnitude of the transient current, which flows for a few cycles after the circuit is energized and is then damped by the resistance inherent in a practical inductor (considered in Art. 8–5). Therefore, in the steady-state condition, the constant of integration is zero; the steady-state condition is the condition under consideration here.

PURE CAPACITIVE CIRCUITS

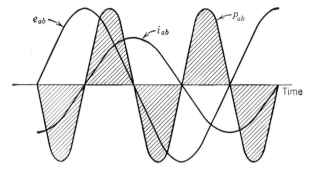

Fig. 8–3. Power versus time in a pure inductive circuit.

The instantaneous power in the pure inductive circuit is given by the relation,

$$p_{ab} = e_{ab}i_{ab} = (E_m \sin \omega t)(-I_m \cos \omega t)$$

$$= (E_m \sin \omega t)I_m \sin\left(\omega t - \frac{\pi}{2}\right) \tag{8-14}$$

The average power in the pure inductive circuit is zero.*

A plot of the instantaneous power reveals, as shown in Fig. 8–3, that the power fluctuates between positive and negative values and that the average power is zero. Thus, the energy stored in the magnetic field while the current is increasing in magnitude is returned to the source while the current is decreasing.

8–4. PURE CAPACITIVE CIRCUITS

If the inductor of Fig. 8–2 is replaced by a capacitor having constant capacitance and negligible resistance and inductance, the current in the circuit is a function of the time rate of change of voltage. Thus,

$$i_{ab} = C \frac{de_{ab}}{dt} \tag{8-15}$$

For the given voltage supply,

$$i_{ab} = C \frac{d(E_m \sin \omega t)}{dt} \tag{8-16}$$

and

$$i_{ab} = \omega C E_m \cos \omega t \tag{8-17}$$

Therefore,

$$I_m = \omega C E_m \tag{8-18}$$

* The student should prove this statement.

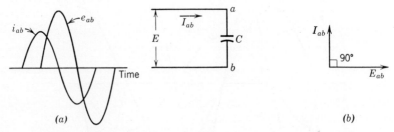

Fig. 8–4. Voltage and current relationships in a pure capacitive circuit.

and, in terms of effective values,

$$I = \omega CE = 2\pi f CE \tag{8-19}$$

Equation 8–19 may be written

$$I = \frac{E}{X_C} \tag{8-20}$$

where

$$X_C = \frac{1}{2\pi f C} \tag{8-21}$$

The factor X_C is called the **capacitive reactance**; it is expressed in ohms when f is in cycles and C is in farads.

The plots of voltage and current versus time for the pure capacitive circuit are shown as Fig. 8–4a, and the corresponding vector diagram is indicated in Fig. 8–4b.

As shown in Fig. 8–4, the current in a pure capacitive circuit leads the applied voltage by one fourth of a cycle or, in terms of the angular position of the vectors, by 90 degrees.

The instantaneous power in the pure capacitive circuit is given by the relation,

$$p_{ab} = e_{ab} i_{ab} = (E_m \sin \omega t)(I_m \cos \omega t) \tag{8-22}$$

$$= (E_m \sin \omega t) I_m \sin \left(\omega t + \frac{\pi}{2} \right) \tag{8-23}$$

The average power in the pure capacitive circuit is zero.

A plot of the instantaneous power versus time for this circuit is similar to that for the inductive circuit in that the power fluctuates between positive and negative values, as shown in Fig. 8–5. One difference between the inductive and capacitive circuits is that energy is stored by the capacitor when the voltage is increasing in magnitude whereas energy is stored in the magnetic field of the inductor when the current is increasing in magnitude.

8–5. PRACTICAL CONSIDERATIONS

The circuit elements considered in the three preceding articles are pure elements; that is, the resistor has negligible inductance and capacitance, the inductor has negligible resistance and capacitance, and the capacitor has negligible resistance and inductance. The assumption of pure elements in an actual circuit may lead to calculated results that are not consistent with the experimental data. Therefore, each circuit component should be analyzed as to the expected departure from the purity assumed.

Whether or not a resistor can be assumed to be a pure element depends upon the construction of the resistor. Resistors wound in the form of coils, as the slide-wire types, may have inductance values that would make the inductive reactance an appreciable quantity, compared to the resistance value. The resistor then has to be considered as a resistor and an inductor in series. Coil-type resistors wound so that the inductance is negligible are obtainable, however. The resistors considered in circuits and discussions in this book are assumed to be pure elements, unless otherwise specified.

The inductance of a given circuit is proportional to the rate of change of flux linkages of the circuit (Chapter 6). Therefore, a circuit component that has inductance must, by nature of its construction, also have resistance. The magnitude of the resistance in comparison to the inductive reactance is, of course, dependent upon the construction of the component and the frequency. At power frequencies, the ratio of resistance to inductive reactance is larger for inductors having only nonmagnetic materials in the magnetic circuit than for those having magnetic materials.

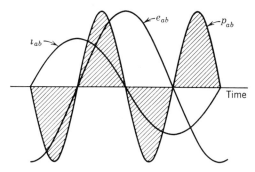

Fig. 8–5. Power versus time in a pure capacitive circuit.

In considering the resistance of an inductor, the nature of the magnetic circuit may be an important factor. At the usual power frequencies, the resistance of inductors having nonmagnetic cores may be taken as the resistance of the conductors of which the inductor is constructed. The resistance of inductors having magnetic cores, however, is somewhat larger than the resistance of the conductor because of the hysteresis and eddy-current losses that occur in the magnetic circuit. These losses must be supplied from the line and, since no average power is consumed in a pure inductance, they make the resistance of the inductor appear larger than the true conductor resistance. A resistance value that includes both the conductor resistance and the additional resistance due to hysteresis and eddy-current losses in the magnetic circuit is called the **effective resistance.**

As indicated in the preceding discussion, a coil connected to an a-c supply may be represented in a circuit diagram as a resistor and an inductor in series.

In general, capacitors may be assumed to be pure elements. Conditions arise, however, in which a capacitor consumes power. The capacitor is then represented as pure capacitance in parallel with pure resistance. The parallel connection of the resistor represents the resistance of the leakage path for current over the surface between the terminals of the capacitor.

8–6. NONLINEAR ELEMENTS IN A-C CIRCUITS

The preceding discussion of the individual circuit elements considers that the values of resistance, inductance, and capacitance are independent of the current. However, a-c circuits may include components that are

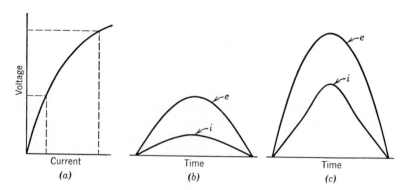

Fig. 8–6. Voltage and current relationships for a nonlinear resistor in an a-c circuit.

SERIES COMBINATIONS OF RESISTANCE, INDUCTANCE, AND CAPACITANCE

nonlinear; that is, the values of resistance, inductance, or capacitance may depend upon the values of voltage, current, or frequency.

The treatment of a-c circuits involving nonlinear components is more complex than that for the equivalent d-c circuit, because the voltage supply is alternating. In general, the effect of a nonlinear element in an a-c circuit with a sinusoidal voltage supply is to distort the wave form of the current so that it is not sinusoidal; therefore, the current cannot be represented by a rotating vector of constant magnitude. Thus, if a resistor having an instantaneous voltage–current characteristic similar to that shown in Fig. 8–6a is connected to a sinusoidal voltage supply, the current wave form depends upon the maximum value of the alternating voltage. If the maximum value of voltage is such as to be on the linear portion of the characteristic, the current wave is sinusoidal, as shown for a half-cycle of voltage in Fig. 8–6b; if the maximum value of voltage is such as to be on the nonlinear portion of the curve, the current wave form is as shown for a half-cycle of voltage in Fig. 8–6c.

Unless the wave form of the current can be expressed mathematically, the effective value of current and the average value of power in the circuit containing the nonlinear resistor of Fig. 8–6 must be determined graphically.

The circuit parameter most frequently found to be nonlinear is inductance. In general, any iron-core inductor is a nonlinear element, since inductance is determined by the expression

$$L = N \frac{d\phi}{di} \times 10^{-8} \qquad (8\text{–}24)$$

and, in an iron-core inductor, the flux does not vary uniformly with current.

The effects of a nonlinear inductor in an a-c circuit are similar to those of a nonlinear resistor. These effects are considered in more detail in a later chapter.

Unless otherwise specified, the circuit elements considered in circuit discussions in this book are assumed to be linear elements.

8–7. SERIES COMBINATIONS OF RESISTANCE, INDUCTANCE, AND CAPACITANCE

The circuit diagram for the series connection of a resistor, an inductor, and a capacitor to an a-c supply of voltage E and frequency f is shown as Fig. 8–7. (The diagram shows the elements to be pure elements although, as previously indicated, the inductor must have resistance also; for the solution of a circuit problem, the resistance of the inductor is added to the resistance of the resistor to form a circuit similar to that shown in Fig. 8–7. However, the actual voltage between the terminals of the inductor includes the voltage drop in the resistance of the inductor as well as the voltage drop due to the inductive reactance. In the discussion that follows, it is assumed that the resistance of the inductor is negligible and that the terminals shown in the circuit of Fig. 8–7 are the actual terminals of the circuit components.)

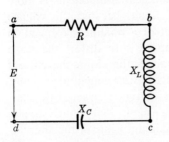

Fig. 8–7. Resistance, inductance, and capacitance in series.

The voltage relation applying to the circuit of Fig. 8–7 is

$$\overline{E_{da}} + \overline{E_{ab}} + \overline{E_{bc}} + \overline{E_{cd}} = 0 \qquad (8\text{--}25)$$

or

$$\overline{E_{ad}} = \overline{E_{ab}} + \overline{E_{bc}} + \overline{E_{cd}} \qquad (8\text{--}26)$$

If the values of the component voltages in terms of the current are substituted,

$$\overline{E_{ad}} = \overline{IR} + \overline{IX_L} + \overline{IX_C} \qquad (8\text{--}27)$$

where

$$I = I_{abcd} \qquad (8\text{--}28)$$

As the bar over the individual voltages indicates, the additions are vectorial.

With the current vector as a reference line for the construction of a vector diagram (since the current is common to all elements of the

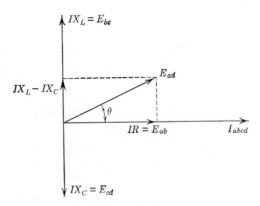

Fig. 8–8. Vector diagram for R, L, and C in series.

SERIES COMBINATIONS OF RESISTANCE, INDUCTANCE, AND CAPACITANCE

circuit), the vector diagram corresponding to Eqs. 8–26 and 8–27 is shown as Fig. 8–8. (It is assumed that the inductive reactance X_L is larger than the capacitive reactance X_C; therefore, IX_L is larger than IX_C.) The phase relations of the voltages across the circuit elements with respect to the current through the elements are determined from relations previously established in this chapter.

The magnitude of the current in the circuit of Fig. 8–7 is determined as follows:

From the vector diagram

$$E_{ad} = \sqrt{(IR)^2 + (IX_L - IX_C)^2} \tag{8-29}$$

or

$$E_{ad} = I\sqrt{R^2 + (X_L - X_C)^2} \tag{8-30}$$

Therefore,

$$I = \frac{E_{ad}}{\sqrt{R^2 + (X_L - X_C)^2}} \tag{8-31}$$

The angular phase difference θ between the total voltage E_{ad} and the total current I_{abcd} may also be determined from the vector diagram. Thus,

$$\cos\theta = \frac{IR}{E_{ad}} = \frac{R}{\sqrt{R^2 + (X_L - X_C)^2}} \tag{8-32}$$

In the circuit of Fig. 8–7, only the resistor consumes average power, since it is assumed that the inductor and the capacitor are pure elements and since, as previously established, pure inductors and capacitors dissipate no average power. Thus,

$$P = I^2 R = IE_{ab} = I(E_{ad} \cos\theta) \tag{8-33}$$

The preceding discussion is concerned with a particular circuit in which the three circuit elements are connected in series and in which the inductive reactance is larger than the capacitive reactance. The relationships developed, however, are applicable to any series combination of the three elements, if the proper terms are omitted. For example, if the circuit consisted of resistance only, Eqs. 8–31, 8–32, and 8–33 would reduce to Eqs. 8–34, 8–35, and 8–36, respectively.

$$I = \frac{E}{\sqrt{R^2}} = \frac{E}{R} \tag{8-34}$$

$$\cos\theta = \frac{R}{\sqrt{R^2}} = 1 \tag{8-35}$$

$$P = EI \cos\theta = EI \tag{8-36}$$

Equations 8–34, 8–35, and 8–36 are identical to those established in Art. 8–2 for a resistive circuit in which the inductance and the capacitance are negligible. (The cosine of the phase angle θ between the voltage and current is unity, indicating that the voltage and current are in phase, as previously established.)

Because the vector for current lags the vector for total voltage in Fig. 8–8, the three-element circuit under consideration is inductive in nature. Therefore, an equivalent circuit for this three-element circuit consists of a pure resistor R in series with a pure inductor having a reactance equal to $X_L - X_C$.

8–8. IMPEDANCE

The **impedance** of a portion of a circuit containing no power source is defined as the ratio of the effective value of the voltage between the terminals of the circuit to the effective value of the current in the circuit. Therefore, the impedance of a circuit is the apparent resistance or opposition of a circuit to the flow of alternating current. Impedance is symbolized by Z. Thus,

$$Z = \frac{E}{I} \tag{8-37}$$

If E is in volts and I is in amperes, Z is in ohms.

The expression for the impedance of the circuit of Fig. 8–7 in terms of the resistance and the reactance is (from Eq. 8–31)

$$Z_{abcd} = \frac{E_{ad}}{I_{abcd}} = \sqrt{R^2 + (X_L - X_C)^2} \tag{8-38}$$

A portion of the vector diagram shown as Fig. 8–8 is redrawn as Fig. 8–9a. If the indicated magnitudes of the vectors are divided by the current I, the result is the triangle shown as Fig. 8–9b. The triangle is called an **impedance triangle**; it is not a vector diagram because resistance and reactance are not vector quantities. The calculation of

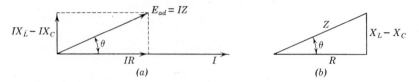

Fig. 8–9. Vector diagram and impedance triangle for R, L, and C in series.

APPARENT POWER AND POWER FACTOR

the circuit impedance from the triangle components agrees with Eq. 8-38.

As for other relations developed for the series circuit of resistance, inductance, and capacitance, Eq. 8-38 is applicable to any series combination of the three circuit elements.

8-9. APPARENT POWER AND POWER FACTOR

If power is associated with the product of voltage and current, the power that the three-element circuit of Fig. 8-7 apparently dissipates is given by $E_{ad}I_{abcd}$. Therefore, the product of the effective voltage and the effective current of a circuit is called the **apparent power** (symbolized by U). The unit for apparent power is the **volt-ampere** (abbreviated **va**).

The ratio of the average power (also called the true power or active power) to the apparent power is the **power factor** of the circuit. For the circuit under consideration,

$$\text{Power factor} = \frac{P}{U} = \frac{E_{ad}I_{abcd}\cos\theta}{E_{ad}I_{abcd}}$$

or

$$\text{Power factor} = \cos\theta \tag{8-39}$$

Thus, the cosine of the phase angle between the voltage and the current of a circuit is the power factor of the circuit.

From Eq. 8-32, the power factor for a series circuit consisting of resistance, inductance, and capacitance is

$$\cos\theta = \frac{R}{\sqrt{R^2 + (X_L - X_C)^2}} = \frac{R}{Z} \tag{8-40}$$

Equation 8-40 is applicable to any series combination of the three circuit elements.

Since the value of the power factor in any circuit equals the cosine of an angle, it can never be greater than unity. In addition to having a numerical value, the power factor of a circuit usually carries a notation that signifies the nature of the circuit: that is, whether the equivalent circuit is inductive or capacitive. Thus, the power factor might be expressed as 0.8 leading or 0.8 lagging. The leading and lagging refer to the phase of the current vector with respect to the voltage vector. Thus, a lagging power factor means that the current lags the voltage and that the circuit is inductive in nature. Power factor may be expressed as a percentage as well as a decimal; thus, the 0.8 lagging power factor might be expressed as 80 per cent lagging.

The value of the power factor of electric circuits is economically important to both the producer and the consumer of electric power. It is important to the producer because it affects the rating of the equipment needed to deliver a given amount of power. For example, if 100 kilowatts of power are to be delivered at a power factor of 1.0, the equipment must be rated at 100 kilovolt-amperes (100,000 volt-amperes). However, if the same power is to be delivered at a power factor of 0.5 and at the same voltage, the equipment must be rated at 200 kilovolt-amperes; therefore, the equipment must carry twice as much current for the 0.5 power-factor load as for the 1.0 power-factor load. The producer of electric power must therefore make a greater capital investment to deliver the 100 kilowatts at a power factor of 0.5 than he would to deliver the same power at a power factor of 1.0. Because of the higher capital investment at the lower power factor, the producer sells the power at different rates so that the consumer having equipment with a low power factor pays a higher rate than the consumer operating at a high power factor. Thus, the value of the power factor is also important to the consumer. (The power factor of most circuits is a lagging one; therefore, some power producers offer a premium to consumers having circuits that operate at a leading power factor.)

8–10. REACTIVE POWER

The relationship between apparent power and average power can also be shown by employing the vector diagram for the three-element series circuit. Figure 8–10a shows the diagram resulting after each of the vectors is multiplied by the effective value of current. The I^2R side represents the average power; the I^2Z side, the apparent power.

Figure 8–10b is the same triangle as that shown in Fig. 8–10a except

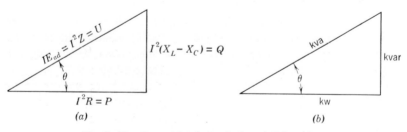

Fig. 8–10. Power triangle for R, L, and C in series.

PARALLEL COMBINATIONS OF RESISTANCE, INDUCTANCE, AND CAPACITANCE

that the sides are labeled in terms of the units for average power, apparent power, and **reactive power** (symbolized by Q). As indicated in the two diagrams of Fig. 8–10, the reactive power in a circuit is given by the expression

$$Q = U \sin \theta \tag{8-41}$$

or, for this circuit,

$$Q = E_{ad} I_{abcd} \sin \theta \tag{8-42}$$

The quantity $\sin \theta$ is called the **reactive factor** of the circuit. As indicated, it is the ratio of the reactive power to the apparent power. The reactive power represents the apparent power taken by the pure reactance of a circuit. For a circuit having only pure reactance, the triangle shown as Fig. 8–10b reduces to a vertical line. In such a circuit the apparent power is all reactive power, the true power being zero, as previously established. Similarly, in a circuit having only resistance, the power triangle reduces to a horizontal line, because the reactive power is zero.

The unit for reactive power is the **var** (var being an abbreviated term for **volt-ampere-reactive**). The power triangle shown as Fig. 8–10b has its sides dimensioned in kilowatts, kilovolt-amperes, and kilovars (1000 vars).

The power triangle can be effectively applied in the solution of some circuit problems, particularly those parallel circuits where power requirements and power factors of individual loads are known and where the total power and power factor are to be determined.

8–11. PARALLEL COMBINATIONS OF RESISTANCE, INDUCTANCE, AND CAPACITANCE

The circuit diagram for the parallel connection of a resistor, an inductor, and a capacitor to an a-c supply is shown as Fig. 8–11. The components are assumed to be pure elements.

The current relation applying to the circuit of Fig. 8–11 is

$$\overline{I_{aa'}} + \overline{I_{b'b}} + \overline{I_{e'e}} + \overline{I_{c'c}} = 0 \tag{8-43}$$

or

$$\overline{I_{aa'}} = \overline{I_{bb'}} + \overline{I_{ee'}} + \overline{I_{cc'}} \tag{8-44}$$

If the values of the component currents in terms of the voltage are substituted,

$$\overline{I_{aa'}} = \frac{\overline{E_{ad}}}{R} + \frac{\overline{E_{ad}}}{X_L} + \frac{\overline{E_{ad}}}{X_C} \tag{8-45}$$

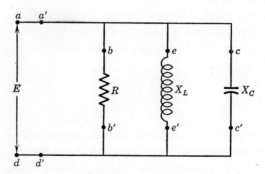

Fig. 8–11. Resistance, inductance, and capacitance in parallel.

With the voltage vector as a reference for the construction of a vector diagram (since the voltage is common to all elements of the circuit), the vector diagram corresponding to Eqs. 8–44 and 8–45 is shown as Fig. 8–12. For purposes of discussion, the inductive reactance is assumed to be larger than the capacitive reactance; therefore, the current through the inductor is less than the current through the capacitor.

From the vector diagram, the magnitude of the total current in this circuit is

$$I_{aa'} = \sqrt{(I_{bb'})^2 + (I_{cc'} - I_{ee'})^2} \tag{8-46}$$

or

$$I_{aa'} = \sqrt{\left(\frac{E_{ad}}{R}\right)^2 + \left(\frac{E_{ab}}{X_C} - \frac{E_{ad}}{X_L}\right)^2} \tag{8-47}$$

or

$$I_{aa'} = E_{ad}\sqrt{\left(\frac{1}{R}\right)^2 + \left(\frac{1}{X_C} - \frac{1}{X_L}\right)^2} \tag{8-48}$$

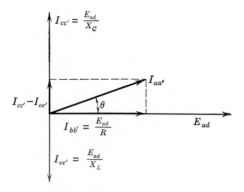

Fig. 8–12. Vector diagram for R, L, and C in parallel.

SERIES RESONANCE

Since the impedance of a circuit is the ratio of the effective voltage to the effective current, the impedance of this parallel circuit is given by the expression

$$Z = \frac{1}{\sqrt{\left(\frac{1}{R}\right)^2 + \left(\frac{1}{X_C} - \frac{1}{X_L}\right)^2}} \tag{8-49}$$

Inspection of Eq. 8–49, which applies to a parallel circuit, and of Eq. 8–38, which applies to a series circuit of the same elements, shows clearly that the impedances are different and that the methods of calculation are different.

In general, the solution of a parallel circuit problem is simplified if a vector diagram, such as that shown in Fig. 8–12, is used as the basis for the solution. Thus, for the circuit given, calculations are made as follows:

Impedance: $\quad Z = \dfrac{E_{ad}}{I_{aa'}} \tag{8-50}$

Power factor: $\quad \cos \theta = \dfrac{I_{bb'}}{I_{aa'}} \tag{8-51}$

Apparent power: $\quad U = E_{ad} I_{aa'} \tag{8-52}$

Power: $\quad P = E_{ad} I_{aa'} \cos \theta \tag{8-53}$

Reactive power: $\quad Q = E_{ab} I_{aa'} \sin \theta \tag{8-54}$

Expansion of the expression for power for this parallel circuit reveals that the only power consumed in the circuit is that dissipated in the resistor. Thus,

$$P = E_{ad} I_{aa'} \cos \theta = E_{ad} I_{aa'} \frac{I_{bb'}}{I_{aa'}} \tag{8-55}$$

$$= E_{ad} I_{bb'} = I_{bb'}^2 R \tag{8-56}$$

One observation of interest, in comparing the series connection of the three elements with the parallel connection of the same elements, is the fact that one connection appears inductive and the other connection appears capacitive.

8–12. SERIES RESONANCE

The current in the series circuit consisting of resistance, inductance, and capacitance is given by the expression

$$I = \frac{E_{ad}}{\sqrt{(R)^2 + (X_L - X_C)^2}} \tag{8-57}$$

When the inductive reactance equals the capacitive reactance, the current in the circuit is a maximum and is in phase with the applied voltage. Under these conditions, the circuit has **series resonance**. Any circuit containing inductance and capacitance connected in series is said to have series resonance when the current in the circuit is in phase with the voltage across the circuit or when the power factor of the circuit is unity.

The effects of series resonance in a circuit depend upon the magnitudes of resistance and reactance. A low value of resistance may result in current that is excessive, since, at resonant conditions, the current in the circuit is limited only by the resistance; large values of reactance may result in excessive voltages appearing across each of the reactive elements.

As an example of series resonance, consider that a 100-volt a-c supply is connected to a series combination consisting of a 10-ohm resistor, a pure inductor having a reactance of 1000 ohms, and a pure capacitor having a reactance of 1000 ohms. The current in this circuit is 10 amperes. Therefore, the inductor and the capacitor each have 10,000 volts between their terminals.

8–13. PARALLEL RESONANCE

The total current in the parallel circuit of Fig. 8–11 is given by the expression

$$I_{aa'} = E_{ad} \sqrt{\left(\frac{1}{R}\right)^2 + \left(\frac{1}{X_C} - \frac{1}{X_L}\right)^2} \tag{8-58}$$

When, for this special parallel case, the inductive reactance is equal to the capacitive reactance, the current entering the circuit is in phase with

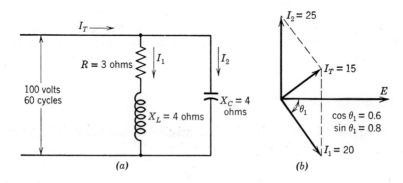

Fig. 8–13. Parallel a-c circuit.

COMPLEX NOTATION

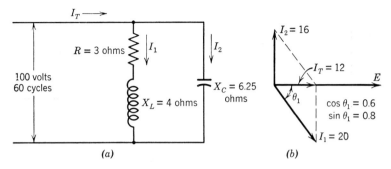

Fig. 8-14. Circuit having parallel resonance.

the applied voltage. Under these conditions, the circuit has **parallel resonance**. Any circuit having inductance and capacitance connected in parallel is said to have parallel resonance when the current entering the circuit is in phase with the voltage across the circuit or when the power factor of the circuit is unity.

In general, the condition for resonance in a parallel circuit is not that the inductive reactance be equal to the capacitive reactance but that the circuit have unity power factor. In the special case, equal reactances resulted in unity power factor, but this may not always be true; an example is the circuit shown as Fig. 8-13a. The reactances in the parallel circuit of Fig. 8-13a are equal, but the circuit does not have parallel resonance. Conversely, the circuit of Fig. 8-14a has reactances that are unequal, but the circuit has parallel resonance because the total power factor is unity.

8-14. COMPLEX NOTATION

One caution that must always be observed in analyzing a-c circuits is that, in general, quantities cannot be considered arithmetically but must take into account the phase relationships that apply. Thus, although the effective value of voltage divided by the absolute magnitude of impedance gives the effective value of current, the nature or make-up of the impedance must be considered in order that the current be placed in its proper position with respect to voltage on the resulting vector diagram. Similarly, an impedance made up of resistance in series with inductive reactance does not have a magnitude given by the arithmetic sum of the two but must be considered to be the hypotenuse of a right triangle that has the values of resistance and inductive reactance as the

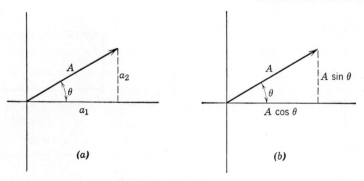

Fig. 8–15. Vector components.

two legs. Because of phase relationships, problems that are just moderately complex often appear impossible to analyze by the fundamental methods considered earlier in this chapter. Although not impossible by such methods, the solution is often more easily found by the application of complex notation, which makes use of the fact that a vector may be completely specified by the complex coordinates, or the complex number, representing the tip of the vector.

The vector of magnitude A shown in Fig. 8–15a is considered to be made up of two components, one horizontal and one vertical, labeled a_1 and a_2. Thus,

$$A = \sqrt{a_1^2 + a_2^2} \tag{8-59}$$

The angle θ that the vector makes with the horizontal axis is found from the relation

$$\tan \theta = \frac{a_2}{a_1} \tag{8-60}$$

As shown in Fig. 8–15b, the sides of the vector could also be labeled in terms of the angle the vector makes with the horizontal. Thus,

$$a_1 = A \cos \theta \tag{8-61}$$
$$a_2 = A \sin \theta \tag{8-62}$$

and

$$A = \sqrt{(A \cos \theta)^2 + (A \sin \theta)^2} \tag{8-63}$$

In addition to Eqs. 8–59 and 8–63, the magnitude of the vector could be expressed by the vector equation

$$\bar{A} = \overline{A \cos \theta} + \overline{A \sin \theta} \tag{8-64}$$

By introducing an operator j that indicates that the vector to which it

COMPLEX NOTATION

is assigned is rotated 90 degrees counterclockwise from the reference axis, the magnitude of the vector can be expressed by a complex number. Thus,

$$\bar{A} = A \cos \theta + jA \sin \theta \qquad (8\text{-}65)$$

Equation 8-65 means that the vector of length A is made up of a component, $A \cos \theta$, along the horizontal, or reference, axis (called the real axis) and a component, $A \sin \theta$, upward along the vertical axis (called the imaginary axis). If the vector had been in the fourth quadrant, with the vertical component downward, the equation would have been written

$$\bar{A} = A \cos \theta - jA \sin \theta \qquad (8\text{-}66)$$

the significance of the $-j$ being that the vector is rotated 90 degrees clockwise from the reference axis.

Another form for the complex representation of a vector can be developed from the exponential expression for sine and cosine functions. Thus,

$$\sin \theta = \frac{1}{2j}(\epsilon^{j\theta} - \epsilon^{-j\theta}) \qquad (8\text{-}67)$$

and

$$\cos \theta = \tfrac{1}{2}(\epsilon^{j\theta} + \epsilon^{-j\theta}) \qquad (8\text{-}68)$$

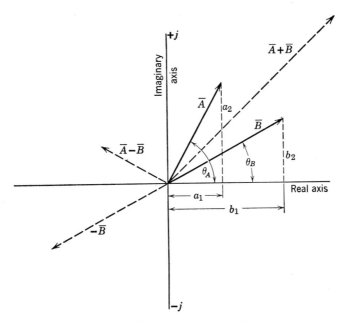

Fig. 8-16. Addition and subtraction of vectors.

then
$$A(\cos\theta + j\sin\theta) = A\epsilon^{j\theta} \tag{8-69}$$

Equation 8-69 in turn can be written as

$$A\cos\theta + jA\sin\theta = A\angle\theta \tag{8-70}$$

to indicate that the vector of magnitude A makes an angle θ with the real axis. (A positive angle means a counterclockwise angle and is denoted by $\angle\theta$. A negative angle is indicated by $\angle\!\!\!\!\diagdown\theta$ or $\angle-\theta$.)

The two vectors shown in Fig. 8-16 may be used to illustrate the application of ordinary algebraic processes to complex numbers. As indicated, the vectors are given by

$$\bar{A} = a_1 + ja_2 = A\epsilon^{j\theta_a} \tag{8-71}$$

$$\bar{B} = b_1 + jb_2 = B\epsilon^{j\theta_b} \tag{8-72}$$

For these vectors, the following identities exist:

$$\begin{aligned}\bar{A} + \bar{B} &= a_1 + ja_2 + b_1 + jb_2 \\ &= (a_1 + b_1) + j(a_2 + b_2)\end{aligned} \tag{8-73}$$

$$\bar{A} - \bar{B} = (a_1 - b_1) + j(a_2 - b_2) \tag{8-74}$$

Also:

$$\begin{aligned}\bar{A}\bar{B} &= A\epsilon^{j\theta_a}B\epsilon^{j\theta_b} \\ &= AB\epsilon^{j(\theta_a+\theta_b)} \quad \text{or} \quad AB\angle(\theta_a + \theta_b)\end{aligned} \tag{8-75}$$

$$\begin{aligned}\frac{\bar{A}}{\bar{B}} &= \frac{A\epsilon^{j\theta_a}}{B\epsilon^{j\theta_b}} \\ &= \frac{A}{B}\epsilon^{j(\theta_a-\theta_b)} \quad \text{or} \quad \frac{A}{B}\angle(\theta_a - \theta_b)\end{aligned} \tag{8-76}$$

In the preceding expressions, the rectangular form of complex notation is employed for addition and subtraction, and the polar form for multiplication and division. It is quite proper to use the rectangular form for multiplication and division, but the polar form is usually more convenient. In applying the rectangular form, however, a j^2 term is encountered, and this term is evaluated by noting that $j = \sqrt{-1}$ and therefore $j^2 = -1$.

8-15. APPLICATION OF COMPLEX NOTATION

Because alternating voltages and currents may be represented by vectors in a single plane, complex notation may be applied to these quantities, and Eqs. 8-73 and 8-74 may be used to solve Kirchhoff

APPLICATION OF COMPLEX NOTATION

law equations involving currents or voltages. However, the operations prescribed by these expressions are identical with the geometric addition and subtraction of the horizontal and vertical components of the corresponding vectors. The application of complex algebra to a-c circuit problems is of more value if impedance, as well as current and voltage, is also expressed as a complex number.

Although impedance is not a vector quantity, it may be represented by complex notation because there are associated with it both a magnitude and an angle (power-factor angle). The convention employed for impedance is that the angle associated with an inductive impedance is positive, and that associated with a capacitive impedance is negative. Thus, the impedance of a pure inductance L is represented by

$$\bar{Z}_L = \omega L \epsilon^{j(\pi/2)} = 0 + j\omega L \tag{8-77}$$

For a pure resistance R,

$$\bar{Z}_R = R\epsilon^{j0} = R + j0 \tag{8-78}$$

For a pure capacitance C,

$$\bar{Z}_C = \frac{1}{\omega C} \epsilon^{-j(\pi/2)} = 0 - j\frac{1}{\omega C} \tag{8-79}$$

By applying the complex notation for current, voltage, and impedance, and the processes of complex algebra (Eqs. 8–73 through 8–76), any a-c circuit problem can be treated exactly as if it were a d-c circuit with resistors replacing the impedances. For example, the net impedance of the series combination of pure resistance R and a pure inductance L is

$$\bar{Z} = R + j\omega L = \sqrt{R^2 + \omega^2 L^2} \; \epsilon^{j \tan^{-1}(\omega L/R)} \tag{8-80}$$

This solution is in agreement with that derived in Art. 8–8.

The circuit of Fig. 8–17 provides an example for the application of complex notation to an a-c circuit problem. In this circuit the constants E, L, C, and R are known, and the magnitude and phase of the current are required. The solution proceeds as follows:

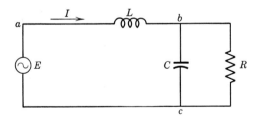

Fig. 8–17.

1. The current is found by dividing the source voltage by the total impedance. Thus,

$$I_{ab} = \frac{E_{ac}}{Z_T} \tag{1}$$

2. The total impedance is the sum of the inductive impedance and the impedance of the parallel combination of R and C. The impedance of the parallel combination is found, as for parallel resistors in a d-c circuit, from the relation

$$\frac{1}{Z_{bc}} = \frac{1}{Z_C} + \frac{1}{Z_R} \tag{2}$$

or

$$Z_{bc} = \frac{Z_C Z_R}{Z_C + Z_R} \tag{3}$$

Substituting the complex notation,

$$\bar{Z}_{bc} = \frac{-j\dfrac{R}{\omega C}}{R - j\dfrac{1}{\omega C}} \tag{4}$$

The complex fraction of Eq. 4 can be rationalized by multiplying both numerator and denominator by $\left(R + j\dfrac{1}{\omega C}\right)$, the **complex conjugate** of the denominator. Thus,

$$\bar{Z}_{bc} = \frac{-j\dfrac{R}{\omega C}\left(R + j\dfrac{1}{\omega C}\right)}{\left(R - j\dfrac{1}{\omega C}\right)\left(R + j\dfrac{1}{\omega C}\right)} = \frac{\dfrac{R}{\omega^2 C^2} - j\dfrac{R^2}{\omega C}}{R^2 + \dfrac{1}{\omega^2 C^2}} \tag{5}$$

Because the denominator of Eq. 5 is real, that equation may be considered to represent an impedance consisting of a resistance, R_s, in series with a capacitive reactance X_{cs} where

$$R_s = \frac{R/\omega^2 C^2}{R^2 + 1/\omega^2 C^2} = \frac{R}{1 + \omega^2 R^2 C^2} \tag{6}$$

and

$$X_{C_s} = \frac{R^2/\omega C}{R^2 + 1/\omega^2 C^2} = \frac{R^2 \omega C}{1 + \omega^2 R^2 C^2} \tag{7}$$

The total impedance of the circuit is the sum of the impedance of this parallel combination and that of the inductance L. Thus,

$$\bar{Z}_T = R_s + j(\omega L - X_{C_s}) \tag{8}$$

HARMONICS

3. The current is, therefore,

$$I_{ab} = \frac{E_{ac}}{\sqrt{R_s^2 + (\omega L - X_{C_s})^2}} \tag{9}$$

and the angle θ between the current and voltage is such that

$$\tan \theta = \frac{\omega L - X_{C_s}}{R_s} \tag{10}$$

Note. Had the voltage vector not been the reference but been such that, in polar form,

$$\bar{E} = E\underline{/\theta_1} \tag{11}$$

then

$$I = \frac{E\underline{/\theta_1}}{\bar{Z}_T\underline{/\theta}} = \frac{E}{Z_T}\underline{/\theta_1 - \theta} \tag{12}$$

where θ is such that Eq. 10 is applicable. Thus, in Eq. 12, the magnitude of the current vector is given by the same expression as Eq. 9, and the angle of the current vector by $(\theta_1 - \theta)$. As previously noted, the angle between the voltage and current vectors is θ.

Had actual magnitudes been given for the circuit constants in the problem, it is doubtful that the application of complex notation would have decreased the labor involved in the solution. However, such application provided a more direct approach to a solution than would have been possible without it. Furthermore, the use of complex notation made possible the derivation of general expressions for the circuit. Thus, Eq. 9 is a general expression for the magnitude of the current in terms of the applied voltage, circuit constants, and frequency.

8–16. HARMONICS

Sinusoidal wave forms of voltage and current have been considered in the preceding discussion of a-c circuits, even though, as previously indicated, the wave forms encountered in practice are not always sinusoidal. One justification for the assumption of sinusoidal quantities is the fact that any periodic wave, regardless of form, can be represented as the summation of a series of sine waves of multiple frequency and of different maximum values and phase positions. Thus, the resultant voltage wave of Fig. 8–18, assumed to be the output voltage of a generator, represents the summation of the voltages of three sine-wave generators in series, when each generator produces a sine-wave component as indicated.

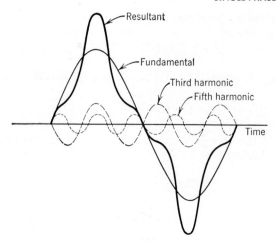

Fig. 8–18. Resultant voltage wave and harmonic components.

The three sine waves that combine to make up the resultant voltage wave in Fig. 8–18 are of different amplitudes and frequencies. One wave has the frequency of the resultant wave and is called the **fundamental**; the other waves have frequencies that are three and five times the frequency of the fundamental and are called the third and fifth **harmonics**, respectively.

The equation for any distorted periodic voltage wave can be expressed by Fourier's series in the form

$$e = E + E_{m1} \sin (\omega t + a_1) + E_{m2} \sin (2\omega t + a_2)$$
$$+ E_{m3} \sin (3\omega t + a_3) + \cdots + E_{mn} \sin (n\omega t + a_n) \quad (8\text{–}81)$$

where E is a direct component (or average value of the voltage), $E_{m1} \sin (\omega t + a_1)$ is the fundamental, and the other terms are the harmonics. Although this series has an infinite number of terms, only a few terms are required to express the essential features of most wave forms found in practical applications.

In general, the voltage wave forms have positive and negative alternations that are equal in magnitude and similar in shape. In such cases, the d-c component and the even harmonics are not present, and the equation involves only the fundamental and the odd harmonics. Thus, the equation for the wave in Fig. 8–18 is given as

$$e = E_{m1} \sin \omega t + E_{m3} \sin (3\omega t + 180°) + E_{m5} \sin 5\omega t \quad (8\text{–}82)$$

The effect of the harmonic content of a given voltage wave form upon a circuit depends upon the amplitude and frequency of the

HARMONICS

sine waves that are assumed to compose the wave form. To determine this effect, the fundamental and each harmonic may be assumed to produce independent currents in the circuit under consideration. Since there is only one value of current at any instant, the wave form of the resultant current is then found by an algebraic summation of the independent currents at each instant. (This method is known as superposition.)

As an example, if the voltage wave expressed by Eq. 8–82 is applied to a pure resistor, the current at any instant is expressed as

$$i = \frac{E_{m1}}{R} \sin \omega t + \frac{E_{m3}}{R} \sin (3\omega t + 180°) + \frac{E_{m5}}{R} \sin 5\omega t \quad (8\text{–}83)$$

or

$$i = I_{m1} \sin \omega t + I_{m3} \sin (3\omega t + 180°) + I_{m5} \sin 5\omega t \quad (8\text{–}84)$$

Comparison of Eqs. 8–82 and 8–84 reveals that the current wave for the resistance circuit has the same form as the voltage wave.

If the voltage wave given by Eq. 8–82 is impressed on a pure inductor of constant inductance, the current at any instant is expressed as

$$i = \frac{E_{m1}}{\omega L} \sin (\omega t - 90°) + \frac{E_{m3}}{3\omega L} \sin (3\omega t + 90°) + \frac{E_{m5}}{5\omega L} \sin (5\omega t - 90°) \quad (8\text{–}85)$$

As indicated in Eq. 8–85, the current contributions of the fundamental and harmonic voltages lag these voltages by 90 degrees. Also, as the higher-order harmonics are reached, the current contributions of these harmonics become more negligible because of the increased reactance at the increased frequency. Thus, if the reactance at the fundamental frequency for the inductor circuit is equal to the resistance in the resistor circuit, the amplitude of the current in the inductor due to the fifth-harmonic voltage is only one fifth of that in the resistor due to the same voltage. Thus, the resultant current in the inductor tends to have a wave form more nearly sinusoidal than the current in the resistor.

If the same voltage wave form is impressed across a pure capacitor of constant capacitance, the current at any instant is expressed as

$$i = \frac{E_{m1}}{1/\omega C} \sin (\omega t + 90°)$$

$$+ \frac{E_{m3}}{1/3\omega C} \sin (3\omega t - 90°) + \frac{E_{m5}}{1/5\omega C} \sin (5\omega t + 90°) \quad (8\text{–}86)$$

As indicated in Eq. 8–86, the current contributions of the fundamental and harmonic voltages lead these voltages by 90 degrees. Also, as the higher-order harmonics are reached, the current contributions of the higher harmonic voltages become proportionately greater because of the decreased reactance at the higher frequency. If the resistance in the resistor circuit is equal to the capacitive reactance at the fundamental frequency in this example, the amplitude of the current in the capacitor due to the fifth-harmonic voltage is five times that in the resistor due to the same voltage. Thus, the resultant current wave in the capacitor circuit is more distorted than the voltage wave that causes it.

These examples are included to give a general picture of the effects of distorted wave forms upon circuits. Calculations involving nonsinusoidal voltages and circuits that consist of combinations of resistance, inductance, and capacitance become rather complex and, by necessity, are beyond the scope of this text.

PROBLEMS

8–1. A voltage of the form $141.4 \sin 377t$ is impressed across a resistor of 10 ohms.
  (a) What is the expression for current?
  (b) What is the average rate of energy dissipation?
  (c) What is the effective value of current?
  (d) What is the frequency of the supply?
  (e) Draw a vector diagram.
  (f) What is the power factor for this circuit?

8–2. The voltage of Prob. 8–1 is impressed across an inductor of 0.02 henry. The inductor is assumed to have negligible resistance.
  (a) What is the expression for current?
  (b) What is the average rate of energy dissipation?
  (c) What is the effective value of current?
  (d) Draw a vector diagram.
  (e) What is the power factor for this circuit?

8–3. The voltage of Prob. 8–1 is impressed across a capacitor of 100 μf. The capacitor is assumed to be perfect.
  (a) What is the expression for current?
  (b) What is the average rate of energy dissipation?
  (c) What is the effective value of current?
  (d) Draw a vector diagram.
  (e) What is the power factor for this circuit?

8–4. An air-core coil is connected to a 120-volt 60-cycle line. The power taken is 60.0 watts, and the current is 1.5 amperes. Compute:
  (a) The apparent power. *Ans.* 180 va.
  (b) The power factor. *Ans.* 0.333 lag.
  (c) The resistance and the inductance of the coil.
                                        Ans. 26.7 ohms; 0.2 henry.

8–5. (a) What would be the power and the current if the inductor of

Prob. 8–4 were connected to a 120-volt 30-cycle supply? (The resistance of the coil is assumed to be constant.) *Ans.* 180 watts; 2.6 amperes.

(b) Repeat (a) for a 120-volt d-c supply. *Ans.* 4.5 amperes; 540 watts.

8–6. (a) What value of capacitance should be connected in series with the inductor of Prob. 8–4 to cause the resulting circuit to have unity power factor when the combination is connected to a 120-volt 60-cycle supply?

Ans. 35.2 µf.

(b) Compute the current and the voltage across the inductor and capacitor for the circuit as connected in (a).

Ans. 4.5 amperes; $E_L = 360$ volts; $E_C = 339$ volts.

8–7. What size (µf and kva) capacitor should be connected in parallel with the inductor of Prob. 8–4 to cause the resulting circuit to appear as a unity power-factor load when connected to a 120-volt 60-cycle supply?

Ans. 31.1 µf; 0.169 kva.

8–8. The coil of Prob. 8–4 is so built that an iron core can be inserted through it. The previous information concerning this coil is for the condition where the iron core is removed. If the coil is connected to the 120-volt 60-cycle supply and the iron core is then inserted, how do the current, resistance, inductance, and power factor compare to the values given or calculated in Prob. 8–4?

8–9. (a) Show that it is entirely appropriate to consider a coil to be made up of a resistor and an inductor in series instead of as a resistor and an inductor in parallel.

(b) Show that it is entirely appropriate to consider a nonpure capacitor to be made up of a resistor and a pure capacitor in parallel instead of as a resistor and a capacitor in series.

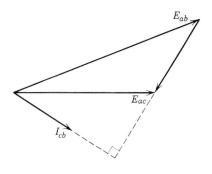

Fig. 8–19.

8–10. Draw the circuit diagram for which the vector diagram shown as Fig. 8–19 is applicable. Label the circuit diagram completely.

8–11. (a) Plot current versus time for a capacitor that has an alternating voltage of triangular wave form applied to it.

(b) Plot voltage versus time for a pure inductor that has an alternating current of triangular wave form flowing through it.

8–12. A resistor of 10 ohms, a pure inductor of 1 henry, and a pure capacitor of 313 μf are to be connected in series. Sketch the approximate curve of impedance versus frequency for this combination, indicating significant values on both the impedance and frequency axes.

8–13. Sketch the curve of impedance versus frequency for each of the following connections:

(a) R and L in series.
(b) R and C in series.
(c) R, L, and C in series.
(d) R and L in parallel.
(e) R and C in parallel.
(f) L and C in parallel.
(g) R, L, and C in parallel.

8–14. It is desired to have the impedance between two points of a circuit as low as possible for a frequency of 60 cycles but as high as possible at all other frequencies. What simple arrangement of elements of pure resistance, inductance, or capacitance would you use? Indicate relative values of the components. At a frequency of 30 cycles, would this impedance appear inductive or capacitive?

8–15. Compute the frequency at which a series circuit of resistance, inductance, and capacitance (all pure elements) is resonant. *Ans.* $f = 1/2\pi\sqrt{LC}$.

8–16. An a-c relay coil is found to have a resistance of 6 ohms and an inductance of 0.0212 henry. The coil is normally connected to a 60-cycle line. In an attempt to improve the performance of the relay, you connect a capacitor in series with it, but find that the current and power remain the same as before the capacitor was added. How much capacitance was added?
Ans. 166 μf.

8–17. (a) For the circuit of Fig. 8–20, what is the magnitude of the voltage E_{ad} when the three-wire supply is a 60-cycle supply having the magnitudes shown, the inductor L has a reactance of 4 ohms, and the resistance R is 3 ohms?
Ans. 150 volts.

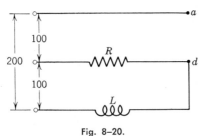

Fig. 8–20.

(b) For the same circuit, what is the magnitude of the voltage E_{ad} when the three-wire supply is a d-c supply having the voltage magnitudes shown?
Ans. 200 volts.

(c) What would be the effect in (a) on the voltage E_{ad} if the inductor is replaced by a capacitor having a reactance of 4 ohms?

(d) For the circuit of (c), what is the magnitude of the voltage E_{ad} when the three-wire supply is a d-c supply?
Ans. 100 volts.

8–18. An unknown circuit element takes 60 watts and 1.5 amperes from a 120-volt 60-cycle a-c source. When a capacitor of capacitance $C = 35.2 \times 10^{-6}$ farad is connected in series with the unknown circuit element, the current increases to 4.5 amperes, while the power increases to 540 watts.

(a) Determine the unknown circuit element.
(b) Is the answer to (a) the only possibility? Explain.

8–19. A load connected to a 220-volt 60-cycle source takes a power of 1760 watts and has a power factor of 0.8 lagging. What capacitance should be connected in series with this load to reduce the current to one-half the original value? *Ans.* 4.95 μf.

8–20. What resistance should be connected in series with the load of Prob. 8–19 to reduce the current to one-half the original value? *Ans.* 24.4 ohms.

8–21. You receive an impedance box to be used in connection with a certain circuit. From 60-cycle current, voltage, and power data you decide that the box has an impedance of 10 ohms and a power factor of 0.6. Upon opening the box, you find that the box contains a resistor connected in parallel with a capacitor. What is the resistance of the resistor and the reactance of the capacitor? *Ans.* $R = 16.6$ ohms; $X_C = 12.5$ ohms.

8–22. (*a*) An a-c source consisting of an emf E and a series impedance Z of power factor angle θ (inductive) is used to supply power to a variable resistance R. For what value of R is the power consumed in R a maximum? *Ans.* $R = Z$.

(*b*) If additional circuit elements (resistance, inductance, capacitance) of suitable magnitudes are available, show how more power can be taken out of the supply source.

8–23. The conductors connecting a load to a 60-cycle single-phase supply have a total resistance of R ohms and a total inductive reactance of X ohms. The rated current of the conductors is I amperes. Draw vector diagrams showing the relation of the load voltage E_l to the supply voltage E_s for each of the following load conditions:

(*a*) A pure-capacitive load of I amperes.
(*b*) A pure-resistive load of I amperes.
(*c*) A pure-inductive load of I amperes.

8–24. (*a*) Prove that multiplication of two vectors, such as given by eqs. 8–71 and 8–72, using the rectangular form gives the same result as using the polar form.

(*b*) Repeat (*a*) for the division of two vectors.

8–25. (*a*) Determine the total impedance and the power factor of the circuit shown as Fig. 8–21. *Ans.* 3.54 ohms; 0.99 lead.

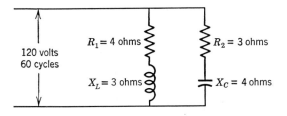

Fig. 8–21.

(*b*) What current would the circuit of Fig. 8–21 take if it were connected to a 120-volt d-c supply? Explain briefly. *Ans.* 30 amperes.

8-26. (*a*) In the circuit of Fig. 8–22, what value of inductance is necessary to cause the complete circuit to act as a unity power factor load at 60 cycles? (The inductor is assumed to have negligible resistance.)

Ans. 0.0159 henry.

(*b*) What current will flow in the circuit of Fig. 8–22 for a 220-volt d-c supply? *Ans.* 17.6 amperes.

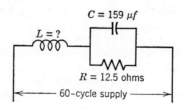

Fig. 8–22.

8-27. (*a*) A 100-turn air-core coil of negligible resistance is connected to a 120-volt 60-cycle supply. What is the maximum value of flux linking the coil? *Ans.* 450,000 lines.

(*b*) What would be the effect, upon the current and maximum flux, of inserting an iron core in the coil of (*a*)?

8-28. The exciting coil of an a-c electromagnet is normally connected to a 120-volt 60-cycle supply. If it is assumed that the wire of the coil has negligible resistance, would a change in the number of turns affect the lifting power of the magnet? *Justify your answer.*

8-29. A 100-turn air-core coil having negligible resistance is connected to a 120-volt 60-cycle supply and takes a current of 1 ampere. What current will the coil take when connected to the same supply if the number of turns is reduced to 50? Explain. *Ans.* 4 amperes.

8-30. (*a*) The two coils shown in Fig. 8–23 are identical in every respect and have a common iron magnetic circuit. Coil 1 draws 1 ampere when connected to a 120-volt 60-cycle supply as shown. What will be the total current taken by the combination if coil 2 is connected in parallel with coil 1 by joining *c* to *a* and *d* to *b*? Explain. (Coil resistance and leakage flux may be considered to be negligible.) *Ans.* 1 ampere.

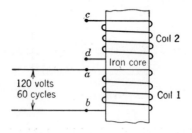

Fig. 8–23.

PROBLEMS 171

(b) What current will flow if d is connected to a and a 240-volt 60-cycle supply is connected between c and b? *Ans.* 0.5 ampere.

8–31. Figure 8–24 shows a circuit consisting of a resistance load connected in series with two identical iron-core coils to an a-c supply. One leg of the iron core has a coil connected to a d-c supply. The current in the load is found to be a function of the current in the d-c coil. Explain.

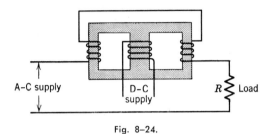

Fig. 8–24.

8–32. Figure 8–25 shows a circuit commonly used to produce a shift in the phase of a voltage without affecting the magnitude of the voltage. The shift in phase is produced by adjustment of the rheostat in series with the capacitor. Using vector diagrams, prove that a change in the rheostat setting does not affect the magnitude of the voltage E_{ab}.

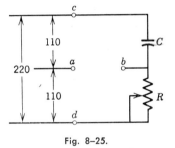

Fig. 8–25.

8–33. For the circuit of Fig. 8–26, determine the magnitude of the voltage E_{ab}. *Ans.* 132 volts.

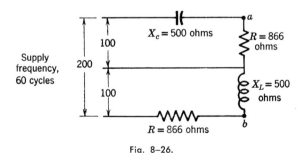

Fig. 8–26.

8–34. The nameplate of a certain device reads as follows: 120 volts, 6 kw, 80% power factor.
(a) What is the rated current of this device? *Ans.* 62.5 amperes.

(b) The power factor of the device could be lagging or leading. How do you prove conclusively what it actually is, if the only instruments available are ammeters? Assume that any other equipment that might prove helpful is available. Show reasoning for your answer.

8–35. A certain device connected to a 120-volt 60-cycle supply takes 10 amperes and 720 watts. Determine:

(a) The series combination of resistance and inductive reactance to which the device is equivalent (at 60 cycles).

Ans. $R = 7.2$ ohms; $X_L = 9.6$ ohms.

(b) The parallel combination of resistance and inductive reactance to which the device is equivalent (at 60 cycles).

Ans. $R = 20$ ohms; $X_L = 15$ ohms.

(c) Whether the two combinations of resistance and inductance in (a) and (b) are equivalent at 120 volts, 30 cycles.

8–36. (a) Assume that you have a radio, a desk lamp, and an electric fan in your room. How are these elements connected: in series or in parallel? Why?

(b) An electric installation receives power from a 117-volt 60-cycle line through cables that are rated at 250 amperes. The installation consists of the following:

Load	Rating	Efficiency	Power Factor
(A) Pump motor	15 hp	85%	80% lagging
(B) Refrigerator motor	5 hp	85%	90% leading

(The ratings of the motor are outputs.)

What is the maximum unity power-factor lighting load, in addition to loads A and B, that could be connected at this installation without exceeding the cable rating? Ans. 10.7 kw.

8–37. An induction motor (which normally has a lagging power factor) takes 24 amperes from a 220-volt 60-cycle line when running at full load. The motor delivers 4.8 hp under these conditions at an efficiency of 84%. It is desired to bring the power factor of the line to 95% lagging by connecting a capacitor in parallel with the motor. What size capacitor must be used (μf and kva)? Ans. 93.3 μf; 1.7 kva.

8–38. A load center of a small industrial plant is to be connected to the 130-volt 60-cycle supply, which is 200 ft away, by means of two No. 0 solid copper conductors which are to be 1 ft apart. (Resistance of No. 0 wire is 0.538 ohm per mile. Inductive reactance per single conductor of No. 0 at 1-ft spacing and at 60 cycles is 0.546 ohm per mile.)

(a) The load at the load center consists of a capacitive impedance of 1.19 ohms, with a power factor of 0.866, connected in parallel with an inductive impedance of 0.685 ohm with a power factor of 0.5. If the voltage at the supply remains constant at 130 volts, what is the voltage at the load?

Ans. 120 volts.

(b) Draw a complete vector diagram that clearly shows all the voltages and currents involved in the connection of the load center to the supply.

PROBLEMS

8–39. For the circuit shown in Fig. 8–27:

(a) Approximately what percentage of the amplitude of the alternating voltage appears across the 5000-ohm load resistor? *Ans.* 75%.

(b) Approximately what percentage of the direct voltage appears across the 5000-ohm load resistor? *Ans.* 91%.

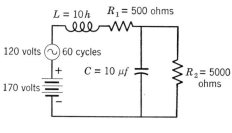

Fig. 8–27.

■ 9 Polyphase A-C Circuits

9–1. INTRODUCTION

The a-c circuits considered previously in this book may be classified as single-phase circuits. A single-phase circuit is "either an a-c circuit which has only two terminals or one which, having more than two terminals, is intended to be so energized that the potential differences between all pairs of terminals are either in phase or differ in phase by 180 degrees." * Schematic diagrams and vector diagrams for some circuits that are classified as single-phase circuits according to the definition are shown as Fig. 9–1. In each example, the load may be considered to consist of one impedance or a network of impedances.

In Fig. 9–1a, the load is supplied by a single a-c generator with only two terminals; this circuit represents the simplest of all single-phase circuits. The circuit of Fig. 9–1b is similar to that of Fig. 9–1a in that the load has only two terminals. However, the source in Fig. 9–1b consists of two a-c generators in series. Although the two generators produce voltages 90 degrees out of phase, the resultant voltage supplied to the load is a single alternating voltage and the circuit is a single-phase circuit. Therefore, the circuit of Fig. 9–1b can be represented by an equivalent circuit similar to that shown in Fig. 9–1a.

The circuit shown in Fig. 9–1c is a three-wire circuit commonly used by power companies for residential power distribution. In the example shown, the two loads might represent two different residences or two blocks of load within an individual residence. As indicated by the

---
* American Standard Definitions of Electrical Terms.

INTRODUCTION

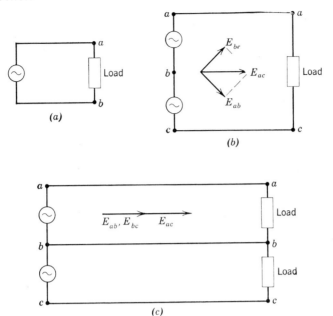

Fig. 9-1. Single-phase circuits.

vector diagram, the two sources are in series and have voltages of the same frequency and in phase. The voltages across the individual loads are therefore in phase or 180 degrees out of phase, depending upon the manner in which the voltages are measured, and the circuit is a single-phase circuit. In practical application, as found in residential distribution circuits, the voltages across the individual loads would be nominally 115 volts and the total voltage nominally 230 volts.

A polyphase circuit may be defined as a circuit with more than two terminals and intended to be so energized that the potential differences between pairs of terminals are of the same frequency but differ in phase. Thus, if the two generators shown in Fig. 9-1b are connected to the loads shown in Fig. 9-1c, the resulting circuit is a polyphase circuit. It would, in fact, be termed a two-phase circuit.

Another definition states that a polyphase circuit is an interconnection of a group of single-phase circuits, the voltages of which are of the same frequency but differ in phase. The polyphase circuit relations developed in subsequent articles are derived from the interconnection of such single-phase circuits.

Although the term polyphase covers circuits of two or more phases, this chapter is devoted only to circuits classified as three-phase, because this is the predominant type of polyphase circuit. Two-phase and six-phase circuits find some application, and these connections are considered when the devices employing them are discussed.

9–2. THREE-PHASE FOUR-WIRE CIRCUIT

The voltage–time diagrams for the three independent single-phase generators of Fig. 9–2a are shown in Fig. 9–2b. As indicated, the

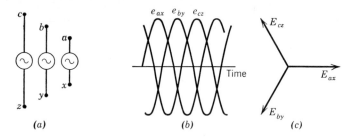

Fig. 9–2. Three single-phase circuits.

voltages have the same magnitude and frequency, but differ in phase by 120 degrees. These phase relationships are shown more clearly in the vector diagram of Fig. 9–2c, which is drawn to relate the voltages in time even though the generators are physically independent.

If a connection is made between terminals x and y of the generators shown in Fig. 9–2a, no current flows in the connection because a circuit has not been completed (assuming that the capacitive reactance of the open circuit between terminals a and b is infinite). Similarly, a connection could be made between y and z without current flow. The connection of these terminals is shown in Fig. 9–3, and, as indicated, the common terminal is labeled n.

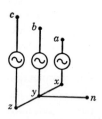

Fig. 9–3. Three-phase source.

Although the connection of terminals x, y, and z does not cause current to flow, it does relate the voltages of the generators physically. Thus, the potential difference between terminals a and b is now directly related to the voltages E_{ax} and E_{by}, whereas before the connection is made the potential difference between these terminals is indeterminate. The voltage between the termi-

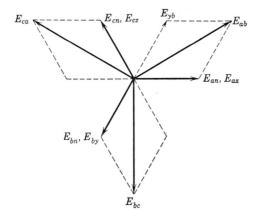

Fig. 9-4. Three-phase voltage relationships.

nals a and b after the connection is made is found from the relation

$$\bar{E}_{ab} = \bar{E}_{ax} + \bar{E}_{yb} \tag{9-1}$$

or

$$\bar{E}_{ab} = \bar{E}_{ax} - \bar{E}_{by} \tag{9-2}$$

and is shown on the vector diagram of Fig. 9-4. The voltages E_{bc} and E_{ca}, found in a similar manner, are also shown on this diagram.

Because the voltage E_{ab} is the vector sum of two equal vectors that are 60 degrees apart, it has a magnitude given by

$$E_{ab} = \sqrt{3}\, E_{ax} \tag{9-3}$$

The interconnection of the three single-phase generators creates a single source having four terminals such that the definition of a polyphase source is satisfied; that is, it has more than two terminals and the potential differences between pairs of terminals are of the same frequency but differ in phase. A three-phase source is characterized by the fact that there are three equal voltages that differ in phase by 120 degrees. This particular connection has two sets of such voltages and is known as a three-phase four-wire source.

In the connection shown in Fig. 9-3, the terminals a, b, and c are called **line terminals,** the terminal n the **neutral terminal.** Therefore, voltages measured between line terminals are called **line voltages;** voltages measured from a line terminal to the neutral terminal are called **line-to-neutral** or **phase, voltages.** Each of the three generators makes up one phase of the three-phase supply.

As indicated in Fig. 9-4, the phase sequence of the supply voltages is

E_{an}, E_{bn}, E_{cn}. Instead of using voltage notation, the usual practice in designating the phase sequence of a three-phase supply is to apply the letters or numbers that refer to the given phases. Thus, the phase sequence of the three-phase supply shown in Fig. 9–3 is given as a–b–c.

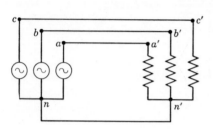

Fig. 9–5. Connection of single-phase loads to three-phase generator.

Three independent single-phase loads may be connected from the line terminals to the neutral terminal, as shown in Fig. 9–5, and remain independent if the impedance of the neutral conductor is negligible. Under these conditions, the neutral terminal of the source is at the same potential as the neutral terminal of the loads, and the line terminals of the source are at the same potentials as the corresponding line terminals at the load. Therefore, the voltages from the lines to neutral at the load equal the voltages from the lines to neutral at the source. A change in the impedance of any or all of the loads results only in a change in the current taken by the altered loads; there is no change in the magnitude or phase of the line-to-neutral voltage of any of the phases (assuming the impedances of lines and generators to be negligible).

At first glance it appears that the neutral conductor must have a current-carrying capacity equal to the sum of the capacities of the three line conductors. However, since the neutral current is the vector sum of the three line currents, the magnitude of the neutral current depends not only on the magnitudes of the line currents but also on their phase relations. In most circuits encountered in practice, the neutral current is equal to or smaller than any one of the line currents and may even be zero. (The exceptions are those circuits having severe load unbalance. A **balanced three-phase load** is one that has the same impedance and power factor in each phase.) Thus, the neutral current for the unbalanced three-phase load shown in Fig. 9–6a is shown on the vector diagram of Fig. 9–6b to be smaller than any one of the line currents. For this connection, the loads on the three phases are assumed as follows:

Phase a: Z_a, lagging power factor.
Phase b: Z_b, unity power factor.
Phase c: Z_c, leading power factor.

The total average power taken by the unbalanced three-phase load of

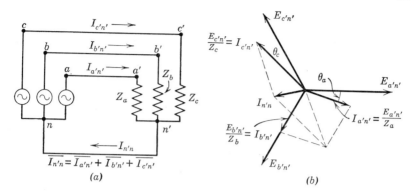

Fig. 9-6. Three-phase four-wire circuit.

Fig. 9-6 is the sum of the powers taken by the individual loads. Thus,

$$P_t = P_a + P_b + P_c \tag{9-4}$$

or

$$P_t = E_{a'n'}I_{a'n'}\cos\theta_a + E_{b'n'}I_{b'n'}\cos\theta_b + E_{c'n'}I_{c'n'}\cos\theta_c \tag{9-5}$$

9-3. BALANCED WYE-CONNECTED LOADS

If a balanced three-phase load is connected to a three-phase four-wire supply, the neutral current is zero. The vector diagram for such a load is shown as Fig. 9-7a. As indicated, the phase currents are equal and 120 degrees apart; therefore, the vector sum of the currents is zero.

Because the neutral current is zero, the neutral conductor may be removed without affecting the remainder of the system. The line voltages are equal and 120 degrees apart, as they were originally, and, since the impedances and power factors of the three phases are identical, the phase voltages remain unchanged. The three-phase load, without the neutral conductor, is shown as Fig. 9-7b; this type of load connection is called a **wye** connection.

Balanced three-phase loads, such as three-phase motors, are often wye-connected in such a manner that the neutral point is not accessible; that is, only three terminals of the load are available. For such connections, all measurements must be made at the line conductors. The relationship between line and phase quantities is easily determined by inspection of the connection and vector diagrams. Thus, for a balanced

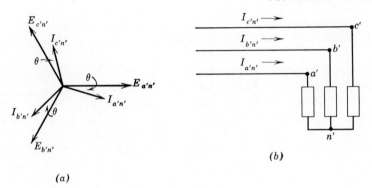

Fig. 9-7. Balanced wye circuit.

wye-connected load,

$$I_{a'n'} = I_{b'n'} = I_{c'n'} = I_p \tag{9-6}$$

$$I_p = I_L \tag{9-7}$$

$$E_{a'n'} = E_{b'n'} = E_{c'n'} = E_p \tag{9-8}$$

$$E_L = \sqrt{3}\, E_p \tag{9-9}$$

where the subscripts p and L denote phase and line quantities, respectively.

The total average power taken by the balanced wye-connected load is the sum of the individual powers, or

$$P_t = E_{a'n'}I_{a'n'} \cos \theta + E_{b'n'}I_{b'n'} \cos \theta + E_{c'n'}I_{c'n'} \cos \theta \tag{9-10}$$

Since the phase voltage, phase current, and phase power factor are the same for the three phases, Eq. 9–10 may be written

$$P_t = 3E_p I_p \cos \theta \tag{9-11}$$

If the voltage and current in Eq. 9–11 are expressed in terms of line quantities,

$$P_t = 3\frac{E_L}{\sqrt{3}} I_L \cos \theta = \sqrt{3}\, E_L I_L \cos \theta \tag{9-12}$$

Equation 9–12 differs from power relations previously discussed in this book in that the power factor in Eq. 9–12 is the cosine of the angle between phase voltage and phase current and not the cosine of the angle between line voltage and line current. Equation 9–12 is derived for the purpose of providing a convenient power relation for a balanced three-phase load.

9-4. POLYPHASE VECTOR DIAGRAMS

Two vector diagrams that apply to the phase and line voltages of four-wire circuits and balanced wye-connected circuits are shown as Figs. 9–8a and 9–8b. The vector diagram of Fig. 9–8a is similar to those previously considered; the diagram of Fig. 9–8b has the same elements but has the line-voltage vectors so located that they complete an equilateral triangle. The formation of the triangle is in agreement with Kirchhoff's law of voltage as applied to the traverse involving the line terminals; that is,

$$\overline{E_{a'b'}} + \overline{E_{b'c'}} + \overline{E_{c'a'}} = 0 \tag{9-13}$$

A variation of the vector diagram of Fig. 9–8b is shown as Fig. 9–8c. In this diagram, the corners of the triangle are labeled with the letters identifying the line terminals; the origin of the phase-voltage vectors is labeled as the neutral terminal. The points on the diagram so labeled are treated as if they were the actual terminals of the circuit. Thus, the distance between any two points on the diagram represents the voltage between those points in the circuit. This type of diagram is known as a topographic vector diagram.

The principal advantage of the topographic diagram is its application to wye-connected loads that are unbalanced. This type of load is not frequently encountered in practice and, when it does occur, is usually the result of a fault on the system.

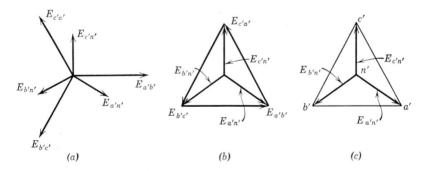

Fig. 9-8. Three-phase voltage diagrams.

9–5. UNBALANCED WYE-CONNECTED LOADS

Fig. 9–9a shows the connection of a balanced wye-connected load to a three-phase supply. The topographic vector diagram applying to this circuit is shown as Fig. 9–9b.

To illustrate the effects of an unbalanced wye-connected load on the three-phase system, it is assumed that the load element connected between terminals c' and n' is short-circuited (as shown by the dashed line in Fig. 9–9a). Terminal c' is then at the same potential as terminal n'; therefore, voltages $E_{a'n'}$ and $E_{b'n'}$ are equal to $E_{a'c'}$ and $E_{b'c'}$, respectively. Since the magnitudes and phase relations of the line voltages are fixed by the source, the voltages of the unfaulted phases of the load must rise until they are equal to the line voltages indicated in the preceding sentence.

The effect of the short circuit is represented on the topographic diagram by the transfer of the neutral point from the center of the triangle to a position coincident with the terminal c', as shown in Fig. 9–9c. Examination of Fig. 9–9c reveals that the diagram portrays the voltage conditions that exist during the presence of the short circuit on the system.

The purpose in using the short-circuited phase as an example of an unbalanced wye-connected load is to demonstrate, as simply as possible, that the potential of the neutral with respect to the line terminals in a wye-connected system is variable and depends upon the impedances of the phases. Thus, in this example of unbalance, the neutral shifted from the normal potential position for a balanced wye-connected load to a position at the vertex of the topographic triangle.

If phase c' is not short-circuited but has a smaller load impedance than the other two phases, the shift in the neutral is not so great but

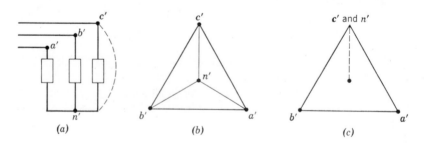

Fig. 9–9. Application of topographic vector diagram.

DELTA-CONNECTED LOADS

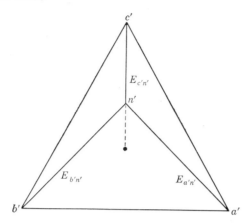

Fig. 9-10. Topographic vector diagram.

is, nevertheless, toward the terminal c', thus changing the magnitudes of the voltages on the other phases. The topographic vector diagram for this condition ($Z_{a'n'}$ equal to $Z_{b'n'}$ and greater than $Z_{c'n'}$; equal phase power factors) is shown as Fig. 9-10.

As indicated, the potential of the neutral with respect to the line terminals in a wye-connected load depends upon the degree of load unbalance. Topographically, the position of the neutral is not limited to the inside of the triangle formed by the line voltages; in some instances of severe load unbalance, it may appear outside this triangle.

The conclusion to be drawn from this discussion of unbalanced wye-connected loads is that, because of the variation in phase voltages on an unbalanced wye-connected system, individual single-phase loads should not be connected in wye unless they form a balanced three-phase load at all times.

9-6. DELTA-CONNECTED LOADS

Figure 9-11 shows the three elements of a three-phase load connected in **delta**. As shown, each phase of the load is connected between a pair of line terminals; therefore, the voltage of each phase of the load is equal to the line voltage.

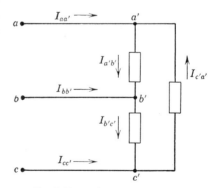

Fig. 9-11. Delta-connected circuit.

The phase currents for the delta-connected load are determined from the voltage of the phase (the line voltage) and the impedance of the phase. The line currents depend upon the currents in adjacent phases and, for the circuit of Fig. 9–11, are found as follows:

$$\overline{I}_{aa'} = \overline{I}_{a'b'} - \overline{I}_{c'a'} \qquad (9\text{--}14)$$

$$\overline{I}_{bb'} = \overline{I}_{b'c'} - \overline{I}_{a'b'} \qquad (9\text{--}15)$$

$$\overline{I}_{cc'} = \overline{I}_{c'a'} - \overline{I}_{b'c'} \qquad (9\text{--}16)$$

An example of a vector diagram for an unbalanced delta-connected load is shown as Fig. 9–12a.

When the delta-connected load is balanced, the phase currents are equal and 120 degrees apart; therefore, the line currents are also equal and 120 degrees apart, as shown in Fig. 9–12b. Each line current is therefore equal to $\sqrt{3}$ times a phase current.

The total average power dissipated by a delta-connected load is the sum of the phase powers. Thus,

$$P_t = E_{a'b'}I_{a'b'}\cos\theta_{a'b'} + E_{b'c'}I_{b'c'}\cos\phi_{b'c'} + E_{c'a'}I_{c'a'}\cos\phi_{c'a'} \qquad (9\text{--}17)$$

Since, for a balanced load, the phase voltages, phase currents, and power factors are equal, Eq. 9–17 may be written

$$P_t = 3E_p I_p \cos\theta \qquad (9\text{--}18)$$

If the voltage and current in Eq. 9–18 are expressed in terms of line quantities,

$$P_t = 3(E_L)\frac{I_L}{\sqrt{3}}\cos\theta = \sqrt{3}\,E_L I_L \cos\theta \qquad (9\text{--}19)$$

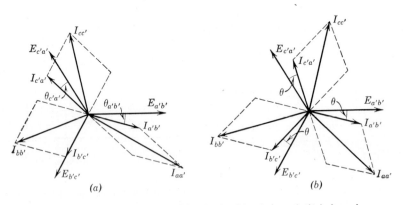

Fig. 9–12. Vector diagrams for delta circuits; (a) unbalanced, (b) balanced.

THREE-PHASE RATINGS

Equation 9–19 is the same as the power expression previously developed for the balanced wye-connected load, and, as in that expression, the power factor is the cosine of the angle between phase voltage and phase current, not the cosine of the angle between line voltage and line current.

9–7. THREE-PHASE RATINGS

The voltage ratings of three-phase supplies are specified in terms of effective line values. Thus, a 220-volt three-phase supply is one having 220 volts between lines. The voltages ratings of three-phase loads, regardless of the type of load connection, are similarly specified.

Three-phase load units are usually balanced; therefore, a single power rating and a single current rating are given for the unit. The current rating is the line current, regardless of the type of load connection.

The power rating of a three-phase load refers to the total power dissipated in the unit. The power dissipated in each phase is one-third the total power.

PROBLEMS

9–1. The three line leads of a 208-volt three-phase four-wire supply are designated as A, B, and C, respectively. The fourth wire, or neutral, is designated as N. The phase sequence is A–B–C. Compute the currents in the four wires when the following loads are connected to this supply:

From A to N: 10 kw, unity power factor
From B to N: 15 kva, 0.866 lag
From C to N: 15 kva, 0.866 lead

Ans. $A = 83.3$ amperes; $B = 125$ amperes; $C = 125$ amperes; $N = 133.2$ amperes.

9–2. Repeat 9–1 for the load from C to N removed.
Ans. $A = 83.3$ amperes; $B = 125$ amperes; $C = 0$; $N = 67.2$ amperes.

9–3. Repeat 9–1 for the loads from C to N and B to N removed.
Ans. $A = 83.3$ amperes; $B = C = 0$; $N = 83.3$ amperes.

9–4. The voltage measured between terminals of a three-phase three-wire generator is recorded as 208 volts. A three-phase load consisting of three 10-ohm resistors in wye is connected to the terminals of the generator.
(*a*) Draw a connection diagram.
(*b*) What should be the voltage rating, the current rating, and the power rating of each load resistor? *Ans.* 120 volts; 12.0 amperes; 1440 watts

(c) If one of the resistors should become open-circuited, what would be the current, the voltage, and the power of the remaining resistors?
Ans. 104 volts; 10.4 amperes; 1081.6 watts.

(d) Using double-subscript notation, draw a complete vector diagram for the original connection (all resistors intact).

(e) Draw a complete vector diagram for (c).

9–5. The resistance of one of the resistors of the three-phase load used in Prob. 9–4 is variable from 0 to infinity (short circuit to open circuit). If this resistor is varied between its limits, what are the limits of the voltage across each of the other resistors? *Ans.* 104 volts; 208 volts.

9–6. The voltage measured between terminals of a three-phase three-wire generator is recorded as 208 volts. A three-phase load consisting of three 10-ohm resistors in delta is connected to the terminals of the generator.

(a) Draw a connection diagram.

(b) What should be the voltage rating, the current rating, and the power rating of each load resistor? *Ans.* 208 volts; 20.8 amperes; 4326 watts.

(c) What are the line currents? *Ans.* 36 amperes.

(d) If one of the resistors should become open-circuited, what would be the current and the voltage of the remaining resistors?
Ans. 20.8 amperes; 208 volts.

(e) Determine the line currents for (d). Indicate the lines in which the currents exist. *Ans.* 20.8 amperes; 20.8 amperes; 36 amperes.

(f) Using double-subscript notation, draw a complete vector diagram for the original connection (all resistors intact).

(g) Draw a complete vector diagram for (d), including the line currents calculated in (e).

9–7. The resistance between two terminals of a three-phase loading unit (assumed balanced) is measured as 12 ohms.

(a) What is the resistance of each phase if the unit is wye-connected?
Ans. 6 ohms.

(b) What is the resistance of each phase if the unit is delta-connected?
Ans. 18 ohms.

9–8. (a) A 440-volt three-phase three-wire line supplies power to a wye-connected induction motor. The motor delivers 10 hp and operates at an efficiency of 85%. The motor power factor is 80%. If the motor is a balanced load, what is the line current? *Ans.* 14.4 amperes.

(b) Repeat (a) for the motor being delta-connected.
Ans. 14.4 amperes.

9–9. Three heaters (pure resistance) are connected in delta to a 220-volt three-phase line. The powers consumed by the individual heaters are, respectively, 11.0 kw, 15 kw, and 17.6 kw. What must be the current-carrying capacity of each of three identical cables supplying this load?
Ans. 128.5 amperes.

9–10. Two three-phase loads are connected to the same 208-volt three-phase supply. Load A is a 10-kw wye-connected heater. Load B is a 15-kw wye-connected heater. Calculate the total current delivered to the point at which the loads are connected. *Ans.* 69.3 amperes.

9–11. Two three-phase loads are connected to the same 208-volt three-phase supply. Load A is a 10-kw wye-connected heater. Load B is a 10-hp

PROBLEMS

85%-efficient wye-connected motor with 80% lagging power factor. Calculate the total current delivered to the point at which the loads are connected. *Ans.* 55.3 amperes.

9-12. The three line leads of a three-phase four-wire generator are designated, respectively, as A, B, and C. The fourth wire, or neutral, is designated as N. The voltage measured from the line leads to neutral are all 120 volts. The line voltages are found to be as follows:

From A to B: 120 volts
From B to C: 120 volts
From C to A: 208 volts

(a) What should be done to balance the generator line voltages?
(b) What is the rated line voltage of the generator? *Ans.* 208 volts.

9-13. For the circuit of Fig. 9-13, determine the magnitude of the voltage E_{ad} when the phase sequence of the balanced 208-volt three-phase 60-cycle supply is a–b–c. *Ans.* 275 volts.

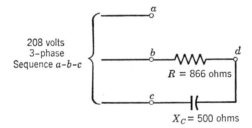

Fig. 9-13.

9-14. For the circuit of Fig. 9-14, determine the magnitude of the voltage E_{xy}. *Ans.* 0 volts.

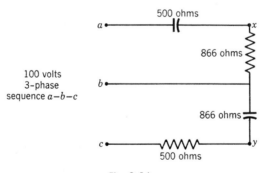

Fig. 9-14.

9-15. For the circuit of Fig. 9-15, determine the current in each line and the neutral. *Ans.* $I_A = 6$; $I_B = 12$; $I_C = 6$; $I_N = 6$.

188   POLYPHASE A-C CIRCUITS

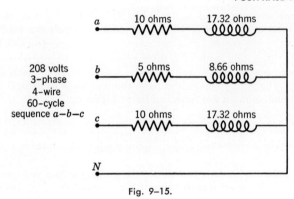

Fig. 9–15.

9–16. For the circuit of Fig. 9–16, determine the magnitude of the voltage E_{xy}.   Ans. 208 volts.

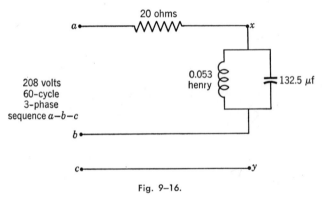

Fig. 9–16.

9–17. For the circuit of Fig. 9–17, determine the magnitude of the voltage E_{MT}.   Ans. 173.2 volts.

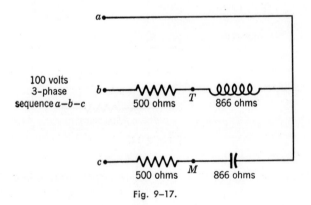

Fig. 9–17.

10 Electrical Instruments and Measurements

10-1. INTRODUCTION

In the discussions of circuits in the preceding chapters, no mention is made of the methods of measuring the electrical quantities involved. It is the purpose of this chapter to consider (*a*) the construction and the theory of operation of various types of instrument mechanisms (or instrument movements), and (*b*) the incorporation of these mechanisms into instruments to measure a desired electrical quantity.

The names of the instruments and meters used to measure some of the various electrical quantities are given in Table 10-1.

TABLE 10-1

Quantity	Measuring Device
Current	Ammeter
Voltage	Voltmeter
Power	Wattmeter
Frequency	Frequency meter
Power factor	Power-factor meter
Energy	Watt-hour meter
Resistance	Ohmmeter

Technically, an instrument is a device for measuring the present value of the quantity under observation; a meter is a device that measures and registers the integral of an electrical quantity with respect to time.*

* American Standard Definitions of Electrical Terms.

According to these definitions, the wattmeter is an instrument, whereas the watt-hour meter is a meter. In a practice, however, the term meter is used interchangeably with the term instrument without strict regard for the technical definitions. Table 10–1 is verification of this fact, in that all the measuring devices listed are called meters, whereas only the watt-hour meter is a meter.

The basic instrument mechanisms are:
(a) The permanent-magnet moving-coil, or D'Arsonval, mechanism.
(b) The electrodynamometer mechanism.
(c) The moving-iron mechanism.

10–2. THE PERMANENT-MAGNET MOVING-COIL MECHANISM

The operation of the permanent-magnet moving-coil mechanism depends on the reaction between the current in a movable coil and the field of a permanent magnet. The basic components of the mechanism are shown schematically as Fig. 10–1a. It is assumed that the coil of Fig. 10–1a is so mounted that it can rotate and that it supports a pointer which has the same angular deflection as the coil.

If, when the coil is energized, the current in the individual conductors is in the direction shown in Fig. 10–1b, the forces on the conductors cause the coil to rotate in a clockwise direction (and the pointer to deflect upscale). If its rotation were unhindered, the coil would rotate until the plane of the coil coincided with axis $A-A$; in this position, the torque tending to cause rotation is zero. Actually, the movement of the coil is restrained by springs, and the angular deflection of the coil

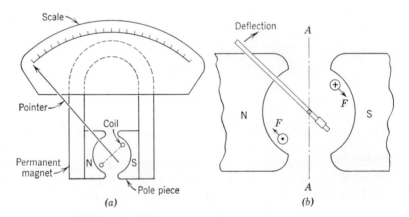

Fig. 10–1. Components of a permanent-magnet moving-coil mechanism.

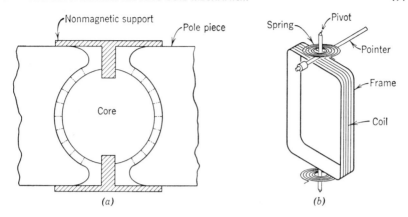

Fig. 10–2. Permanent-magnet moving-coil mechanism details.

and pointer depends upon the value of current in the coil. The maximum permissible deflection of the coil shown in Fig. 10–1a is approximately 90 degrees, as indicated by the instrument scale.

In actual construction, the magnetic circuit of the mechanism does not have the large air gap shown in Fig. 10–1; in one design, it is completed by a soft-iron core, as shown in Fig. 10–2a. The addition of the core results in a decreased reluctance of the magnetic circuit and in a uniform, radial flux in the air gap. The core is held in position between the soft-iron pole pieces by two nonmagnetic supports.

Instead of being a single-turn coil, as shown in Fig. 10–1b, the coil of the mechanism has many turns and, in one design, is wound on a light, rectangular, aluminum frame having centrally located pivots at each end, as shown in Fig. 10–2b. The pointer is attached to one of these pivots, and each pivot has a spiral spring which not only provides the restraining torque of the mechanism but also provides an electric connection to the coil. The springs are wound in opposite directions to compensate for changes in temperature. The moving system is statically balanced in all positions by counterweights, only one of which is shown in Fig. 10–2b.

The magnetic and electric components shown in Fig. 10–2 are combined in the instrument as indicated in Fig. 10–3. The pivots of the moving system are mounted in jewel bearings which are supported by a nonmagnetic strap connecting the two pole pieces. As indicated, the rotation of the moving system is limited not only by the restraining torque of the springs but also by the supports for the core. The force on a conductor carrying current in a transverse magnetic field

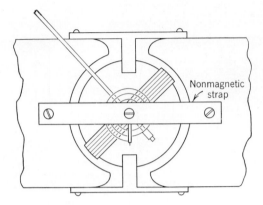

Fig. 10-3. Permanent-magnet moving-coil mechanism.

is proportional to the current in the conductor and the flux density of the field and acts in a direction perpendicular to both the direction of the current and the direction of the field. Since the flux in the air gap of the mechanism may be assumed to be uniform and radial, the force, and therefore the torque, on the conductors of the moving element is directly proportional to the current. Thus, if the current is constant,

$$T_D = KI \tag{10-1}$$

where T_D is the torque tending to cause deflection of the moving element. The deflection of the moving element is opposed by the restraining torque developed in the spiral springs. This opposing torque is proportional to the angle of deflection. Thus,

$$T_R = K'\alpha \tag{10-2}$$

where T_R is the restraining torque, and α is the angle of deflection.

Since, for equilibrium, the restraining torque must be equal to the deflecting torque, Eqs. 10-1 and 10-2 are combined to give

$$K'\alpha = KI \tag{10-3}$$

Thus, the angular deflection of the pointer is directly proportional to the current, and the scale for the instrument is uniform.

If the current through the coil is not constant but is varying at a low rate, the pointer may follow the variations, depending on the inertia of the movement. If the frequency of the variations is greater than 30 cycles, however, the pointer takes a position corresponding to the average developed torque and, therefore, to the average current.

The magnitude of the current required to produce full-scale deflection of the pointer depends upon the make and grade of instrument and is usually of the order of a few milliamperes. The resistance of the current-carrying coil is of the order of 1 ohm.

Damping of the moving system is provided by the aluminum frame. As the frame moves, voltages are induced in its sides; the resulting current reacts with the field to develop a force opposing rotation, thus providing damping action.

10–3. THE ELECTRODYNAMOMETER MECHANISM

The principle of operation of the electrodynamometer mechanism is the same as that for the permanent-magnet moving-coil mechanism. In construction, however, the two differ radically.

The magnetic circuit of the electrodynamometer mechanism is devoid of magnetic material; it is composed entirely of nonmagnetic materials, principally air. The magnetic field is provided by two fixed coils, connected in series, which carry a current dependent upon the application of the instrument. The moving system is similar to that of the permanent-magnet moving-coil mechanism in that electric connections to the moving coil are made through spiral springs, which also provide the restraining torque of the movement. The schematic construction of the mechanism is shown as Fig. 10–4. The electric connections to the fixed coils and the movable coil depend upon the application of the mechanism. In some applications the fixed coils and the movable coil are connected in series; in others, they are energized separately. Regardless of connection, the force on the conductors of the moving coil is proportional to the flux produced by the fixed coils and to the current in the moving coil. Thus, if the flux and current are

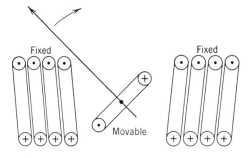

Fig. 10–4. Schematic construction of the electrodynamometer mechanism.

constant,

$$F_d = K\phi_f I_m \tag{10-4}$$

where F_d is the force developed on the conductors, ϕ_f is the flux produced by the fixed coils, I_m is the current in the movable coil, and K is a constant depending upon the construction of the moving element and the system of units. Since the flux produced by the fixed coils is directly proportional to the current in the fixed coils, Eq. 10-4 may be converted to

$$F_d = K' I_f I_m \tag{10-5}$$

where I_f is the current in the fixed coils.

The torque developed on the moving system depends not only upon the force developed on the conductors but also upon the position of the moving system with respect to the field, since the force developed is always mutually perpendicular to the field and the current in the conductors of the moving system. The space relations among the flux, the force, and the moving system are shown schematically as Fig. 10-5.

The torque developed on the moving system of Fig. 10-5 may be expressed by the relation

$$T_d = K'' I_f I_m d \sin \alpha = K''' I_f I_m \sin \alpha \tag{10-6}$$

where d is the distance from the center of the conductors to the center of the moving system, and α is the angle of deflection of the moving coil from the axis mutually perpendicular to the field and the direction of current.

Inspection of Fig. 10-4 reveals that the force on the moving system is still such as to cause upscale deflection if the currents in both the fixed and moving coils are reversed. Thus, the mechanism is

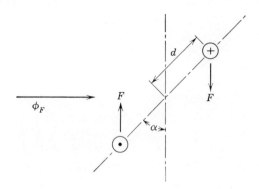

Fig. 10-5. Force on current-carrying coil.

THE MOVING-IRON MECHANISM

applicable to circuits in which both coils are energized from a-c supplies. When each coil carries an alternating current, the force (therefore, the torque) pulsates, and the pointer takes a position corresponding to the average torque. The relations that apply for the a-c connection are similar to those developed for the constant-current application but are modified to take into account that the torque is pulsating.

In order to develop relations that are perfectly general for the usual a-c case, it is assumed that the currents in the fixed and moving coils are expressed by

$$i_f = I_{m(f)} \sin \omega t \tag{10-7}$$

and

$$i_m = I_{m(m)} \sin (\omega t \pm \beta) \tag{10-8}$$

The instantaneous developed force on the conductor is given by the relation

$$F_d \text{ (inst)} = K' i_f i_m \tag{10-9}$$

By integration, the average force is found to be

$$F_d \text{ (avg)} = K'' I_f I_m \cos \beta \tag{10-10}$$

where I_f is the effective value of current in the fixed coils, I_m is the effective value of current in the moving coil, and β is the phase angle between the two currents. Therefore, the average developed torque on the movement is given by

$$T_d \text{ (avg)} = (K'' I_f I_m \cos \beta)(d \sin \alpha) \tag{10-11}$$

or

$$T_d \text{ (avg)} = K''' I_f I_m \cos \beta \sin \alpha \tag{10-12}$$

Magnetic materials are not included in the magnetic circuit of the electrodynamometer mechanism because the principal application of the mechanism is in a-c instruments (hysteresis and eddy-current losses are thus avoided). The increased reluctance of the magnetic circuit, however, necessitates that the current required for full-scale deflection be considerably greater than that for the permanent-magnet moving-coil mechanism.

10-4. THE MOVING-IRON MECHANISM

The basis of operation of one form of the moving-iron mechanism is that like magnetic poles have a repelling force developed between them. One construction based on this principle is shown schematically as Fig. 10–6a.

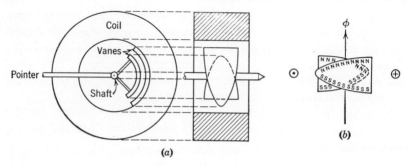

Fig. 10–6. Moving-iron mechanism.

The basic components of the mechanism illustrated in Fig. 10–6a are a fixed coil and two soft-iron pennant-shaped vanes, one of which is attached to the pointer shaft, the other to the coil. The retarding torque on the moving system is provided by a spiral spring which, in this instance, does not have to provide the electric connection. The only electric connection in the mechanism is that to the fixed coil.

When the coil is energized, the field produced by the current in the coil causes magnetic poles to be formed on the edges of the iron vanes, as shown in Fig. 10–6b. Adjacent edges of the vanes are of similar polarity, and, therefore, a repelling force exists between vanes. The vanes are so designed that there is a resultant tangential force, causing the movable vane to move away from the fixed vane and resulting in an upscale deflection of the pointer. Inspection of Fig. 10–6b reveals that the pointer also deflects upscale when the current in the coil is reversed. Therefore, the mechanism is applicable to a-c instruments.

The force of repulsion between the vanes is a function of the product of the strengths of the magnetic poles formed at the edges of the vanes, in addition to other factors. The pole strengths, in turn, are a function of the current flowing in the coil. The force of repulsion, therefore, is a function of the square of the current in the coil. Since the developed torque in the mechanism is a function of the developed force, the torque is a function of the current squared.

10–5. D-C VOLTMETERS

Although any of the mechanisms described in the preceding articles may be applied in a d-c voltmeter, only the permanent-magnet mov-

D-C VOLTMETERS

ing-coil mechanism is commonly used. This article considers only the application of the permanent-magnet moving-coil mechanism, but the principles discussed are also applicable to the other mechanisms.

To measure a voltage in a circuit, a voltmeter is connected in parallel with the portion of the circuit in which the voltage is of interest. The current taken by the voltmeter should be small so that the operation of the circuit in which the voltmeter is connected is not appreciably affected (the voltmeter current causes an additional voltage drop in the circuit from the supply to the point at which the voltage is being measured).

Because the permanent-magnet moving-coil mechanism has a resistance of the order of 1 ohm and requires a current of the order of a few milliamperes for full-scale deflection, a resistor must be connected in series with the mechanism if it is to be used to measure any voltage larger than a few millivolts; any larger voltage causes the pointer to deflect off-scale and may damage the moving coil. The series-connected resistor must have a value such that full-scale deflection occurs when the rated voltage of the voltmeter is applied to the terminals of the series combination. For example, to construct a 150-volt d-c voltmeter incorporating a mechanism having a resistance of 1 ohm and requiring 10 milliamperes for full-scale deflection necessitates a series-connected resistor of 14,999 ohms. The connection of the components of the voltmeter is shown schematically as Fig. 10–7a.

The terminals of the voltmeter of Fig. 10–7a are marked 150 and + (− is sometimes used). The 150 signifies that the range of this instrument is from 0 to 150 volts. As indicated in the discussion of the mechanism, the scale is approximately linear. The + (or −) indicates the terminal that must be positive (or negative) for an upscale deflection of the pointer.

Voltmeters are often multirange; that is, the voltmeter is so con-

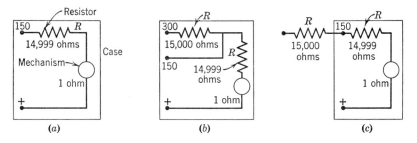

Fig. 10–7. Voltmeter details.

structed that full-scale deflection is obtained with any of several voltages applied to appropriate terminals of the instrument. The connection shown in Fig. 10–7b is an example of a multirange d-c voltmeter. As indicated, the instrument has both a 0- to 150-volt range and a 0- to 300-volt range. Thus, each graduation of the scale has two values; one applies when the instrument is connected for the 150-volt range; the other applies when connected for the 300-volt range.

The multirange instrument shown as Fig. 10–7b is complete in itself. The 150-volt instrument shown as Fig. 10–7a can be made multirange by connecting an external resistor in series with it, as shown in Fig. 10–7c. A resistor thus employed is called a **multiplier resistor.** Since the scale of the original 150-volt instrument is presumably graduated from 0 to 150, a multiplier of 2 must now be applied to the scale.

10–6. D-C AMMETERS

An ammeter is connected in series with the device or circuit in which current is to be measured and should therefore have as low a resistance as possible so that the voltage drop across it does not appreciably affect the operation of the remainder of the circuit. The ammeter should, of course, be capable of carrying the current expected in the circuit.

Any of the three mechanisms may be employed in ammeters to measure a direct current but, as in the case of d-c voltmeters, only the permanent-magnet moving-coil mechanism is commonly used.

Since the permanent-magnet moving-coil mechanism requires such a small current for full-scale deflection, a major portion of the current to be measured in most circuits cannot be passed through the instrument and must therefore be diverted. Thus, if the mechanism is paralleled with a low-valued resistor, a definite portion of the current in the circuit is shunted around the mechanism, the amount depending on the relative resistances of the parallel resistor (called the **shunt**) and the mechanism. The connection of the mechanism and shunt is shown in Fig. 10–8a.

Instead of having a resistance rating, shunts are rated in terms of current and voltage (usually amperes and millivolts). When rated current flows through the shunt, the rated voltage appears across it. Thus, to construct a 5-ampere ammeter incorporating a 10-milliampere 1-ohm permanent-magnet moving-coil mechanism requires a

D-C AMMETERS

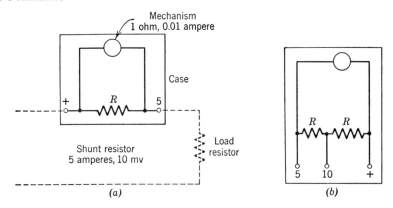

Fig. 10–8. Ammeter details.

5-ampere 10-millivolt shunt, since 5 amperes through the shunt produces a 10 millivolt drop across the shunt and full-scale deflection of the mechanism pointer. Such an ammeter is shown in Fig. 10–8a.

When 5 amperes flows through the shunt of the ammeter shown in Fig. 10–8a, the mechanism has full-scale pointer deflection. The full-scale graduation is then labeled as 5 amperes, even though the actual current in the line is 5.01 amperes; the current taken by the mechanism is neglected, and the line current is assumed to be the same as the shunt current.

Multirange ammeters are constructed essentially as shown in Fig. 10–8b. The mechanism is connected across a shunt so that, in this example, full-scale deflection occurs when 5 amperes flows through the shunt. To make the instrument multirange, a tap is made on the shunt so that full-scale deflection occurs when 10 amperes flows through only a portion of the original shunt.

The instruments shown in Fig. 10–8 are self-contained. In practice, measurements of direct current are often made with millivoltmeters and external shunts, particularly when the currents are in excess of 25 amperes. The scale of a millivoltmeter to be used with a particular external shunt is usually graduated in amperes applying to the shunt. Often, millivoltmeters and shunts are interchangeable, in which case the millivoltmeters may have scales graduated in millivolts, amperes, or arbitrary divisions. Thus, the use of a millivoltmeter and an external shunt may involve a multiplier to convert the scale reading to amperes. For example, if a 50-millivolt instrument having a scale graduated from 0 to 100 is connected to a shunt having a rating of 50 amperes 50 millivolts, the scale reading of the millivoltmeter must be multiplied by 0.5 to convert it to amperes.

10-7. A-C VOLTMETERS

Both a-c and d-c voltmeters have a high-valued resistor in series with a mechanism. The discussion of a-c voltmeters is therefore limited to a consideration of the various mechanisms.

If an alternating voltage having a frequency greater than 30 cycles is impressed on a permanent-magnet moving-coil mechanism, the pointer of the mechanism does not deflect, since the average current and the average torque are zero. Therefore, this mechanism is not directly applicable as an a-c voltmeter.

Electrodynamometer Voltmeters. The electrodynamometer mechanism, with its fixed and movable coils connected in series, is the type commonly used in a-c voltmeters. The average developed torque in the mechanism (according to Eq. 10–12) is given by

$$T_d \text{ (avg)} = K''' I_f I_m \cos \beta \sin \alpha \qquad (10\text{–}13)$$

Since the coils are in series and therefore carry the same current, Eq. 10–13 reduces to

$$T_d \text{ (avg)} = K''' I^2 \sin \alpha \qquad (10\text{–}14)$$

where I is the effective value of current through the coils. The current in the coils is proportional to the voltage applied and, therefore, Eq. 10–14 may be written

$$T_d \text{ (avg)} = K''' \frac{E^2}{R^2} \sin \alpha \qquad (10\text{–}15)$$

where E is the effective value of voltage applied to the voltmeter, and R is the total resistance of the mechanism and the series resistor. (It is assumed that the reactance of the coils is negligible in comparison to the resistance introduced by the series resistor.) Although the pointer deflects according to the square of the effective value of voltage applied, the nonlinear scale of the mechanism is graduated in terms of the effective value.

The electrodynamometer type of voltmeter can be used to measure a direct voltage, since the value of a direct voltage is the effective value. When precise measurement of a direct voltage is required, it is common practice to measure the voltage twice, the second measurement being taken with the terminals of the voltmeter reversed, and then

A-C VOLTMETERS

to average the readings. This precaution is taken to eliminate the effects of stray magnetic fields.

Moving-Iron Voltmeters. The current through the coil of a moving-iron voltmeter is proportional to the voltage applied. Since the torque developed in the mechanism is a function of the current squared, the torque is therefore a function of the voltage squared. When connected to an a-c supply, therefore, the average torque is a function of the effective value of the voltage applied. The pointer deflection is dependent upon the average torque, and therefore upon the effective value of voltage applied.

The moving-iron voltmeter is also applicable for direct-voltage measurements, with the same precaution as indicated for the electrodynamometer type.

Rectifier-Type Voltmeters. A *rectifier* is a device that permits an appreciable flow of current in only one direction. Thus, if a rectifier is connected to an a-c supply, current flows during one half-cycle but not the other, as shown by the current–time diagram in Fig. 10–9b, and the average value of current in the circuit is not zero. The circuit diagram of Fig. 10–9a shows the rectifier connected in series with a resistor and a permanent-magnet moving-coil mechanism to form an a-c voltmeter. (The symbol for the rectifier indicates the direction in which current flows.) A more common connection of the rectifier for instrument purposes, however, is the bridge circuit shown as Fig. 10–9c; with such a circuit, current flows through the mechanism during the entire cycle, as indicated in Fig. 10–9d. (A more complete discussion of rectifiers is taken up in Chapter 23.)

The mechanism shown in Fig. 10–9c deflects according to the aver-

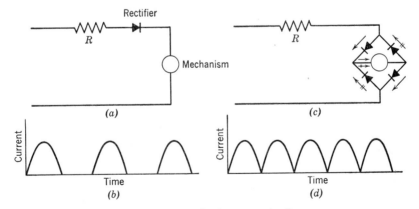

Fig. 10–9. Rectifier instrument details.

age value of the current through it. Thus, if the supply voltage is sinusoidal in form, the instrument has a deflection corresponding to 0.637 times the peak value of current, which, because of the series resistance, is proportional to the peak value of voltage. The scale, however, is graduated in terms of the effective value; therefore, a factor of 1.11 (the ratio of effective value to average value of a rectified sine wave) is introduced by the scale. This fact is particularly important when the voltage to be measured is nonsinusoidal. Thus, if this instrument is applied to measure a direct voltage, the reading on the scale is 1.11 times the actual value of voltage, since the average value of a direct voltage is the same as its effective value.

10-8. A-C AMMETERS

The mechanism of a-c ammeters is usually the moving-iron type. The electrodynamometer mechanism is not employed except for the measurement of very small currents, owing to the limitation on the current that a practical moving coil can carry. A shunt is not practical for the electrodynamometer mechanism, since the division of the current between the shunt and the coils varies with the frequency (because of the reactance of the coils). Therefore, the instrument would be accurate at only one frequency, and then only if the current were sinusoidal in wave form.

The coil of a moving-iron ammeter is constructed of a few turns of large wire so that it can carry all the current to be measured. As in the case of the moving-iron voltmeter, the ammeter has a pointer deflection corresponding to the effective value of current through it, and the instrument scale is so graduated.

Rectifier-type ammeters are available for the measurement of currents but are limited to the current that the rectifier and the mechanism can safely carry.

10-9. WATTMETERS

The basic element of the wattmeter is the electrodynamometer mechanism. The fixed and movable coils of the mechanism are connected separately: the fixed coil so that it carries the current of the circuit in which the power is to be measured, the movable coil so that it carries a current proportional to the voltage of the circuit. One method of connecting the two coils is shown schematically in

WATTMETERS

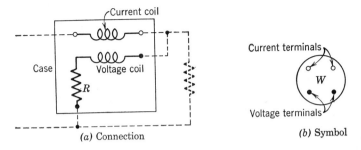

(a) Connection (b) Symbol

Fig. 10-10. Wattmeter details.

Fig. 10-10. As indicated, the movable coil has a resistor connected in series with it to limit the current.

The average torque developed in the mechanism is given by

$$T_d \text{ (avg)} = K''' I_f I_m \cos \beta \sin \alpha \qquad (10\text{-}16)$$

For the application as a wattmeter, the fixed-coil current I_f is the circuit current I; the movable-coil current I_m is proportional to the circuit voltage E; that is,

$$I_m = \frac{E}{R} \qquad (10\text{-}17)$$

where R is the total resistance in the movable-coil circuit. Therefore, Eq. 10-16 may be written

$$T_d \text{ (avg)} = K'''' EI \cos \beta \sin \alpha \qquad (10\text{-}18)$$

For this application, the angle β between the voltage and current is the power-factor angle θ of the circuit; therefore, Eq. 10-18 may be written

$$T_d \text{ (avg)} = K''(EI \cos \theta) \sin \alpha \qquad (10\text{-}19)$$

Thus, the average developed torque and the subsequent deflection of the instrument are functions of the average power of the circuit.

For the connection shown in Fig. 10-10, the wattmeter measures the power in the circuit; this is not true of all wattmeter applications. In some connections of the wattmeter, the angle β between the fixed-coil current and the movable-coil voltage is not the power-factor angle, and therefore the wattmeter does not measure power. In such applications, however, the angle β is often a function of the power-factor angle, and the deflection of the wattmeter is related to the power in the circuit.

Because a downscale deflection may occur if the wattmeter is not

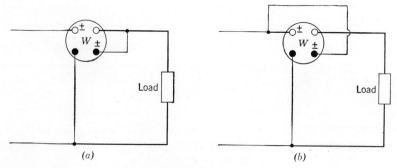

Fig. 10–11. Wattmeter connections.

properly connected, the wattmeter terminals that should have the same polarity to result in an upscale deflection are labeled ±. Two possible correct connections of the wattmeter are shown as Figs. 10–11a and 10–11b.

For either connection shown in Fig. 10–11, the wattmeter indicates a power greater than that actually taken by the load. In Fig. 10–11a, the current in the current coil (fixed coil) is greater than the load current by an amount equal to the current taken by the voltage coil (movable coil); the power measurement is therefore too large by the amount of power taken by the voltage coil. In Fig. 10–11b, the voltage coil measures a voltage that is larger than the load voltage by an amount equal to the voltage drop across the current coil; the power measurement is therefore too large by the amount of power taken by the current coil. Of the two sources of error, the power loss in the voltage coil is more readily calculated, since the voltage-coil circuit is essentially pure resistance. Therefore, if the wattmeter is connected as shown in Fig. 10–11a, the actual power dissipated in the load is given by

$$P_L = P_W - \frac{E^2}{R_m} \qquad (10\text{--}20)$$

where P_L is the power in the load, P_W is the reading of the wattmeter, E is the voltage across the voltage coil, and R_m is the resistance of the voltage-coil circuit.

The calculation indicated by Eq. 10–20 is not necessary when a compensated wattmeter is employed. In this instrument (connected like the wattmeter in Fig. 10–11a), the current in the voltage coil is passed through a supplementary fixed coil so located that the additional magnetic effect of the voltage-coil current in the current coil is canceled. The magnetic field causing deflection is then due only to the load current, and a true indication of the power in the circuit results.

The ratings of a wattmeter are given in terms of power, voltage, and current. Since the power in a circuit is the product of the voltage, current, and power factor, the instrument may have an on-scale deflection but, because of a low power factor, have its voltage coil or current coil overloaded. The voltage and current ratings are therefore the most important of the three ratings given. When power is being measured with a wattmeter, the circuit should also contain a voltmeter and ammeter so that a check can be made to see that the voltage and current ratings of the wattmeter are not exceeded.

10–10. MEASUREMENT OF POWER IN THREE-PHASE CIRCUITS

The total power in a three-phase circuit is the sum of the powers in the individual phases and, therefore, can be measured by the proper connection of a wattmeter in each of the phases. Figures 10–12a and 10–12b show such a connection in three-phase wye and delta circuits, respectively. (As indicated in Fig. 10–12a, the connection also applies to a three-phase four-wire circuit.)

In many three-phase circuits the methods of power measurement indicated in Fig. 10–12 are not applicable because of the construction of the devices in the circuit; that is, the neutral terminal of a wye-connected device is often not available for connection to the wattmeter voltage coil, or the phases of a delta-connected device cannot be opened for the insertion of the wattmeter current coils. In such

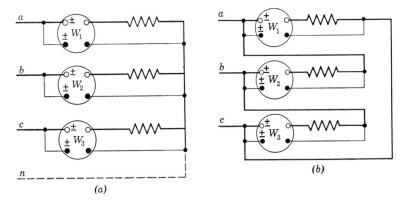

Fig. 10–12. Power measurement by the three-wattmeter method.

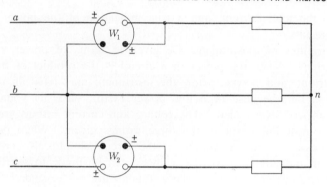

Fig. 10–13. Power measurement by the two-wattmeter method.

circuits, the total power in the circuit is usually measured by the **two-wattmeter method**.

In the two-wattmeter method, each wattmeter is so connected that its current coil carries a line current, and its voltage coil is connected to a line voltage. The line voltage involved is between the line in which the current coil of the wattmeter is connected and the line in which there is no current coil, as shown in Fig. 10–13. (The load shown in Fig. 10–13 is wye-connected, but the method is equally applicable to delta-connected loads.)

In the two-wattmeter method, the sum of the readings of the two wattmeters is the total power. This fact is proved as follows:

(*a*) For the circuit of Fig. 10–13, the instantaneous power measured by wattmeter 1 is given by

$$p_1 = e_{ab}i_{an} \tag{10-21}$$

Similarly, for wattmeter 2,

$$p_2 = e_{cb}i_{cn} \tag{10-22}$$

(It is assumed that the power loss in the wattmeter is negligible.)

(*b*) Substituting, in Eqs. 10–21 and 10–22, the phase voltages that combine to produce the line voltages gives

$$p_1 = e_{ab}i_{an} = e_{an}i_{an} + e_{nb}i_{an} \tag{10-23}$$

and

$$p_2 = e_{cb}i_{cn} = e_{cn}i_{cn} + e_{nb}i_{cn} \tag{10-24}$$

(*c*) Adding the instantaneous powers yields

$$p_1 + p_2 = e_{an}i_{an} + e_{nb}(i_{an} + i_{cn}) + e_{cn}i_{cn} \tag{10-25}$$

(*d*) Since

$$i_{an} + i_{cn} = -i_{bn} \tag{10-26}$$

MEASUREMENT OF POWER IN THREE-PHASE CIRCUITS

Eq. 10–25 may be written

$$p_1 + p_2 = e_{an}i_{an} - e_{nb}i_{bn} + e_{cn}i_{cn} \tag{10-27}$$

(e) Since

$$e_{bn} = -e_{nb} \tag{10-28}$$

Eq. 10–27 may be written

$$p_1 + p_2 = e_{an}i_{an} + e_{bn}i_{bn} + e_{cn}i_{cn} \tag{10-29}$$

Equation 10–29 indicates that the sum of the instantaneous powers measured by the wattmeters is equal to the sum of the instantaneous phase powers. The sum of the average powers, which are the values indicated by the wattmeter deflections, is therefore equal to the total average power in the circuit.

For the preceding proof, no restrictions are placed on the balance of the load; the proof holds for both balanced and unbalanced loads. Although the proof is for a wye-connected circuit, a similar analysis of a delta-connected circuit produces the same result.

A proof of the validity of the two-wattmeter method for measuring the power in balanced loads reveals an aspect of the method that is not apparent in the preceding proof. The proof for a balanced wye-connected load proceeds as follows:

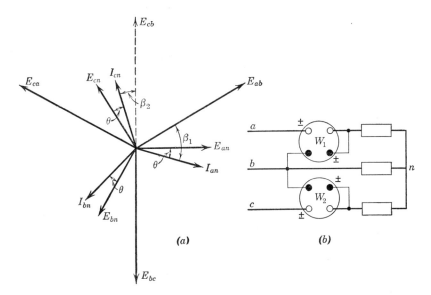

Fig. 10–14. Power measurement in a balanced three-phase circuit by the two-wattmeter method.

(a) The connection diagram for the load is shown in Fig. 10–14b; the vector diagram is shown in Fig. 10–14a.

(b) The reading of wattmeter 1 is given by

$$P_1 = E_{ab}I_{an} \cos \beta_1 \qquad (10\text{–}30)$$

where E_{ab} and I_{an} are the effective values of voltage and current, respectively, and β_1 is the phase angle between this voltage and current. Similarly, the reading of wattmeter 2 is given by

$$P_2 = E_{cb}I_{cn} \cos \beta_2 \qquad (10\text{–}31)$$

(c) Inspection of the vector diagram reveals that Eqs. 10–30 and 10–31 can be rewritten

$$P_1 = E_{ab}I_{an} \cos(30° + \theta) \qquad (10\text{–}32)$$

and

$$P_2 = E_{cb}I_{cn} \cos(30° - \theta) \qquad (10\text{–}33)$$

By substitution of the proper trigonometric identities, Eqs. 10–32 and 10–33 become

$$P_1 = E_{ab}I_{an} \cos 30° \cos \theta - E_{ab}I_{an} \sin 30° \sin \theta \qquad (10\text{–}34)$$

and

$$P_2 = E_{cb}I_{cn} \cos 30° \cos \theta + E_{cb}I_{cn} \sin 30° \sin \theta \qquad (10\text{–}35)$$

(d) Adding Eqs. 10–34 and 10–35 gives

$$P_1 + P_2 = (E_{ab}I_{an} + E_{cb}I_{cn}) \cos 30° \cos \theta \qquad (10\text{–}36)$$

(e) Because the load is balanced,

$$E_{ab} = E_{cb} = E_L \qquad (10\text{–}37)$$

and

$$I_{an} = I_{cn} = I_L \qquad (10\text{–}38)$$

Therefore, Eq. 10–36 can be written

$$P_1 + P_2 = 2E_L I_L \cos 30° \cos \theta \qquad (10\text{–}39)$$

or

$$P_1 + P_2 = \sqrt{3}\, E_L I_L \cos \theta \qquad (10\text{–}40)$$

Thus, the sum of the readings of the two wattmeters is equal to the total power in the circuit.

Equation 10–32 indicates that wattmeter 1 reads negatively, or downscale, if the power-factor angle of the circuit is greater than 60 degrees. (If the load had been capacitive, the sign of the power-factor angle would be reversed, and wattmeter 2 would deflect downscale.) Under these conditions, the current coil of the wattmeter should be reversed to produce an upscale deflection; the arithmetic difference in the readings is then the total power. Thus, when the wattmeters are properly connected, the algebraic sum of the wattmeter readings in the two-wattmeter method is the total power. (A downscale deflection means that the algebraic sign of that measurement is negative.)

The algebraic summation of the two wattmeter readings is accomplished directly in the **polyphase wattmeter,** which incorporates two sets of wattmeter elements having a common shaft. The wattmeter elements are connected according to the two-wattmeter method, and the resulting deflection of the instrument pointer is a function of the algebraic summation of the torques produced by the individual elements. The deflection of the instrument pointer is therefore a function of the total power.

10–11. OHMMETERS

One instrument for measuring resistance directly is called an ohmmeter. The circuit for a simple ohmmeter consists of a battery, two resistors, and a permanent-magnet moving-coil mechanism, connected as shown in Fig. 10–15. The resistance to be measured is connected between the open terminals of the instrument.

When the terminals of the instrument are open (indicating infinite resistance between the terminals), no current flows in the mechanism, and there is no deflection. Therefore, in terms of resistance, the normal zero point on the scale of the instrument is marked ∞. If the terminals are short-circuited (indicating zero resistance between the terminals), maximum current flows. The internal resistance of the instrument, therefore, is such that maximum deflection occurs when the terminals

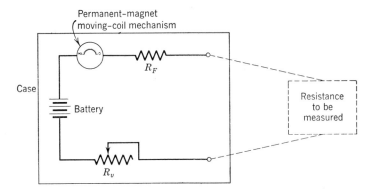

Fig. 10–15. Elementary ohmmeter circuit.

are short-circuited. In terms of resistance, the point of maximum deflection is marked 0. Other points on the scale are marked according to the deflection that occurs when given resistors are connected between the terminals. Because of its limits, the scale must have nonlinear graduations.

In normal use, the ohmmeter is first tested for full-scale deflection with the terminals short-circuited; any necessary correction is made by adjusting the variable resistor R_v. The purpose of the variable resistor is to change the internal resistance of the ohmmeter as the battery ages and becomes weaker. This means that the ohmmeter is less accurate as the battery ages, since the scale reading is dependent upon the internal resistance of the instrument. The ohmmeter, however, is not a precision instrument; if accurate measurement of resistance is required, other methods must be applied.

The preceding discussion is concerned with the simplest type of ohmmeter. Actually, ohmmeter circuits are more complex than that shown, but, basically, these complex circuits operate on the principles discussed.

10–12. WATT-HOUR METERS

The construction of a watt-hour meter is dependent upon whether the meter is to be applied to a-c or to d-c circuits. In general, types designed for d-c circuits can be applied to a-c circuits, but the reverse is not true. Although the detailed theory of operation of both types is best understood after d-c and a-c motor theory is studied, it is possible to consider here the general theory of operation applying to both types.

Figure 10–16 shows, schematically, the general construction of the watt-hour meter. As indicated, the operating mechanism has a voltage coil and a current coil, connected in the same manner as the coils of a wattmeter. Since there is no retarding spring, the shaft rotates owing to the motor action produced when the coils are energized. The number of revolutions of the shaft is a measure of the electric energy passing through the meter and is recorded on dials which are geared to the shaft.

The aluminum disk connected to the shaft provides the retarding torque of the mechanism. As the disk rotates, voltages are induced in the disk because it cuts the flux produced by the permanent magnets. The direction of the resulting current in the disk is such that the force developed because of the reaction between the permanent-

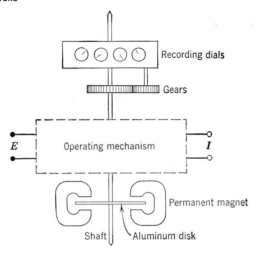

Fig. 10-16. Watt-hour meter construction.

magnet field and the current opposes the rotation. Since the induced voltage (and therefore the current) in the disk is proportional to the rotational velocity of the disk, the retarding torque is also proportional to the disk velocity. Thus,

$$T_R = K'n \qquad (10\text{--}41)$$

where T_R is the retarding torque, and n is the speed in revolutions per unit time.

The average torque developed to cause rotation of the disk is proportional to the average power passing through the operating mechanism. Thus,

$$T_d = KEI \cos \theta = KP \qquad (10\text{--}42)$$

For equilibrium, the retarding torque must be equal to the developed torque; thus,

$$K'n = KP \qquad (10\text{--}43)$$

If both sides of Eq. 10-43 are multiplied by time t,

$$K'nt = KPt \qquad (10\text{--}44)$$

or

$$nt = K''Pt \qquad (10\text{--}45)$$

Since the product nt represents the number of revolutions of the disk in time t, and the product Pt represents the energy passing through the meter in time t, the number of revolutions of the disk is proportional to the energy passing through the meter.

The nameplate of a watt-hour meter usually provides a watt-hour-meter constant, in addition to the other essential information. This constant, which may be a whole number or a fraction, gives the amount of energy passing through the meter for each revolution of the disk. Thus, if the watt-hour-meter constant K_h is ⅓, then ⅓ of a watt-hour is recorded for each revolution of the shaft.

10–13. MEASUREMENT OF NONELECTRICAL QUANTITIES

Often a nonelectrical quantity can be converted into an electrical quantity that is proportional to the nonelectrical quantity; the non-electrical quantity then becomes measurable by electrical methods. Some examples are temperature, displacement, speed, light, pressure, humidity, and sound.

The electrical measurement of some nonelectrical quantities can be made by means of the instrument mechanisms already discussed; others require electronic instrumentation which is discussed later. The detailed discussion of these measurements is too lengthy to be undertaken in this book, but a general consideration of the electrical measurement of a few of the examples provides a glimpse of the convertibility and measurement of nonelectrical quantities.

The measurement of temperature by measuring the change in resistance of a temperature-sensitive resistor is discussed in Chapter 2. Another electrical method having wide application employs the thermocouple. A thermocouple consists of a pair of electric conductors of dissimilar metals so joined as to produce a thermal emf when the junctions are at different temperatures. A simple thermocouple circuit consisting of copper and Constantan (an alloy) conductors and a permanent-magnet moving-coil mechanism is shown as Fig. 10–17.

A difference in temperature between the junctions of the thermocouple circuit results in a direct emf in the circuit. The emf is a function of the temperature difference (for copper–Constantan ther-

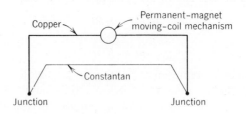

Fig. 10–17. Copper–Constantan thermocouple circuit.

mocouples, the calibration is approximately 1 millivolt per 25° C difference in temperature between the junctions) and is readily calculated if the current and the resistance of the circuit are known. The mechanism is included in the circuit so that the current may be determined. The copper–Constantan combination can be employed to measure temperatures in the range from $-190°$ C to $350°$ C; other combinations of metals and alloys are available to extend the range to about $1500°$ C.

All three electric circuit parameters—resistance, capacitance, and inductance—are employed for measuring mechanical displacement. If the movable contact of a rheostat having one fixed contact and one movable contact is attached to a body having mechanical displacement, the position of the movable contact—and therefore the resistance between the fixed and the movable contacts—is a function of the mechanical displacement. The displacement can therefore be indicated and measured by including the rheostat in the appropriate electric circuit.

Methods of measurement of mechanical displacement employing capacitance and inductance are similar to that described for resistance. The displacement is used to alter the position of one plate of a capacitor or the position of an iron core in an inductor. The resulting change in capacitance or inductance is a function of the displacement and can be determined electrically.

The fact that a voltage is induced in a coil as it revolves in a magnetic field is the basis for one method of electrically measuring the speed of a rotating device. The revolving coil is incorporated in d-c and a-c tachometer generators, which have a voltage output that is a function of the speed of rotation. If the relationship between voltage and speed is known, the speed may be determined by measuring the voltage with an appropriate voltmeter. Since the frequency of the alternating voltage induced in the revolving coil is also a function of the speed, the speed may also be determined by a measurement of the frequency of the voltage.

The examples discussed are only a few of the many applications of electric circuits and mechanisms to the measurement and control of nonelectrical quantities.

PROBLEMS

10–1. A d-c voltmeter that has a resistance of 3000 ohms gives full-scale deflection with a potential difference of 150 volts across its terminals.

(a) How much current passes through the instrument?

(b) Show, by a diagram of connections, how the range of the voltmeter can be extended to give full-scale deflection when connection is made across 600 volts. Include values.

(c) What would have to be done to have the instrument indicate full-scale deflection with 15 volts applied? Indicate values.

10–2. Two 150-volt d-c voltmeters, having resistances of 15,000 and 5000 ohms, respectively, are connected in series to a 150-volt d-c source.

(a) What should be the reading of each instrument?

Ans. 112.5 volts; 37.5 volts.

(b) What is the maximum voltage that can safely be measured with this combination? *Ans.* 200 volts.

10–3. A 150-volt voltmeter, having a resistance of 15,000 ohms and connected to a d-c supply, indicates 122 volts. The voltmeter is then connected in series with an unknown resistance to the same supply and indicates 14 volts. Determine the value of the unknown resistance. *Ans.* 115,800 ohms.

10–4. A 50-mv 5-ohm d-c millivoltmeter is to be used with the following shunts to measure current: (1) 15 amperes, 50 mv; (2) 25 amperes, 100 mv; (3) 50 amperes, 25 mv. The millivoltmeter has a 0–50 scale, and none of the equipment should be overloaded.

(a) What currents can be measured employing each of the shunts listed above? *Ans.* (1) 15 amperes; (2) 12.5 amperes; (3) 50 amperes.

(b) What is the multiplier to convert the scale reading to amperes for each case? *Ans.* (1) 0.3; (2) 0.25; (3) 2.0.

10–5. The resistance of a resistor is computed from data taken from a voltmeter and an ammeter without allowing for the effect of instrument resistance. If the true value of resistance is larger than the computed value, how were the instruments connected and which instrument reading was misleading? Explain briefly.

10–6. (a) Prove that the deflection of an electrodynamometer mechanism having its coils in series is a function of the effective value of the current passing through the coils regardless of the wave form of the current.

(b) Repeat (a) for a moving-iron mechanism.

10–7. Why do rectifier-type a-c instruments contain permanent-magnet moving-coil mechanisms instead of electrodynamometer or moving-iron mechanisms?

10–8. What might be the effect of the presence of a 60-cycle alternating magnetic field, near the instrument, on the deflections of the following voltmeters connected to the supply indicated:

(a) An electrodynamometer type connected to a 60-cycle a-c supply?

(b) A moving-iron type connected to a d-c supply?

(c) A rectifier type connected to a d-c supply?

10–9. Explain the effect of the presence of a permanent magnet, near the instrument, on the deflections of the following voltmeters connected to the supply indicated:

(a) An electrodynamometer type connected to an a-c supply.

(b) A moving-iron type connected to an a-c supply.

(c) A permanent-magnet moving-coil type connected to a d-c supply.

PROBLEMS

10–10. The voltages listed below are to be measured by the following types of voltmeters: (1) permanent-magnet, (2) electrodynamometer, (3) moving-iron, (4) rectifier. For each of the voltages, determine the value indicated on each of the instruments.

(a) A direct voltage of 120 volts.

Ans. (1) 120; (2) 120; (3) 120; (4) 133.2.

(b) A 60-cycle sinusoidal voltage having a maximum value of 169 volts.

Ans. (1) 0; (2) 120; (3) 120; (4) 120.

(c) A voltage that varies as shown in Fig. 10–18.

Ans. (1) 0; (2) 104; (3) 104; (4) 99.9.

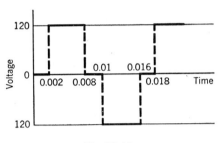

Fig. 10–18.

(d) A voltage that varies as shown in Fig. 10–19.

Ans. (1) 90; (2) 104; (3) 104; (4) 99.9.

10–11. If an accuracy of 1% is desired, what type or types of voltmeters —permanent-magnet, electrodynamometer, moving-iron, or rectifier—could be used to measure the effective value of a voltage having the wave form shown in Fig. 10–20? Assume that the instruments are perfect.

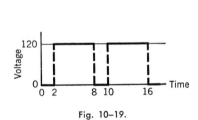

Fig. 10–19.

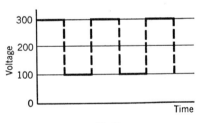

Fig. 10–20.

10–12. A 220-volt d-c supply is connected in series with a 120-volt 60-cycle supply. What value will each of the following types of voltmeter record, if each voltmeter is connected to the output terminals of the supply combination?

(a) Permanent-magnet moving-coil.   *Ans.* 220.
(b) Electrodynamometer.   *Ans.* 251.
(c) Moving-iron.   *Ans.* 251.
(d) Rectifier.   *Ans.* 244.

10–13. A balanced three-phase load of 30 kva at 0.4 lagging power factor is connected to a 208-volt three-phase supply, as shown in Fig. 10–21. The phase sequence of the supply is a–c–b. If the effect of instrument resistance is negligible, what values are recorded on wattmeters W_1 and W_2 (assuming that the wattmeters are zero-center instruments)?

Ans. $W_1 = (+)\ 13{,}930$ watts; $W_2 = (-)\ 1930$ watts.

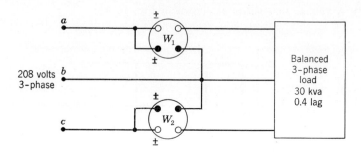

Fig. 10–21.

10–14. Three loads are connected in delta to a 200-volt three-phase supply as shown in Fig. 10–22. The phase sequence of the supply is a–b–c. If the effect of instrument resistance is negligible, what values are recorded on wattmeters W_1 and W_2? *Ans.* $W_1 = 15{,}000$ watts; $W_2 = 26{,}000$ watts.

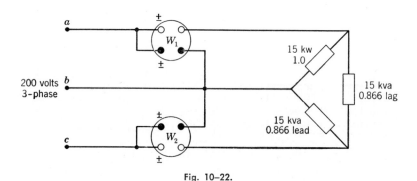

Fig. 10–22.

10–15. Figure 10–23 shows the connection of a three-phase wye-connected heater (pure resistance). Connected as shown, wattmeters 1 and 2 each indicate 1800 watts. If the fuse in line b blows, opening that line, what does each wattmeter read? *Ans.* 900 watts.

PROBLEMS

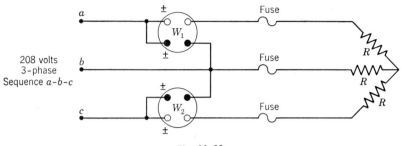

Fig. 10–23.

10–16. Figure 10–24 shows the connection of a balanced three-phase delta-connected heater (pure resistance). Connected as shown, wattmeters 1 and 2 each indicate 6480 watts. If fuse 2 blows, opening that line, what will each wattmeter read?   *Ans.* $W_1 = 6480$ watts; $W_2 = 2160$ watts.

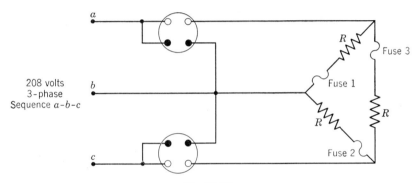

Fig. 10–24.

10–17. A wattmeter is connected as shown in Fig. 10–25. The phase sequence of the supply is a–b–c. The load is a balanced three-phase load having a power-factor angle θ. What does the wattmeter indicate, and what is the practical significance of this deflection?   *Ans.* $E_L I_L \sin \theta$.

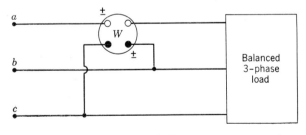

Fig. 10–25.

10–18. Figure 10–26 shows the schematic diagram of a simple ohmmeter. The permanent-magnet moving-coil mechanism requires 0.010 ampere for full-scale deflection.

(a) What value of resistance should be marked on the mid-scale graduation?

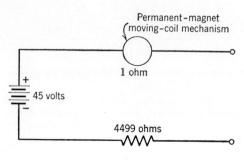

Fig. 10–26.

(b) Repeat (a) for the ¼-scale graduation.
(c) Repeat (a) for the ¾-scale graduation.

Ans. (a) 4500; (b) 13,500; (c) 1500.

10–19. Show how a 0.01-ampere 1-ohm permanent-magnet moving-coil mechanism can be used as:
(a) A 200-volt voltmeter.
(b) A 5-ampere ammeter.
(c) An ohmmeter having a mid-scale value of 5000 ohms.

10–20. The following data are for a circuit containing an ammeter, a watt-hour meter, a wattmeter, and a load:

>Ammeter reading: 5 amperes
>Wattmeter reading: 551 watts
>Revolutions of disk: 27
>Time: 1 min
>Watt-hour constant K_h: ⅓ watt-hr per revolution

If the measurements are correct, what is the accuracy of the watt-hour meter for the data taken? *Ans.* 98%.

10–21. The copper–Constantan thermocouple circuit of Fig. 10–27 is to be used to determine the temperature of a small oven. After the instrument deflection becomes steady, the following data are taken:

R_x, ohms	Instrument Reading, millivolts	Cold Junction Temperature, °F
0	6	77
4	3	77

What is the temperature of the oven as indicated by the thermocouple circuit?
Ans. 325° C.

PROBLEMS

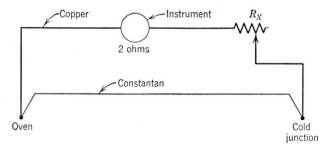

Fig. 10-27.

10-22. It is desired to use a d-c tachometer generator and voltmeter to measure the speed of an automobile. The following facts are known:

> Effective diameter of wheel: 28 in.
> Output voltage of generator: $0.006n$, where n = rpm
> Voltmeter range: 0/3/150
> Voltmeter resistance: 3-volt range, 600 ohms
>   150-volt range, 30,000 ohms
> Gear ratio, axle to generator shaft: 1 to 1

Design a circuit that permits the voltmeter to indicate exactly full scale when the automobile is traveling at the rate of 100 miles per hr.

■ 11 The D-C Machine

11-1. INTRODUCTION

Although a far greater percentage of the electric machines in service are a-c machines, the d-c machine is of considerable industrial importance. (A primary reason for the wide application of a-c machines is the prevalence of alternating-voltage sources. The a-c machine has advantages, however, being generally of lower cost and lower weight compared to d-c machines of the same rating.) The principal advantage of the d-c machine, specifically the d-c motor, is its adaptability to different methods of speed control to provide a wide range of speeds and good speed regulation. The principal industrial application of the d-c machine is therefore in those processes that require fine speed control. In terms of the number of units in service, the automobile industry represents a major consumer of d-c machines, since practically all automobiles are equipped with d-c generators and d-c starting motors.

11-2. D-C MACHINE CLASSIFICATION

Direct-current machines are classified generally as either generators or motors. By definition, an **electric generator** is a rotating machine that transforms mechanical energy into electric energy; an **electric motor** is a rotating machine that transforms electric energy into mechanical energy.

GENERATOR ACTION

The operating characteristics of both motors and generators depend upon the following facts:

(a) A conductor moved through a transverse magnetic field has a voltage induced in it. The magnitude of the voltage is dependent upon the strength of the field and the velocity of the conductor with respect to the field.

(b) A current-carrying conductor in a transverse magnetic field has a force exerted upon it. The magnitude of the force is dependent upon the strength of the field and the magnitude of the current.

Furthermore, standard d-c generators and motors are identical in construction. Therefore, this chapter is concerned with basic generator and motor action and with those details of general construction necessary to an understanding of d-c machine operation. Details of operation that specifically apply to generators or motors are considered in chapters that follow.

11–3. GENERATOR ACTION

Figure 11–1a shows, in schematic form, the fundamental parts of a d-c machine. As indicated, a coil is mounted on a cylinder of magnetic material, called the **armature,** which can be rotated in the

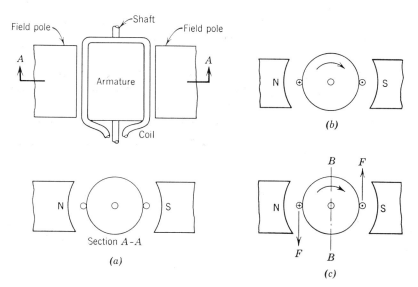

Fig. 11–1. Details of elementary generator.

magnetic field produced by two permanent-magnet **field poles**. It is assumed, although not shown, that provision is made for connecting a resistor to the open ends of the coil.

If the armature is rotated by some external means, a voltage is induced in the coil. When the rotation is in the direction indicated in Fig. 11–1b, the polarity of the induced voltage is such as to tend to cause current to flow as indicated by the cross and dot in the conductor cross section. When the coil is open, however, no current flows in the coil even though a voltage is induced. Under these conditions, therefore, the power output of the device causing rotation is only that necessary to overcome friction and to supply the hysteresis and eddy-current losses that occur as a result of the rotation of the armature in the magnetic field.

If a resistor is connected to the free ends of the rotating coil, current flows as indicated in Fig. 11–1c. Because the conductors carry current while in a magnetic field, a force is exerted on the conductors in such a direction as to oppose the rotation of the armature. Therefore, if the armature is to be rotated at the same speed as before the resistor was connected, the power output of the device causing rotation must be increased. Thus, the addition of the resistor results in the acquisition of electric energy at the cost of increased mechanical energy at the shaft. The action described therefore represents that of an elementary electric generator.

11–4. A-C TO D-C CONVERSION

Inspection of Fig. 11–1c reveals that the polarity of the voltage induced in the coil is changed when the coil conductors cross the axis B–B. Thus, the generator is, in reality, an elementary a-c generator. The wave form of the alternating voltage produced depends upon the pattern of the magnetic field in the air gap. If the field is assumed

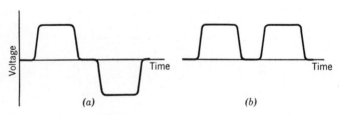

Fig. 11–2. Conversion of alternating voltage to direct voltage.

A-C TO D-C CONVERSION

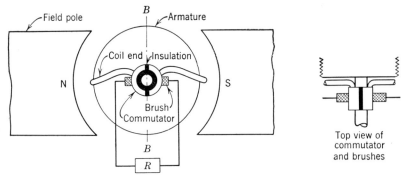

Fig. 11-3. D-c generator elements.

to be uniform and radial, a voltage wave form approximating that of Fig. 11–2a results.

If the alternating voltage of Fig. 11–2a is rectified so that the voltage appearing at the terminals of the machine varies as shown in Fig. 11–2b, the generator can be classified as an elementary d-c generator. This rectification is obtained by means of a segmented connection ring called the **commutator**.

The commutator for the elementary machine under consideration is shown schematically in Fig. 11–3. As indicated, the ends of the coil are connected to the segments of a split copper ring. The segments are so connected to the shaft of the armature that they are insulated from each other and from the shaft. Fixed carbon contacts, called **brushes,** provide the external electric connection to the segments. When the conductors of the machine shown in Fig. 11–3 cross the axis B–B, the polarity of the induced voltage reverses; at the same time, the brush contacts interchange segments. Thus, each brush retains the same polarity throughout a complete revolution of the coil. The voltage

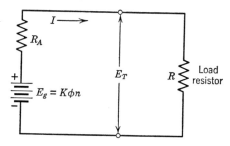

Fig. 11-4. Circuit diagram for elementary d-c generator.

output of the generator is therefore of the pulsating form shown in Fig. 11–2b.

The equivalent circuit for the elementary d-c generator with a resistor connected to its terminals is shown as Fig. 11–4. The battery in this figure is labeled with a magnitude E_g to represent the voltage generated in the coil. The resistor R_A represents the resistance of the coil; the resistor R, the resistance of the load connected to the generator terminals.

An inspection of Fig. 11–4 indicates that the terminal voltage of the generator E_T is given by the relation

$$E_T = E_g - IR_A \tag{11-1}$$

The generated voltage depends upon the rate of change of flux linkage of the coil and, instantaneously, is given by the relation

$$e_g = N\frac{d\phi}{dt} \times 10^{-8} \tag{11-2}$$

The time variation of generated voltage is the pulsating wave form shown as Fig. 11–2b. In representing the generated voltage as a battery in the circuit diagram, however, the average, rather than the instantaneous, value is used. Thus,

$$E_g = \frac{N \Delta\phi}{\Delta t} \times 10^{-8} \tag{11-3}$$

where E_g is the average generated voltage, N is the number of turns, $\Delta\phi$ is the change in flux linking the coil, and Δt is the time of change of flux linkage. If n represents the number of revolutions per second, and ϕ represents the flux per pole, then the average voltage E_g, calculated upon the basis of one-half a revolution, is given by

$$E_g = \frac{N(2\phi)}{(1/2n)} \times 10^{-8} = 4N\phi n \times 10^{-8} \tag{11-4}$$

or, since all factors except the flux and the speed are constants for the given machine,

$$E_g = K\phi n \tag{11-5}$$

Thus, as indicated previously, the voltage induced, or the voltage generated, in the revolving coil depends upon the magnitude of the flux and the velocity of rotation. The battery in Fig. 11–4 is indicated as having the voltage value given by Eq. 11–5.

MOTOR ACTION 225

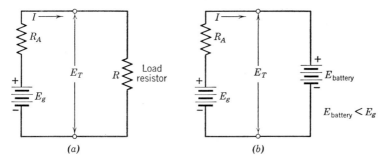

Fig. 11-5. Generator circuit diagrams.

11-5. MOTOR ACTION

The circuit diagram of Fig. 11-4 is shown again as Fig. 11-5a. As previously indicated, the voltage across the load resistor is less than the generated voltage by the amount of the IR drop across the coil resistance.

If the load resistor of Fig. 11-5a is replaced by a battery that has the same voltage as the terminal voltage of the generator and that has negligible internal resistance, as shown in Fig. 11-5b, the current in the circuit is the same as before the resistor was replaced. Therefore, the battery that is added is being charged.

If the power input to the device causing rotation of the generator is reduced, the speed of the generator is reduced. This reduction in speed causes a reduction in the generated voltage and, therefore, a decrease in the charging current to the battery. If the speed reduction is such as to cause the generated voltage of the generator to be

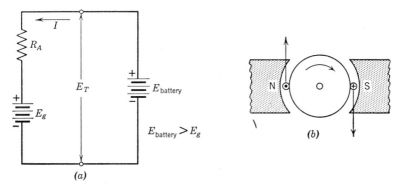

Fig. 11-6. Details of elementary motor.

equal to the battery voltage, no current flows in the circuit. A further reduction in speed causes the voltage of the generator to be less than that of the battery; therefore, current flows in the circuit in a direction opposite to the original direction. The battery is then supplying power to the "generator," as shown in Fig. 11–6a.

For the conditions shown in Fig. 11–6a, the current flow in the coil conductors is as indicated in Fig. 11–6b. The force developed on the conductors (because they carry current while in a magnetic field) is such as to maintain the original direction of rotation; thus the generated voltage maintains its original polarity. Since the electric-energy input to the "generator" assists in the rotation of the machine and is thus converted to mechanical energy, the machine is, by definition, an elementary electric motor.

An inspection of Fig. 11–6a indicates that the voltage relation applying to the elementary motor is given as

$$E_T = E_g + IR_A = K\phi n + IR_A \qquad (11\text{–}6)$$

If a load (such as a cable supporting a weight) is now placed on the shaft of the motor, a rotation-opposing force is exerted, and the speed of rotation decreases. The generated voltage therefore decreases, and the current increases, resulting in an increased force in the direction of rotation. The machine therefore continues to decrease in speed until the increased torque in the direction of rotation is equal to the opposing torque due to the load. Thus, the acquisition of mechanical energy at the shaft is accompanied by an increase in the electric-energy input.

From the preceding analyses of generator and motor action, it should be apparent that the two types of d-c machine are similar in action, that both have a voltage generated, and that both have a force, and therefore a torque, developed on the conductors that form a fundamental part of their construction. The two types of machine differ in the magnitude of the voltage generated with respect to the terminal voltage, and in the magnitude and direction of the torque developed with respect to the speed and direction of rotation. These magnitudes are considered in greater detail in later chapters but are included here to show the basic similarity of the machines.

11–6. MACHINE CONSTRUCTION

In actual construction, the d-c machine differs from the elementary machine of the preceding articles in the following ways:

(a) Except for the air gap between the armature and the field

MACHINE CONSTRUCTION

poles, the magnetic circuit is completed with a magnetic material, usually steel.

(b) The armature is not a smooth, solid cylinder but a laminated body with a tooth-and-slot surface.

(c) The **armature winding** is not a single-turn coil but consists of many coils, usually of more than one turn, which fill the slots on the armature surface.

(d) The commutator has more than two segments, the actual number depending upon the number of coils and the type of winding employed on the armature.

(e) Except for very small machines, such as tachometer generators and magnetos, the magnetic field of d-c machines is not provided by permanent magnets but by field poles that have windings producing the necessary magnetomotive force. Furthermore, practical machines usually have more than two field poles.

Each of the items in the preceding list is considered in more detail in this or succeeding articles.

The complete magnetic circuit of a four-pole d-c machine is shown in section A–A of Fig. 11–7. In addition to indicating the flux paths in the magnetic circuit, the sketch is labeled with some of the terms encountered in the discussion of d-c machine construction. The field yoke, or **frame,** of a d-c machine is usually made of cast or rolled steel, rolled steel being preferable because of its higher permeability. The

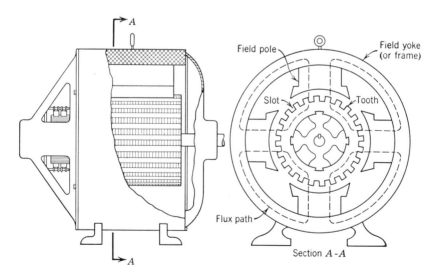

Fig. 11–7. Magnetic circuit of a d-c machine.

field poles are usually constructed of laminated steel and are bolted to the frame.

The tooth-and-slot surface of the armature offers two advantages over the smooth surface: first, it permits the design of a magnetic circuit having a minimum air gap between the armature and the pole face; second, it provides a firm foundation for the conductors, which might otherwise be loosened by the tangential forces to which they are subjected.

The rotating armature is the location of hysteresis and eddy-current losses, called **core losses**. To reduce the eddy-current losses, the armature is built up of laminations, usually insulated from each other by a layer of varnish.

11-7. ARMATURE WINDINGS

The armature winding of the elementary generator considered earlier consisted of a one-turn coil. Such a generator would require an impractically large field flux and speed of rotation to produce a voltage of practical magnitude at its terminals. For this reason, the armature surface is slotted to carry many coils, and therefore many conductors, which are connected in series–parallel combinations so that the average voltage between machine terminals is many times the average voltage per conductor. By employing this type of armature-winding construction, the field flux and speed are kept within

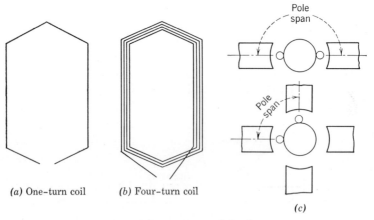

(a) One-turn coil   (b) Four-turn coil
(c)

Fig. 11-8. Armature coil details.

ARMATURE WINDINGS

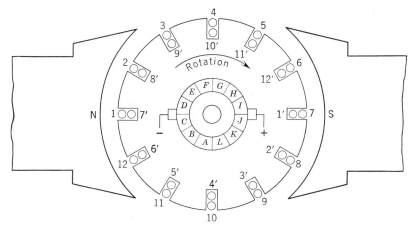

Fig. 11-9. Armature winding details.

practical limits, and the voltage fluctuations are so reduced that the voltage output can be considered constant.

The basic component of all types of armature windings is the armature coil. The coil previously considered was one-turn, as shown schematically in Fig. 11-8a. In practical armature windings, however, the coils usually have more than one turn. If there is more than one turn, the additional conductors are in series, and the voltage at the coil terminals is increased in proportion to the increase in the number of turns. Thus, a coil having four turns, as shown schematically in Fig. 11-8b, has four times the voltage of a one-turn coil for the same conditions of flux and speed. For a given size of slot, however, the increase in voltage gained by increasing the number of turns is offset by a decrease in the current-carrying capacity of the coil, since the coil conductors must be smaller. The number of turns in the armature coils is therefore dependent upon other design considerations.

In many machines, the coil sides are placed a pole span apart. A pole span is the peripheral distance from the center line of one main pole to the center line of the next main pole. As shown in Fig. 11-8c, a pole span is 180 degrees for a two-pole machine, and 90 degrees for a four-pole machine. Many armature windings have coils with sides slightly more or less than a pole span apart, but these coils are not considered in this book.

The analysis of armature windings is simplified if made for a machine specifically designed for the purpose of the analysis. Such a machine is shown schematically as Fig. 11–9. With reference to this machine and the subsequent analysis, it is assumed that the coils are single-turn, with coil sides a pole span apart, and that there are 12 slots on the armature surface.

The conductor combinations that form coils are indicated as 1 and 1', 2 and 2', 3 and 3', etc. There are two coil sides in each slot (the usual practice in winding armatures); one side of each coil is at the bottom of a slot, and the other side is at the top. The connections of the coils to the commutator segments are not shown in Fig. 11–9, in order to simplify the diagram as much as possible.

The ends of the individual coils of the machine of Fig. 11–9 are connected to adjacent commutator bars. Thus, if the armature and commutator surfaces of the machine are developed (laid out in one plane), the coils and their connections to the commutator segments appear as shown in Fig. 11–10. In this figure, the broken lines represent the conductors that are at the bottom of slots; the solid lines, those at the top of slots. The numbers beside the lines correspond to the conductor numbers shown in Fig. 11–9. The polarity marks beside the lines indicate the polarity of the voltage induced in the conductors when the armature is in the position indicated in Fig. 11–9. The squares marked N and S in Fig. 11–10 represent the shadows of the field poles; that is, it is assumed that the field poles are above the developed plan, toward the observer. (One of the coils in Fig. 11–10

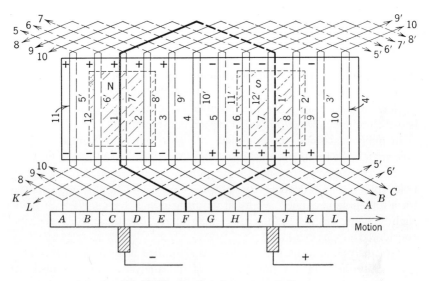

Fig. 11–10. Developed armature winding.

ARMATURE RESISTANCE 231

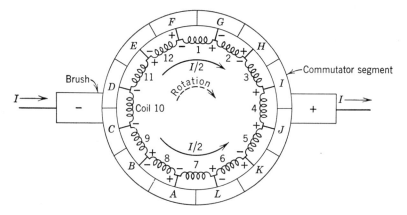

Fig. 11-11. Schematic diagram of armature winding for two-pole machine.

is represented by heavy lines to emphasize the connection to adjacent commutator segments.)

A complete traverse of the armature winding of Fig. 11-10 indicates that the armature coils are connected in series to form a closed loop, and that the algebraic summation of the voltages induced in the coils is zero. The armature winding can therefore be represented schematically as shown in Fig. 11-11. The polarities of the coils in Fig. 11-11 are for the armature in the position shown in Fig. 11-9.

In order to have the maximum voltage between brushes, the brushes must be located so that they make contact at the segments where the polarities of the coils reverse. They are so located in Figs. 11-10 and 11-11. Figure 11-11 indicates that the armature circuit then consists of two parallel paths, each path containing one-half the total number of coils.

There are many variations of armature windings, and the discussion here is limited to the simplest kind. The possible variations come about principally for machines that have more than two poles, but variations are possible even for two-pole machines. The discussion, however, is adequate for a basic understanding of the phenomena that occur in a d-c machine armature and that is its sole purpose.

11-8. ARMATURE RESISTANCE

The resistance of the armature circuit is the resistance between the armature terminals and includes the brushes. In this book, this resistance is symbolized as R_A.

The magnitude of the armature resistance is dependent upon the construction of the machine; except for small machines, it is a low value, less than 1 ohm. Factors governing the resistance value are the number, size, and connection of the armature coils and the contact resistance between the carbon brush and the copper commutator.

11-9. COMMUTATION

As indicated in Fig. 11-11, each coil of the winding carries one-half the total current entering or leaving the armature circuit. However, as the armature rotates, the polarity of the voltage induced in a given coil reverses after the coil makes contact with a brush, and the current through the coil reverses at the same time. Thus, when the coil approaches the contact with the brush, the current through the coil is in one direction; when the coil leaves the contact with the brush, the current has been reversed. The reversal of current in the coil, called **commutation**, takes place while the coil is short-circuited by the brush and is an important factor in the rating and operation of d-c machines.

In the first analysis of commutation, it is assumed: (a) that the insulation between commutator segments is of negligible thickness compared to the width of the commutator segment, (b) that the resistance of an armature coil is negligible compared to the resistance of the carbon brush contact with the copper segments, and (c) that the armature coil has negligible inductance.

The starting point for this analysis of commutation is indicated by the partial schematic diagram shown as Fig. 11-12a, with the coil marked A the coil under consideration. As indicated in the figure, the width of the brush is equal to the width of a commutator segment, and the total armature current is 40 amperes. At the time represented by Fig. 11-12a, the brush is entirely on segment 1, and therefore the current in coil A is 20 amperes, or one-half the total armature current. As the armature rotates, the brush short-circuits coil A, and there are parallel current paths into the brush as long as the short circuit exists. Thus, when the commutator has moved such that the brush is one fourth on segment 2 and three fourths on segment 1, as shown in Fig. 11-12b, the equivalent electric circuit is as shown in Fig. 11-12c, where R_1 and R_2 represent the brush contact resistances on segments 1 and 2, respectively. A resistor is not shown for coil A, since it is assumed that the coil resistance is negligible compared to the brush contact resistance. The values of current in the parallel paths of the equivalent electric circuit are determined by the respective resistances of

COMMUTATION

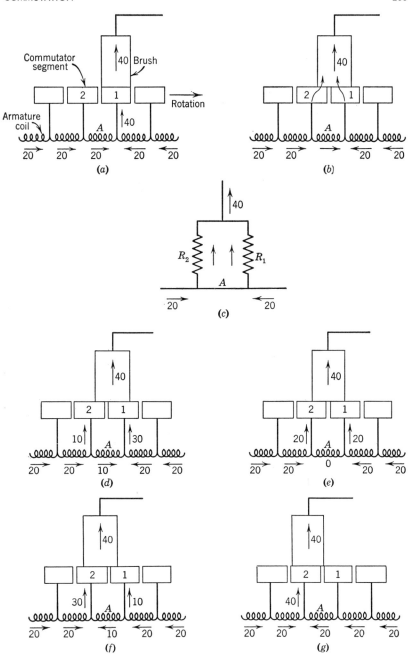

Fig. 11-12. Coil undergoing commutation.

the paths. For the condition shown in Fig. 11–12b, resistor R_2 has three times the resistance of resistor R_1 (assuming that the contact resistance varies inversely as the area of contact), and the current distribution is as shown in Fig. 11–12d.

When the brush is one half on segment 2 and one half on segment 1, the brush contact resistances in the two paths are equal, and the current distribution is as shown in Fig. 11–12e.

The current distribution for two later times in the commutation period, the period when the coil is short-circuited by the brush, is shown in Figs. 11–12f and 11–12g. Figure 11–12f shows the distribution when the brush is three fourths on segment 2 and one fourth on segment 1; Fig. 11–12g, the distribution at the instant the brush is about to break contact with segment 1.

A plot of the value of current in coil A versus time appears as shown in Fig. 11–13 (the original direction of current in the coil is assumed to be positive). The uniform reversal of current in the coil undergoing commutation is a necessary condition for ideal commutation.

Ideal commutation results in the preceding discussion because it is assumed that the armature coil has negligible inductance, which is not true in actual machines. The effect of coil inductance is a delay in the reversal of the coil current, so that, instead of the uniform reversal shown in Fig. 11–13, the current reverses as shown in Fig. 11–14a. The delay in current reversal means that the current in the coil is not completely reversed at the instant the brush breaks contact with the commutator segment, as indicated in Fig. 11–14b, with the result that an arc is formed between the brush and the segment. The effect of the arcing is pitting of the commutator segments and, possibly, eventual failure of the machine.

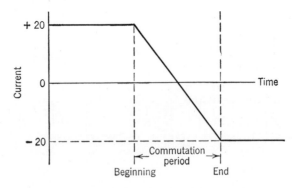

Fig. 11–13. Current versus time for ideal commutation.

FIELD WINDINGS

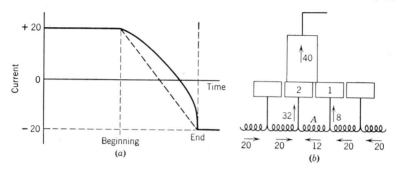

Fig. 11-14. Effect of inductance upon commutation.

The effects of inductance upon commutation are offset in most d-c machines by the installation of a **commutating winding.** The purpose of this winding is to induce a voltage in the coil undergoing commutation so that the voltage of self-induction caused by the reversal of the coil current is neutralized, thus allowing the current to reverse uniformly. (The commutating winding is considered in more detail in Art. 11–11.)

The relative thickness of the insulation between commutator segments, the resistance of the armature coil, and the relative width of the brush also affect the reversal of current, but an analysis of their effects is not necessary to this discussion.

11-10. FIELD WINDINGS

The windings for producing the main magnetic field in a d-c machine are wound on the field poles as shown schematically in Fig. 11–15. Each pole, regardless of the number of poles, has a winding on it, and the windings are connected in series to form a field winding. Field windings are generally classified according to the method of connection, that is, **shunt, series,** or **separately excited.**

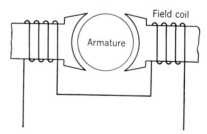

Fig. 11-15. Field coil connections.

The shunt field winding is designed to be connected in parallel with the armature, as shown schematically in Fig. 11–16a. The resistance of the winding is large so that the current in the winding is small

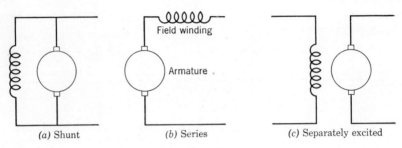

Fig. 11–16. D-c machine connections.

compared to the rated armature current of the machine. (The current taken by shunt windings is usually of the order of 2 to 5 per cent of the rated current of the machine.) To produce the necessary mmf, the winding consists of many turns.

The series field winding is designed to be connected in series with the armature, as shown schematically in Fig. 11–16b. The resistance of the winding is made as low as possible so that the voltage drop across the winding is the minimum. (The actual resistance of a series winding approximates that of the armature circuit.) To produce the same flux at rated conditions, the turns of a series winding are fewer than those of a shunt winding, since the series winding carries the rated current of the machine.

The separately excited field winding is energized from a separate voltage supply, as shown schematically in Fig. 11–16c. The characteristics of the winding depend upon the application of the machine, and they may range from those of a series winding to those of a shunt winding.

Many d-c machines have both a shunt-type and a series-type winding on their field poles, as shown schematically in Fig. 11–17a. Such

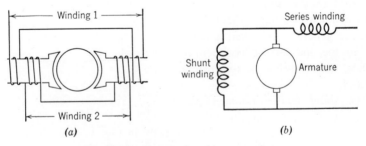

Fig. 11–17. Compound-machine connections.

COMMUTATING WINDINGS

machines are called **compound machines;** they are designed to have both windings energized, as shown in Fig. 11–17b.

When the mmf's of the field windings of a compound machine are additive, the machine is **cumulatively compounded;** when the mmf's are subtractive, the machine is **differentially compounded.**

11–11. COMMUTATING WINDINGS

The purpose of the commutating winding is to induce a voltage in the coils undergoing commutation so that the voltage of self-induction is neutralized, thus producing ideal commutation. Since the coils undergoing commutation are in a plane perpendicular to the axis of the main field poles, as shown in Fig. 11–18, the commutating-winding field must be perpendicular to the main field. The installation of commutating windings therefore necessitates the addition of poles having an axis perpendicular to the main field axis, as shown in Fig. 11–18. The poles upon which the commutating windings are wound are called **commutating poles.**

Since the voltage of self-induction caused by the reversal of the coil current is a function of the magnitude of the armature current, the commutating field strength should also be a function of the armature current and is made so by permanently connecting the commutating winding in series with the armature circuit.

Except for the commutating winding, machine windings are usually

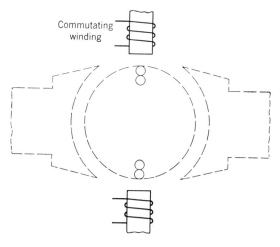

Fig. 11–18. D-c machine having commutating windings.

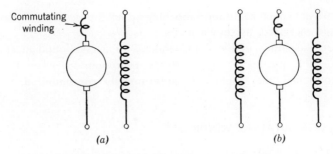

Fig. 11-19. Terminal possibilities for d-c machines.

unconnected when the machine is delivered by the manufacturer. Thus, as shown in Fig. 11-19, a machine designed to be shunt-connected, series-connected, or separately excited has four terminals; a machine designed for compound operation has six terminals. The proper interconnection of the terminals is left to the customer.

11-12. MACHINE RATINGS

The rating of an electric machine is based upon the temperature to which the various parts of the machine are subjected when the machine is operated. Instead of providing complete temperature data, however, the nameplate of a machine specifies the rating in terms of electric or mechanical output, voltage, speed, current, and temperature rise above a given ambient (surrounding) temperature.

The load on a machine is the power that it delivers. **Full load** is the rated power output of the machine. The nameplate ratings are full-load ratings; that is, the values of current and speed are the values of these quantities expected when the machine is operating at the rated voltage and is delivering rated power. Under full-load conditions, the value of temperature rise is the value indicated on the nameplate.

In effect, the nameplate is the manufacturer's guarantee; the manufacturer thus implies that the machine will perform satisfactorily during a long machine life if operated under the limits and conditions specified on the nameplate. It must not be inferred that the machine cannot be operated under conditions different from those on the nameplate. However, operation that exceeds any of the nameplate ratings may result in the premature failure of the machine.

PROBLEMS

11-1. D-c motors and generators are said to be converters of energy. Explain, for each machine, how the rate of conversion is increased when the power output of the machine is increased.

11-2. The armature winding of an eight-pole d-c generator has been completely destroyed. You have the choice of having the armature wound with two paths in the armature circuit or with eight paths in the armature circuit. Compare the voltage, the current, and the power ratings for the two connections if the same total number of coils is used for each.

11-3. A d-c machine is found to have six terminals. The resistances, measured by ohmmeter, between the pairs of terminals found to complete circuits are as follows:

Between terminals 1 and 2: 150 ohms.
Between terminals 3 and 4: less than 1 ohm.
Between terminals 5 and 6: less than 1 ohm.

(a) Identify the terminals.

(b) For those terminals that are not specifically identified in (a), explain how they might be so identified.

11-4. A two-pole 1500-rpm d-c machine has 80 commutator segments. One brush of the machine covers two segments. Determine the time for the current to be reversed in a coil undergoing commutation. *Ans.* 0.001 sec.

11-5. The commutator of a two-pole machine contains 100 segments. A brush covers two segments. The armature circuit delivers a full-load current of 100 amperes.

(a) If the speed of rotation is 600 rpm, what is the average rate of change of current in an armature coil undergoing commutation?
Ans. 50,000 amperes per sec.

(b) If the inductance in each armature coil is 4×10^{-4} henry, what is the average emf of self-induction in each coil during commutation? (*Note:* When this value exceeds about 3 volts, commutating poles are considered necessary.) *Ans.* 20 volts.

11-6. At rated conditions, the shunt winding and the series winding of a certain cumulatively compounded machine have the same mmf. When only the shunt winding is energized at rated conditions, the flux is 80% of that produced when both the shunt and series are connected.

(a) Why does the addition of the series winding increase the flux only 25%?

(b) How much flux, relatively, does the series winding produce if energized alone under rated conditions?

11-7. (a) Is a motor rated in kilowatts or in horsepower?

(b) Is a generator rated in kilowatts or in horsepower?

■ 12 D-C Generators

12–1. D-C GENERATOR RATING

The nameplate of a d-c generator usually contains the following ratings:

 (a) Power   (c) Speed
 (b) Voltage   (d) Temperature rise

The power rating, given in kilowatts, is the rated full-load output of the generator when the terminal voltage is that specified on the nameplate. A current rating is not usually included on the nameplate but can be calculated from the power and voltage ratings. The current rating thus calculated is the current in the lines to the load.

The speed rating of a generator is the speed at which the machine should be operated in order to deliver rated power at rated voltage without exceeding the specified temperature rise. Operation of the machine at a speed below the rated value may result in overheating, not only because of the increased field current necessary to produce rated voltage but also because of the decrease in fanning action due to the decrease in speed. Operation at speeds higher than rated may produce strains in the rotating armature that ultimately cause machine failure.

The temperature rise specified on the nameplate is usually that to be expected if the machine is operated continuously at rated conditions. This temperature rise is based upon a standard ambient temperature, which establishes the permissible ultimate temperature. The standard ambient temperature is dependent on the cooling media and is 40° C

GENERATED VOLTAGE

for air-cooled machines. If the ambient temperature is lower than the standard value, the machine rating may be exceeded somewhat without the allowable ultimate temperature being exceeded. When the machine is so operated, however, provision should be made for measuring the temperature, to make certain that the allowable ultimate is not exceeded.

12-2. GENERATED VOLTAGE

The voltage generated in the elementary d-c generator considered in Chapter 11 has a wave form that is rectangular and that has zero values for an appreciable time; the voltage output of the generator is therefore taken to be the average value of this pulsating wave. In the actual d-c generator, each conductor has such a pulsating voltage generated in it. The voltage between brushes, however, is the summation of the voltages generated in the conductors that form the circuit between the brushes. The position of these conductors on the surface of the armature is such that, at any instant, the circuit between brushes contains conductors that are undergoing every phase of voltage generation; that is, this circuit contains conductors that have zero generated voltage as well as those that have the maximum. The resulting generated voltage is therefore essentially constant, although, if it is viewed on an oscilloscope, a slight rippling effect may be noticeable.

The average voltage generated in the one-turn coil of the elementary generator is given by Eq. 11–4:

$$E_g = 4\phi n \times 10^{-8} \qquad (12\text{–}1)$$

The voltage per coil side, or conductor, is therefore

$$E_g \text{ per conductor} = 2\phi n \times 10^{-8} \qquad (12\text{–}2)$$

The machine for which Eq. 12–1 is developed has two poles. A similar calculation for a four-pole machine results in an average voltage per conductor that is twice the average per conductor for a two-pole machine. The average voltage per armature conductor is therefore

$$E_g \text{ per conductor} = P\phi n \times 10^{-8} \qquad (12\text{–}3)$$

where P represents the number of poles.

The average generated voltage between brushes depends upon the number of conductors in the circuit between the brushes. Therefore, if Z is the total number of conductors on the armature surface and p is the number of paths in the armature circuit, the average generated voltage between brushes is

$$E_g = \frac{Z}{p} P\phi n \times 10^{-8} \quad (12\text{--}4)$$

For a given machine, Eq. 12–4 may be written

$$E_g = K\phi n \quad (12\text{--}5)$$

12–3. FACTORS AFFECTING GENERATED VOLTAGE

It is apparent from Eq. 12–5 that, for a given machine, the only factors affecting the generated voltage are the field flux and speed. These factors affect not only the magnitude of the generated voltage but also its polarity. A reversal of either the field flux or the direction of rotation causes a reversal of the polarity; a reversal of both the field flux and the direction of rotation results in no change in the polarity.

The speed and direction of rotation of the generator depend upon the prime mover causing the rotation. Although it is true that the speed of the generator may decrease when the generator output is increased, the decrease in speed is due to the load-speed characteristic of the prime mover; that is, the speed of the prime mover decreases as the load connected to the prime mover increases, unless the prime mover has a governor adequate to maintain the speed. Therefore, to accomplish a change in generated voltage by a change in speed requires a change in the controls of the prime mover.

The field flux of a given machine depends upon the current in the field winding and upon the condition of the magnetic circuit. A change in generated voltage can therefore be accomplished by changing the field current. Two widely used methods for changing the field current and, therefore, the generated voltage are illustrated in Fig. 12–1. Figure

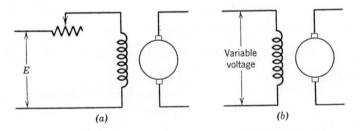

Fig. 12–1. Methods for changing field flux.

FACTORS AFFECTING GENERATED VOLTAGE

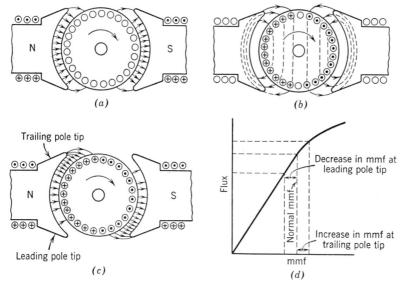

Fig. 12-2. Armature reaction details.

12-1a shows a rheostat connected in series with a field winding energized from a constant-voltage source; Fig. 12-1b shows a field winding energized from a variable-voltage source.

It might be assumed that the field flux remains constant if the current in the field winding is maintained at a constant value. Actually, the flux may change owing to a change in the condition of the magnetic circuit. Such a change in magnetic-circuit conditions may occur when the armature current changes and may result in a change in the field flux. Because the effect upon the field flux depends upon the armature current, the effect is called **armature reaction.**

Armature reaction is the effect, upon the magnetic circuit, of the mmf produced by the current flowing in the armature conductors. This effect can be determined by first considering the field patterns produced by the field and the armature mmf's acting independently and then combining the patterns. Thus, Fig. 12-2a schematically shows the field pattern produced by the field winding acting alone; Fig. 12-2b, that produced by the armature mmf alone; and Fig. 12-2c, the combination of the two patterns.

Figure 12-2 shows that the two mmf's act at right angles to each other except in the region at the pole tips. At the one pole tip, the mmf's are additive; at the other pole tip, the mmf's are subtractive. At first glance, it appears that the effect of the addition and subtraction is to distort the field so that the increase in flux at one pole tip is equal to the decrease at the other, with no change in the total flux from the pole. Actually, because of its small cross-sectional area, the pole tip having the addition of the mmf's becomes saturated so that the increase in flux is not so great at that tip as the decrease at the other tip. This effect is evident in Fig. 12–2d, which shows a schematic magnetization curve for the combined steel-and-air magnetic circuit.

The over-all effect of armature reaction, therefore, is a decrease in the total flux from the pole compared to that expected for the given field winding current. The amount of the decrease depends upon the current in the armature but is not a linear function of this current. In actual value, the amount of the decrease, for rated current in the armature circuit, may range from 1 to 5 per cent of the total flux.

12–4. THE SATURATION CURVE

Although the generated voltage is a linear function of flux, the variation in the generated voltage with a change in the current in the field winding depends upon the characteristics of the magnetic circuit, since flux is not a linear function of field current. The plot that shows the variation in voltage with field current is called the **saturation curve.**

The data for a no-load saturation curve are obtained experimentally by operating the machine at zero load output and constant speed and recording the change in terminal voltage as the field current is varied. The circuit for obtaining the data is shown as Fig. 12–3a; the plot of the data is shown as Fig. 12–3b.

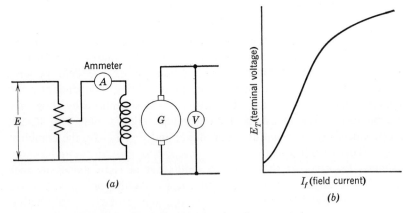

Fig. 12–3. Circuit for obtaining data for saturation curve.

Since the terminal voltage is the difference between the generated voltage and the armature-resistance drop, or

$$E_T = E_g - I_A R_A \qquad (12\text{-}6)$$

and since the armature current in the circuit of Fig. 12-3a is negligible at no load (being only the voltmeter current), the no-load terminal voltage is equal to the generated voltage. The plot shown as Fig. 12-3b is therefore a plot of generated voltage versus field current.

The shape of the saturation curve is similar to that of a magnetization curve. The curve is, in fact, a magnetization curve for the combined magnetic circuit. The ordinate is generated voltage and, since the speed is held constant, is directly proportional to the flux. The abscissa is field current and is proportional to the mmf. Thus, the curve is a plot of flux versus mmf for the magnetic circuit. The small voltage generated when the field current is zero is due to the residual magnetism in the magnetic circuit.

The no-load saturation curve for one speed can be made to serve as the basis for curves applying to any other speed without additional data. Since, at no load, the flux is a function of the field current (there being no armature reaction), the generated voltage for a given field current is directly proportional to the speed. Thus, for a given field current, the generated voltage E_g' at some speed n' is given by

$$E_g' = \frac{n'}{n} E_g \qquad (12\text{-}7)$$

where E_g is the generated voltage at speed n for the same field current. If calculations are made for a sufficient number of different field currents, the data for a saturation curve at speed n' are obtained.

12-5. SHUNT GENERATOR BUILD-UP

The data for the saturation curve of Fig. 12-3b are obtained using a separately excited generator. The source of the excitation could be another separately excited generator, but somewhere there must be a source that is not a separately excited generator. This source could be a battery but is usually a shunt generator.

A shunt generator derives its excitation from its own armature and is connected as shown in Fig. 12-4a. As indicated, a field rheostat is included in the field circuit for controlling the generated voltage. The field switch and the schematic diagram of Fig. 12-4b are included for purposes of the analysis to follow. As shown in Fig. 12-4b, the

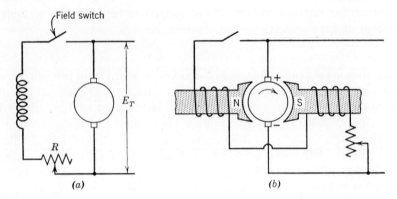

Fig. 12-4. Shunt generator details.

field winding is so connected that, when the field switch is closed, the field current due to the residual voltage in the armature flows to strengthen the residual flux.

In addition to the proper connection shown in Fig. 12–4b, the analysis of shunt generator action requires two different characteristic curves. The first is the no-load saturation curve, which gives the variation in generated voltage with field current, as shown in Fig. 12–5a; the second is the field-resistance characteristic, which indicates how the field current varies with the voltage applied to the field winding, as shown in Fig. 12–5b.

The curves of Figs. 12–5a and 12–5b have their ordinates labeled as the terminal voltage. This designation is correct at all times for the field-resistance characteristic, since the voltage applied to the field

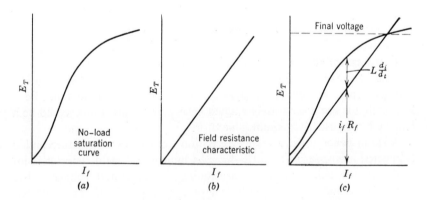

Fig. 12-5. Build-up of a shunt generator.

SHUNT GENERATOR BUILD-UP

when the switch is closed is the terminal voltage. For the no-load saturation curve, the ordinate is actually the generated voltage, which differs from the terminal voltage when the switch is closed because of the armature-resistance drop caused by the field current flowing through the armature. Since this field current is only a small fraction of the rated armature current, the armature-resistance drop due to the field current flowing in the armature may be assumed to be negligible. Thus, for purposes of this analysis, the no-load saturation curve is considered to represent the variation in terminal voltage with field current. This assumption makes the axes of the two curves the same and permits the combination of the curves shown as Fig. 12–5c.

When the field switch is closed, the voltage due to residual magnetism is impressed on the field winding and causes current to flow. This current increases the field flux, causing a greater generated voltage and, therefore, a greater field current. This cycle continues until the voltage applied to the field winding produces the field current necessary to generate that voltage. The end of the cycle is represented in Fig. 12–5c as the intersection of the saturation curve and the field-resistance characteristic. The increase in voltage when the field switch is closed is called the build-up of a shunt generator.

The preceding analysis gives little indication of how the build-up actually occurs. Since the field circuit is inductive, there is a delay in the increase in current upon closing the switch; the rate at which the current increases depends upon the voltage available for increasing it. The equation that applies to the field circuit is

$$E_T = iR + L\frac{di}{dt} \qquad (12\text{–}8)$$

where R is the total resistance of the field circuit, and L is the inductance of the field circuit. Equation 12–8 indicates that the difference between the voltage applied to the field circuit and the iR drop across the field circuit is the voltage available for increasing the current; this difference is shown graphically in Fig. 12–5c.

When the terminal voltage is equal to the resistance drop in the field winding, no voltage is available for increasing the current, which is true at the point at which the curves intersect; therefore, the current becomes constant. The intersection of the two curves therefore indicates the voltage to which the machine builds up. The rate of build-up is governed by the difference between the curves, the rate being greatest where the difference is greatest (assuming constant inductance).

The voltage of a shunt generator is controlled by adjustment of the field-circuit resistance. Thus, a decrease in the resistance results in a resistance characteristic of less slope, and, therefore, the two curves intersect at a higher value of voltage; an increase in resistance causes the voltage to be decreased. Factors affecting the build-up of a shunt generator are the field-circuit resistance, the field-winding connection, the rotation, and the polarity of the residual magnetism. (It is left to the student to determine the effects of each of these factors.)

12-6. GENERATOR OPERATING CHARACTERISTICS

In the selection of a generator for a particular duty, the external characteristic, the relationship between terminal voltage and load current, is probably the most important of the characteristics. When plotted, the external characteristic curve indicates how the terminal voltage varies as the load current changes, the field-circuit conditions and speed remaining constant.

The basis for the development of the external characteristic is the two relations previously developed for the generator, namely,

$$E_T = E_g - I_A R_A \qquad (12\text{--}9)$$

and

$$E_g = K\phi n \qquad (12\text{--}10)$$

which may be combined to give

$$E_T = K\phi n - I_A R_A \qquad (12\text{--}11)$$

If the flux and, therefore, the generated voltage of a separately excited generator are assumed to remain constant (since the field current is to be held constant), the terminal voltage varies linearly with load current, as

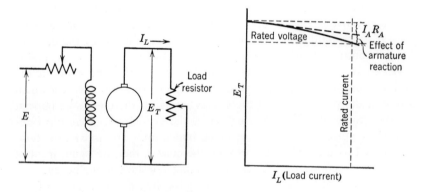

Fig. 12-6. External characteristic of a separately excited generator.

GENERATOR OPERATING CHARACTERISTICS

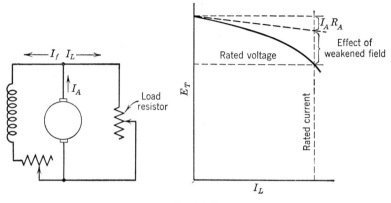

Fig. 12-7.

indicated by the dashed line in Fig. 12-6; the difference between the terminal voltage under load and the terminal voltage at no load is the $I_A R_A$ drop. Actually, as the load current increases, the flux and, therefore, the generated voltage decrease, because of armature reaction, and the characteristic curve droops away from the linear. Therefore, for a separately excited generator, the difference between the terminal voltage under load and the terminal voltage at no load is caused by the $I_A R_A$ drop and the effect of armature reaction.

If the field current is neglected and it is assumed that the load current and armature current are equal, the terminal voltage of a shunt generator decreases as the load current increases because of the increased armature-resistance drop and the effect of armature reaction upon the field flux. In addition, there is a still further decrease in flux because the field current decreases as the terminal voltage decreases. The result is the widely drooping curve in Fig. 12-7, which shows the difference between the terminal voltage under load and the terminal voltage at no load to be due to the $I_A R_A$ drop and the effect of the weakened field.

The external characteristic of a cumulatively connected compound generator depends upon the relative strengths of the shunt and series field windings. As the load current increases, the series field mmf increases and tends to increase the flux and, therefore, the generated voltage. In some generators, the increase in generated voltage is greater than the $I_A R_A$ drop, so that, instead of decreasing, the terminal voltage increases, as shown by curve A in Fig. 12-8. Generators having this characteristic are called **over-compounded generators**. The eventual droop in the characteristic curve occurs because of saturation of the magnetic circuit. A **flat-compounded generator** is one having a full-

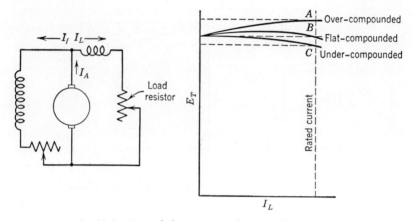

Fig. 12–8. External characteristics of compound generators.

load terminal voltage equal to the no-load terminal voltage, as indicated by the curve labeled *B* in Fig. 12–8. The series winding of this machine is weaker than the one in the over-compounded machine and, therefore, does not increase the flux as much for a given armature current. A

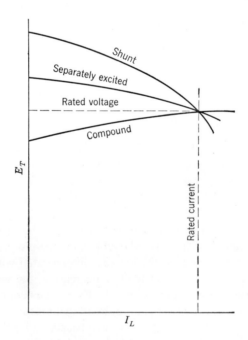

Fig. 12–9. Generator voltage characteristics.

still weaker series winding gives the machine the characteristic labeled C in Fig. 12–8. Since the full-load terminal voltage is less than that at no load, the generator is said to be **under-compounded.**

If it is desired to select a generator of a given full-load rating, the selection depends upon the characteristic desired at some load other than full load. To assist in the selection, the external characteristics of the shunt, separately excited, and compound generators may be plotted on the same axes, as shown in Fig. 12–9. As indicated, all characteristics pass through the same point at full load, since the three generators have the same rating.

12–7. VOLTAGE REGULATION

The change in terminal voltage of a generator between full load and no load is called the **voltage regulation,** usually expressed as a percentage of the voltage at full load. Thus, if the voltage regulation of a generator is given as 10 per cent, it means that the terminal voltage increases 10 per cent as the load is changed from full load to no load. Expressed mathematically,

$$\% \text{ voltage regulation} = \frac{E_{NL} - E_{FL}}{E_{FL}} \times 100 \qquad (12\text{--}12)$$

where E_{NL} is the voltage at no load, and E_{FL} is the voltage at full load. The voltage regulation of a generator is determined with field-circuit conditions and speed held constant.

12–8. GENERATOR PRIME MOVERS

The development of the external characteristics of generators and the expression of voltage regulation are both based upon the assumption that the generator is operating at constant speed. As previously indicated, the speed at which the generator operates depends upon the prime mover. If the prime mover does not have a governor mechanism that maintains the speed approximately constant, the external characteristic and the voltage regulation of the generator are affected. Thus, in selecting a generator for a specific duty, consideration must also be given to the characteristics of the prime mover that is to drive it.

The output rating of the prime mover to drive a generator depends upon the rated load of the generator and the losses associated with the

operation of the generator. The generator losses which must be supplied are:
(a) Friction (brush and bearing) and windage losses.
(b) Copper losses (armature and, if shunt-connected, field winding).
(c) Core losses (hysteresis and eddy-current).

PROBLEMS

12–1. The nameplate of a generator reads as follows: 120 volts, 15 kw, 1200 rpm. The resistance of the armature circuit is found to be 0.05 ohm.
(a) What is the rated current of this machine?   *Ans.* 125 amperes.
(b) When it is operated at constant speed, as a separately excited generator, the following data are taken:

Line current	0	62.5	125
Terminal volts	131.5	126.4	120

Calculate the generated voltage for each of the three loads.
Ans. 131.5; 129.53; 126.25.
(c) In (b), why does the generated voltage decrease as the load increases?
(d) If there were no armature reaction, what would be the terminal voltage for the three loads?   *Ans.* 131.5; 128.37; 125.25.
(e) What is the percentage reduction in flux due to armature reaction at a load of 62.5 amperes and at a load of 125 amperes?   *Ans.* 1.5%; 4%.

12–2. The no-load saturation curve for a generator operating at 1800 rpm is plotted as Fig. 12–10.
(a) Plot the no-load saturation curve for 1500 rpm.
(b) Calculate the generated voltage when the generator is operating at no load with a field current of 4.6 amperes and at a speed of 1000 rpm.
Ans. 144.5 volts.
(c) What field current is required to generate 120 volts at no load when the generator is operating at 900 rpm?   *Ans.* 3.7 amperes.
(d) This machine is operated as a separately excited generator at 1800 rpm with a field current of 4.5 amperes. If a field rheostat having a resistance 0.25 times that of the field circuit is connected in series with the field, what is the no-load voltage when the generator is operating at 1500 rpm?   *Ans.* 197.5 volts.
(e) When the generator is operating at no load as a shunt generator at 1800 rpm, the field current is 4.5 amperes. What is the no-load voltage at 1500 rpm?   *Ans.* 187 volts.
(f) If a field rheostat having a resistance 0.25 times that of the field circuit is connected in series with the field in part (e), what is the no-load voltage at 1800 rpm? at 1500 rpm?   *Ans.* 210 volts; 19 volts.

12–3. A shunt generator has its field winding so connected that the generator builds up to rated voltage with the correct polarity. How does the

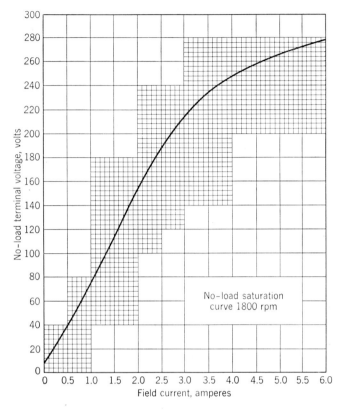

Fig. 12-10.

reversal of the factors listed below affect the polarity and the build-up. assuming that each reversal applies to the original condition?

    (a) Residual flux. (c) Field-winding connection.
    (b) Rotation.   (d) Residual flux and rotation.
    (e) Rotation and field-winding connection.
    (f) Residual flux and field-winding connection.
    (g) Residual flux, rotation, and field-winding connection.

12-4. The first time four shunt generators driven by the same prime mover are operated at rated speed, they fail to build up to rated voltage of 120 volts. The indications of the four voltmeters connected to the armature terminals are listed in the following table; the indications are given for the field switches both open and closed. What should be done to make each generator build up to rated voltage with the correct polarity? (Correct polarity results when the voltmeter reads upscale.)

Generator	Voltage (field switch open)	Voltage (field switch closed)
1	+6	+8
2	Downscale	Closer to zero
3	Downscale	Further downscale
4	+6	+6

12.–5. Plot and briefly explain the terminal voltage versus armature current characteristic of a differentially compounded generator that has its shunt field winding separately excited.

12–6. Plot and briefly explain the terminal voltage versus armature current characteristic of a series generator.

12–7. The generator of Prob. 12–2 has the following nameplate rating: 250 volts, 50 kw, 1800 rpm. The resistance of the field winding is 40 ohms; the resistance of the armature circuit is 0.05 ohm. If this generator has a variable field rheostat with a total resistance of 30 ohms, what is the minimum speed at which this generator could be operated as a shunt machine to have a no-load voltage of 220 volts? Show clearly how you arrived at your answer.   *Ans.* 1456 rpm.

12–8. If the generator of Prob. 12–7 is connected as a separately excited generator, with the field winding connected to a 125-volt supply with no field rheostat, at what speed should the machine be operated to have a terminal voltage of 250 volts when delivering rated current? (Assume a 3% reduction in flux from no load to full load due to armature reaction.)
Ans. 2210 rpm.

12–9. The generator of Prob. 12–7 is separately excited from a 250-volt source, and it is operating at a constant speed of 1800 rpm. If armature reaction is neglected, what should be the rating (resistance and current) of a field rheostat that, if adjusted as the load changes, permits the voltage to be maintained constant at the rated value from no load to full load?
Ans. 20.3 ohms; 4.6 amperes.

12–10. The generator of Prob. 12–7 is to be operated as a shunt generator delivering rated current at rated terminal voltage. At what speed should the machine be operated if the total field circuit resistance is 63 ohms?
Ans. 1910 rpm.

12–11. A cumulatively connected compound generator with its shunt field separately excited is operating at full load. What is the final effect on the voltage, polarity, and compounding of the generator of slowly decreasing the shunt field current and then reversing it so that it has the same value, in the reversed direction, as before the operation started?

12–12. A variable-resistance load is to be connected to a generator through a line that has a total resistance of 0.1 ohm. The load voltage is to be maintained constant at 125 volts from zero load current to 100-ampere load current. Specify the voltage, the power rating, the type of connection, and the voltage regulation of the generator that should be ordered for this installation. (The machine is to be driven at constant speed.)
Ans. 135 volts; 13.5 kw; cum. compd.; −7.4%.

12–13. A certain 125-volt separately excited generator has a voltage regulation of 5% when driven at constant speed at the rated value. If this generator is connected to a Diesel engine that has a speed regulation of 3%, and the Diesel engine operates at the rated speed of the generator at full load, what is the no-load voltage of the generator? (Speed regulation is defined in a manner similar to that applied in defining voltage regulation.)

Ans. 135.19 volts.

■ 13 D-C Motors

13–1. MOTOR RATING

The rating of a d-c motor is specified in terms of the mechanical output under definite operating conditions. As with the rating of the d-c generator, this rating is based upon the temperature rise, above a standard ambient temperature, of certain parts of the machine.

The nameplate data of a d-c motor are:

- (a) Horsepower
- (b) Voltage
- (c) Current
- (d) Speed
- (e) Temperature rise

By the nameplate, the manufacturer, in effect, guarantees that the motor will deliver the rated horsepower without exceeding the rated temperature rise, if it is operated at rated voltage and speed; under these conditions, the line current taken by the motor is that specified on the nameplate.

Since neither horsepower nor torque output is conveniently measurable in most motor applications, the magnitude of the armature current is the best indication of the amount of load being delivered. For this reason, the characteristic curves of motors are often plotted in terms of armature current rather than of output horsepower.

13–2. MOTOR TORQUE

In the discussion of motor action in Chapter 11, it is indicated that the operation of a d-c motor depends upon the fact that a current-carry-

MOTOR TORQUE

ing conductor in a transverse magnetic field has a force exerted upon it. The rotation of the armature, however, depends not only upon the force on the conductor but also upon the resulting developed torque.

The torque developed by a motor is the summation of the torques developed by the forces on the individual armature conductors. For a given conductor, the torque developed is a function of the force, the direction of the force, and the normal distance from the axis of rotation to a line representing the direction of the force. These factors vary as the armature revolves; therefore, the torque varies as the armature revolves. The total torque developed by the motor armature is therefore equal to the product of the average developed torque per armature conductor and the number of such conductors.

The force on a current-carrying conductor in a magnetic field is given by

$$F = CBIl \qquad (13\text{-}1)$$

where B is the flux density, I is the current, l is the length of the conductor in the field, and C is a constant depending upon the system of units. For a given machine this relation reduces to

$$F = C'BI \qquad (13\text{-}2)$$

Investigation of the total torque developed in a given motor reveals that the torque may be expressed as

$$T = K'\phi I_A \qquad (13\text{-}3)$$

where ϕ is the flux per pole, I_A is the total armature current, and K' is a constant which includes the constant of Eq. 13–2, the radius of the armature, the number of paths, and the number of conductors, all of which are constants for a given machine. (The constant in Eq. 13–3 is primed to indicate that it has a different value from the constant previously developed for generated voltage.) Thus, according to Eq. 13–3, the torque developed in a given motor is dependent only upon the field flux and the armature current.

The discussion of motor torque has thus far been concerned with the developed torque, the torque developed within the motor to cause rotation of the armature. Opposing the developed torque, and therefore opposing the rotation, is the resisting torque provided by the load connected to the shaft and by the losses associated with the rotation of the motor armature. For equilibrium conditions, the developed torque is equal to the resisting torque. If the developed torque is greater than the resisting torque, the armature accelerates; if the developed torque is less than the resisting torque, the armature must decelerate. The speed at which a motor operates, therefore, depends upon both the developed torque and the resisting torque.

13-3. STARTING A D-C MOTOR

The discussion of the elementary motor of Chapter 11 is introduced by considering the effect upon a battery-charging generator of a decrease in the power input to the prime mover of the generator. The decrease in power input results in a decrease in the speed and generated voltage of the generator and, subsequently, in a reversal of the current and power flow of the machine. The voltage equation developed for the elementary motor states that

$$E_T = E_g + I_A R_A \qquad (13\text{-}4)$$

Since the elementary motor differs from an actual machine only in the magnitudes of the factors involved, Eq. 13-4 is applicable to any d-c motor.

Although the conversion-of-a-generator-into-a-motor type of analysis is convenient to the introduction of motor theory, the procedure involved is not practical for bringing a motor into operation. In practice, the motor must be started from rest without the aid of a prime mover, which means that the motor must be connected to a voltage supply.

At standstill no voltage is generated in the motor armature; therefore, according to Eq. 13-4, the current that flows is limited only by the armature resistance. Except for very small machines, application of rated voltage to the motor produces an armature current far in excess of the rated value, and, unless the circuit has adequate fuses or circuit breakers, extensive damage may result. As an example, a 10-horsepower 120-volt d-c motor has a rated current of approximately 70 amperes and an armature resistance of approximately 0.1 ohm. Rated voltage applied to this machine would produce a current of 1200 amperes, which would undoubtedly damage the machine unless there were protective devices in the circuit.

A resistor is usually connected in series with the armature circuit to limit the amount of current taken by the motor upon starting. The amount of resistance necessary depends upon the starting-torque requirements, but the resistance is usually such that the magnitude of the current upon starting is in the range from 150 to 200 per cent of the rated value. This greater-than-rated value of current does not harm the motor in the short time necessary to start it, but it does provide the excess torque necessary for acceleration. The added resistor is usually

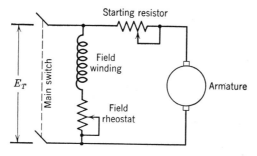

Fig. 13-1. Circuit for starting a shunt motor.

completely variable or variable in steps so that sections of the total resistance can be removed during the starting process. The connections of such a resistor for starting a shunt motor are shown schematically in Fig. 13-1.

For the circuit of Fig. 13-1, the starting process is initiated by closing the main switch. It is assumed that the torque developed upon starting is greater than the resisting torque, and, therefore, the motor accelerates. As the motor accelerates, the generated voltage increases, causing a reduction in armature current. The decrease in current results in a decrease in the difference between the developed torque and the resisting torque and, therefore, a decrease in the acceleration. If, as the motor accelerates, the rheostat is adjusted so as to maintain the armature current at the original value, the developed torque remains constant until the rheostat is completely out of the circuit. At this time, the current decreases until it reaches the value required to produce torque equilibrium, a condition that depends upon the load connected to the shaft.

13-4. MOTOR OPERATION

The steady-state operation of a motor is entirely dependent upon torque equilibrium. Torque equilibrium, in turn, is dependent upon the armature current and field flux (the developed torque) and the shaft load (the resisting torque). A change in any one of these factors disturbs the equilibrium and causes the motor to react accordingly.

An increase in the load connected to the shaft of a motor causes the motor to decrease in speed. As a result of the decrease in speed, the generated voltage decreases, and the armature current increases. If the field flux is assumed to be constant, the speed continues to decrease until the armature current reaches the value necessary to again make the developed torque equal to the resisting torque.

If, while the motor is driving a constant-torque load, a resistor is inserted in the armature circuit, the armature current decreases. If the field flux is constant, the developed torque then decreases, and the motor decelerates, resulting in a decrease in the generated voltage and an increase in the armature current. As before, the motor continues to decelerate until the developed torque is again equal to the resisting torque.

For a consideration of the effect of a change in field flux upon motor operation, it is necessary first to consider the relative magnitudes of the generated voltage and the $I_A R_A$ drop. Thes values vary with machine design and operation, but an approximation of their relative magnitudes is satisfactory. Thus, if the terminal voltage is 100 per cent, the generated voltage under full-load conditions is of the order of 95 per cent; the $I_A R_A$ drop, 5 per cent.

On the basis of the assumed values of generated voltage and $I_A R_A$ drop, an instantaneous 10 per cent decrease in flux causes the generated voltage to drop to a relative value of 85.5 per cent (assuming that the speed cannot change instantaneously because of inertia). The $I_A R_A$ drop must therefore increase to a relative value of 14.5 per cent. The increase in $I_A R_A$ drop represents an increase in armature current of 190 per cent. The developed torque, in terms of the original values, is, therefore,

$$T' = K'(0.9\phi)(2.9I_A) = 2.61 K' \phi I_A$$

Since, after the decrease in flux, the developed torque is 261 per cent of the original torque, the motor accelerates and the armature current decreases. The motor continues to accelerate until the developed torque is again equal to the resisting torque.

In the preceding discussion of the effects of a change in field flux, it is assumed that the field flux is decreased instantaneously. This assumption is made only for the purposes of the discussion; any decrease in field flux, regardless of how made, causes acceleration of the motor. By similar reasoning, it may be seen that an increase in field flux causes deceleration.

13–5. MOTOR POWER REQUIREMENTS

The torque developed in a motor with no load connected to its shaft is only that necessary to equal the resisting torque caused by the friction, windage, and core losses. The current required to produce this torque

SPEED-LOAD CHARACTERISTICS

is only a small fraction of the rated current. The power input to the armature circuit at no load, therefore, is small and represents all loss, since the mechanical output of the motor is zero. This power input is the product of the terminal voltage and armature current. Since

$$E_T = E_g + I_A R_A \tag{13-5}$$

the power input to the armature is given by

$$P = I_A E_T = I_A E_g + I_A{}^2 R_A \tag{13-6}$$

Since $I_A{}^2 R_A$ represents the copper loss in the armature circuit, and, since the entire power input is lost, $I_A E_g$ must represent the total of the friction, windage, and core losses.

When a load is connected to the shaft of the motor, the armature current and, therefore, the power input increase. According to Eq. 13-6, $I_A E_g$ under load conditions must therefore represent the power output plus the friction, windage, and core losses.

The friction and windage losses of a motor depend upon speed; the core losses, upon flux and speed. In some applications it is reasonable to assume that the total of these losses remains constant even though the speed may change. In such applications, the power output of the motor under load conditions may be obtained by subtracting the product of the armature current and generated voltage at no load from a similar product obtained under load conditions. It must be recognized that this method of determining power output is applicable only when the speed and flux change moderately from no-load to load conditions.

13-6. SPEED-LOAD CHARACTERISTICS

The speed–load characteristic is often a deciding factor in the selection of a motor for a particular application. When plotted, this characteristic shows how the speed of the motor changes as load is applied, the terminal voltage and field-circuit resistance remaining constant.

In the development of the speed-load characteristics that follow, it is assumed that the armature current represents the load; this assumption makes for a simpler analysis. It is also assumed that the motor is operating at full load and rated speed and that the load is then removed.

The voltage equation that applies to a motor may be written

$$E_T = K\phi n + I_A R_A \tag{13-7}$$

Solving Eq. 13-7 for speed gives

$$n = \frac{E_T - I_A R_A}{K\phi} \tag{13-8}$$

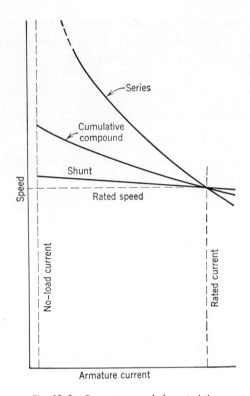

Fig. 13–2. D-c motor speed characteristics.

Equation 13–8 is applied to develop the speed–load characteristics in the discussions that follow.

As the load is removed from a shunt motor, the armature current decreases, and, if the flux is assumed to remain constant, the speed increases. According to Eq. 13–8, the speed–armature-current characteristic is linear, as shown in Fig. 13–2. As indicated in the figure, the characteristic does not have a point at zero armature current, since a small value of current is necessary to maintain rotation at no load.

Removal of the load of a cumulative-compound motor causes a decrease in both the armature current and the flux. The resulting characteristic is therefore as shown in Fig. 13–2. The magnitude of the increase in speed depends upon the compounding; a motor with a relatively large series-field mmf at full load has a greater increase in speed than one with a weak series-field mmf.

Removal of the load of a series motor also effects both the armature current and the flux. The flux is the most important factor in the de-

TORQUE-CURRENT CHARACTERISTICS

termination of the series-motor characteristic since, as the armature current approaches the no-load value, the flux becomes very small. From Eq. 13–8, the speed of the motor then becomes very large, as shown in Fig. 13–2. Because of the excessive speed at no load, series motors are usually directly coupled, rather than belted, to the load to reduce the possibility of the motor's becoming accidentally unloaded.

The speed–load curves shown in Fig. 13–2 are for shunt, compound, and series motors having the same nameplate rating.

The **speed regulation** of a motor is the change in speed from full load to no load and is expressed as a percentage of the speed at full load. Expressed mathematically,

$$\% \text{ speed regulation} = \frac{n_{\text{NL}} - n_{\text{FL}}}{n_{\text{FL}}} \times 100 \qquad (13\text{-}9)$$

where n_{NL} is the speed at no load and n_{FL} is the speed at full load.

13–7. TORQUE–CURRENT CHARACTERISTICS

Another motor characteristic of interest is the relationship between developed torque and armature current. Since the developed torque is proportional to field flux and armature current, the torque–current characteristic depends upon the variation of flux with load, or armature, current. Thus, if armature reaction is neglected, the torque–current characteristic of a shunt motor is linear, as shown in Fig. 13–3. Also shown in this figure is a plot of pulley torque versus armature current.

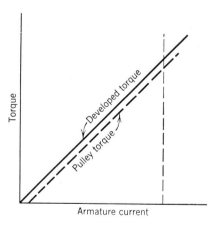

Fig. 13–3. Torque versus current characteristics of a shunt motor.

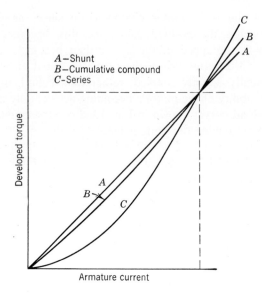

Fig. 13–4. D-c motor torque characteristics.

(Pulley torque is the torque available for driving the external load; therefore it is the developed torque less the torque necessary to drive the armature alone.)

If the field flux of a series motor is assumed to vary linearly with armature current, the developed torque may be expressed as

$$T = K'' I_A^2 \tag{13-10}$$

After the magnetic circuit reaches saturation, however, the flux remains essentially constant, and the torque varies approximately linearly with armature current. The torque–current characteristic of a series motor is shown in Fig. 13–4.

The field flux of a cumulatively connected compound motor increases with an increase in armature current, but it has a no-load value that is a large percentage of the full-load value. The torque–current characteristic of a compound motor therefore has a form in between that of the shunt motor and the series motor, as shown in Fig. 13–4.

The curves of Fig. 13–4 provide an indication of the relative starting torques of the three types of motors (which are assumed to be of the same nameplate rating). Thus, if the starting current is 150 per cent of the rated current, the curves indicate that the series motor provides the greatest starting torque, the shunt motor the least, when motors of

TORQUE-CURRENT CHARACTERISTICS

the same full-load rating are considered. Starting torque is often the deciding factor in the selection of a motor for a particular application.

PROBLEMS

13-1. The nameplate of a d-c shunt motor provides the following information: 10 hp, 120 volts, 71.8 amperes, 1180 rpm. Under normal operating conditions, the resistance of the field circuit is 66.7 ohms, and the resistance of the armature circuit is 0.1 ohm.

(a) What is the nameplate efficiency of this motor?   *Ans.* 86:6%.
(b) Determine the full-load armature current.   *Ans.* 70 amperes.
(c) What percentage of the terminal voltage is the armature-resistance drop at full load?   *Ans.* 5.8%.
(d) What external armature resistance is necessary to produce 150% of rated torque at starting?   *Ans.* 1.04 ohms.
(e) If a constant-torque load equal to the full-load value is connected to the motor when started as in (d), what speed does the motor reach if the external resistance is not removed?   *Ans.* 418 rpm.
(f) What is the rated output torque of this motor, in pound-feet?
   Ans. 44.5 lb-ft.

13-2. A certain 115-volt shunt motor has an armature resistance of 0.1 ohm. When the motor delivers full-load torque with normal field excitation, the armature current is 50 amperes, and the speed is 1100 rpm. (Assume that the resisting torque of the load remains constant and that torque losses may be neglected.)

(a) What is the generated voltage if the field flux is instantaneously reduced to 50% of the normal value? (Assume that speed momentarily remains unchanged.)   *Ans.* 55 volts.
(b) What is the armature current immediately after the flux has been reduced?   *Ans.* 600 amperes.
(c) Determine the developed torque, as a function of normal full-load torque, immediately after the flux has been reduced.   *Ans.* $T_2 = 6T_1$.
(d) Explain, in detail, what happens while conditions are becoming steady. Show that the armature takes 100 amperes and runs at 2100 rpm after conditions have become steady.

13-3. A shunt motor at a certain load takes an armature current of 40 amperes from a 220-volt line and runs at a speed of 500 rpm. The armature resistance is 0.5 ohm. The resisting torque of the load remains constant. Determine the new armature current and speed for each of the following independent cases (neglecting torque losses in the motor):

(a) 2.5 ohms inserted in the armature circuit.
   Ans. 40 amperes; 250 rpm.
(b) Field flux reduced to 80% of original value.
   Ans. 50 amperes; 610 rpm.

13-4. The generator of Prob. 12-6 is connected as a shunt motor to a 240-

volt supply. Driving a constant-torque load, and with its total field-circuit resistance at 40 ohms, the motor draws a line current of 166 amperes and operates at 1500 rpm. What will be the line current and speed if the supply voltage drops to 160 volts? *Ans.* 183 amperes; 1092 rpm.

13–5. The relations for developed torque and generated voltage are, respectively, $T = K'\phi I_a$ and $E_g = K\phi n$. Determine the value of K' in terms of K. *Ans.* $K' = 7.04K$.

13–6. The 10-hp motor of Prob. 13–1 has a full-load armature copper loss of 490 watts. Another 10-hp motor of the same full-load efficiency has an armature copper loss of 750 watts. Which motor should be selected for operation at an average load of 5 hp? Why?

13–7. You are to select a prime mover for a 25-kw 125-volt 1750-rpm shunt generator. The resistance of the armature circuit is 0.05 ohm, and the resistance of the field circuit when the generator is operating at rated conditions is 25 ohms. The only other information available is this memo: "The generator was operated as a motor at no load and rated speed by connecting it to a 240-volt supply with an external resistance in the armature circuit and with the field circuit resistance adjusted to 48 ohms. The line current under these conditions was 16 amperes." Is it possible to calculate the horsepower output required of the prime mover when the generator is operating at rated conditions, from the data given (and neglecting armature reaction)? If so, what is the horsepower requirement? If not, precisely why not?

13–8. The nameplate of a d-c shunt motor supplies the following information: 60 hp, 250 volts, 204 amperes, 1740 rpm. At rated conditions, the resistance of the field circuit is 62.5 ohms. The resistance of the armature circuit is 0.05 ohm.

Calculate the approximate no-load line current of this motor when it is energized at 250 volts with the field-circuit resistance at 62.5 ohms.

Ans. 17 amperes.

13–9. It is desired that the motor of Prob. 13–8 be operated as a separately excited generator delivering rated armature current at rated terminal voltage. At what speed should the machine be driven to produce the desired results if the 62.5-ohm field circuit is energized at 250 volts?

Ans. 1885 rpm.

13–10. Assume that it is desired that the motor of Prob. 13–8 be operated as a separately excited generator delivering rated armature current at rated speed and with its 62.5-ohm field circuit energized from a 250-volt source. With the data given, is it possible to calculate the horsepower output required of the necessary prime mover to produce the desired results? If so, what is the horsepower requirement? If not, why not?

13–11. Assume that the motor of Prob. 13–8 is operating at rated conditions driving a constant-torque load and that the armature resistance is then increased to 2 ohms. Draw and explain the approximate curves of armature current and speed versus time for the period from just prior to the increase in armature resistance to the time when conditions are steady.

13–12. Assume that the motor of Prob. 13–8 is operating at rated conditions driving a constant-torque load and that the field resistance is then increased. Draw approximate curves of armature current and speed versus time for the period from just prior to the increase in field resistance to the time when conditions are steady. Explain briefly.

13-13. A d-c motor operating at no load has its field circuit energized from a 120-volt supply and has its armature circuit energized from a 600-volt supply. When the rheostat in the field circuit is accidentally short-circuited, a large current flows in the armature circuit and causes the circuit breaker in the armature circuit to operate. Briefly explain why the large current flows in the armature circuit under these conditions.

13-14. Two identical machines are connected as shown in Fig. 13-5. When the motor is operating at rated voltage at no load, the opening of the switch S in the field circuit of the generator causes the fuse in the armature circuit to blow. Explain completely.

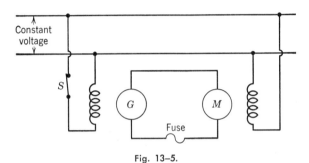

Fig. 13-5.

13-15. The circuit arrangement shown in Fig. 13-6 has been proposed for a bar mill in order that two standard d-c motors that are available might be utilized. The bar mill requires essentially constant torque. Each motor is rated at 40 hp, 1150 rpm, and 230 volts; each has a rated armature-resistance drop of 12 volts. As shown in Fig. 13-6, the motor armature windings are connected in series across a 230-volt line, but the shunt fields are separately excited. The two armatures are coupled to a common gear which drives the load.

(a) What is the motor speed when the motors carry rated current?

Ans. 543 rpm.

(b) Explain what happens to the motor speeds when the coupling between the gear and motor A is broken.

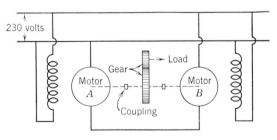

Fig. 13-6.

■ 14 D-C Motor Control

14-1. INTRODUCTION

The operation of a motor may require the control of any of the following factors: (*a*) starting, (*b*) speed, (*c*) direction of rotation, and (*d*) braking. This chapter considers the principles involved in the control of these factors in d-c motor operation and some practical applications of these principles.

14-2. FUNDAMENTALS OF MOTOR STARTING

In starting a motor, it is necessary only that the developed torque be greater than the resisting torque for the armature to accelerate. However, the starting operation is usually such as to produce the greatest developed torque permissible. Thus, if the field flux is a maximum and the armature current is the largest permissible, the developed torque is a maximum. Maximum field flux is obtained by having the resistance of the field circuit a minimum. Any external field-circuit resistance, such as that provided by a field rheostat, is therefore removed before starting. In many applications, however, the field rheostat is preset at a value that allows the motor to operate at a desired speed after the starting operation is completed; therefore, the torque developed at starting may not be the greatest possible with a given current in the armature circuit.

The starting operation consists in (*a*) the insertion of external resist-

FUNDAMENTALS OF MOTOR STARTING

ance into the armature circuit to limit the starting current taken by the motor and (b) the removal of this resistance, in steps, as the motor accelerates. The insertion and removal may be done manually or automatically; both methods are considered in this chapter.

As previously indicated, the amount of resistance inserted into the armature circuit depends upon the current required to accelerate the motor, but it is usually not less than the value that gives a current in the range from 150 to 200 per cent of the rated armature current. The external resistor is such that portions of it may be removed as the motor accelerates; the magnitude of the resistance removed depends upon the type of starter and upon the limits established for the starting current.

The starting current in some automatic starters has two limits: an upper limit and a lower limit. The upper limit is that value established as the maximum permissible for the motor; the lower limit is the value set as a minimum for the starting operation.

The diagram shown as Fig. 14–1 schematically represents an automatic-starter circuit. For an upper limit of 200 per cent of rated armature current and a lower limit of 100 per cent, the steps in the starting of the motor are as follows:

(a) The main switch is closed, energizing the field and armature circuits. The starting resistors limit the armature current to 200 per cent of rated current.

(b) As the motor accelerates, the generated voltage increases; therefore, the armature current decreases.

(c) When the armature current reaches 100 per cent of rated current, switch 1 is closed, short-circuiting resistance R_1. Since the speed and generated voltage momentarily remain the same, the armature current increases. The amount of resistance removed is such that the armature current again equals 200 per cent of rated current.

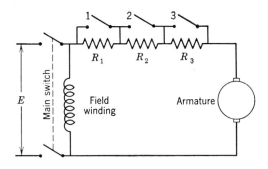

Fig. 14–1. Motor-starting circuit.

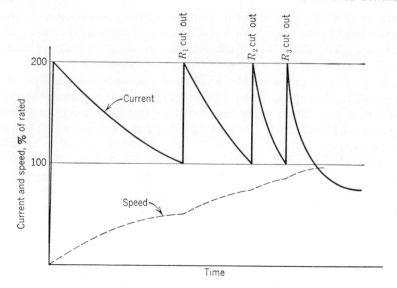

Fig. 14–2. Current and speed versus time during starting operation.

(d) The acceleration continues, the generated voltage increases, and the current again decreases. When the current falls to 100 per cent, switch 2 is closed, short-circuiting resistance R_2. The current again rises to 200 per cent, and the cycle repeats until resistance R_3 is short-circuited.

(e) After reaching the 200 per cent value upon the closing of switch 3, the armature current decreases until it reaches the value necessary to produce torque equilibrium.

A plot of armature current versus time for the starting operation described in the preceding paragraph is shown in Fig. 14–2. A plot of motor speed is also included in Fig. 14–2.

14–3. RATING OF STARTERS

Starters are rated according to the rating of the motor with which they are to be used. Thus, a 10-horsepower 120-volt starter is designed to be used with a 10-horsepower 120-volt motor. The resistance of the starter is such as to permit a starting current in the range of 150 to 200 per cent of the rated armature current, the actual magnitude being dependent upon the particular design.

The current rating of a starter resistor depends upon the time the

MANUAL STARTERS

resistor is expected to be in the circuit, since the temperature rise is dependent upon time. The resistors of most starters are in the circuit for a relatively short time; therefore, resistors of low current rating are employed. Thus, a starter that permits a starting current of 200 per cent of rated armature current may have resistors that are rated at 50 per cent of rated armature current. Such resistors should not be kept in the circuit continuously, and most starters are so designed that it is not possible to keep the resistors in the circuit continuously if the starter is operated normally.

14-4. MANUAL STARTERS

The starting operation described in Art. 14-2 is considered to be that of an automatic starter. Such an operation is desirable in all starters, but for manual starters the operation depends entirely upon the operator. If the resistors are removed too quickly, the current may be excessive, resulting in the operation of the circuit protective device or the damaging of the machine or the starting resistors; if the resistors are removed too slowly, the resistors may overheat, resulting in a shortened life.

Two types of manual starters are the **three-terminal** and the **four-terminal starters.** The schematic connection diagram of the three-terminal starter is shown as Fig. 14-3.

For a three-terminal starter, the starting operation is initiated by moving the operating handle to the first contact point, thus energizing the field circuit and the armature circuit, the latter with external resist-

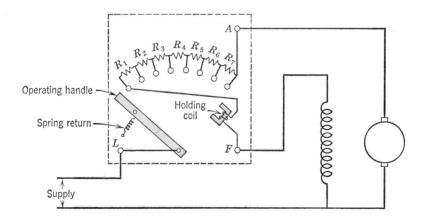

Fig. 14-3. Connection of three-terminal manual starter.

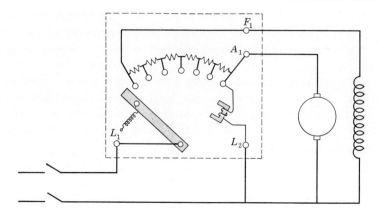

Fig. 14-4. Connection of four-terminal manual starter.

ance in series. Continued movement of the operating handle removes resistance from the armature circuit until the last contact point is reached. At this point, the operating handle, which has a spring return, is held in place by the holding magnet. The coil of the holding magnet is energized by the current in the field circuit.

As resistance is removed from the armature circuit during the starting operation, the resistance is added to the field circuit. The resistance of the field circuit is much larger than that of the starting resistors, so that the additional resistance has little effect upon the field circuit.

One advantage in having the holding magnet coil in series with the field circuit is that the coil becomes de-energized if the field circuit is opened. This allows the operating handle to return to its starting point, de-energizing the motor, in the event that the field circuit is accidentally opened. One disadvantage is that the magnet may become too weak to hold the operating handle if the field current is decreased for speed-control purposes; thus, the motor may be de-energized unnecessarily.

The four-terminal starter is similar in operation to the three-terminal starter but differs in that the holding coil is energized directly by the supply voltage. The schematic connection diagram of the four-terminal starter is shown as Fig. 14-4.

Motors started by means of manual starters are stopped by the opening of the line switch to the supply; thus, the motor is de-energized, and the operating handle returns to its starting point.

14-5. AUTOMATIC-STARTER COMPONENTS

The difference between a manual starter and an automatic starter is in the method of inserting armature resistance and then removing it. The components that make an automatic-starting operation possible are also applied in general control circuits.

The **control circuit** may be considered to be the "heart" of an automatic operation. This circuit incorporates all the elements necessary to initiate and terminate the operation. The connection of the basic elements of a control circuit is shown schematically in Fig. 14–5.

The *start* element is a push-button contact, or switch, held open by a spring. When the button is pressed, the contact closes, but it opens again as soon as the pressure is released.

The *stop* element is a push-button contact held closed by a spring. Pressure on the button opens the contact; upon release of the button, the contact closes again.

The element labeled M is a coil which, when energized, causes contacts M_1 and M_2 to close. The physical arrangement of the coil and one of the contacts is shown schematically as Fig. 14–5b.

The operation of the basic control circuit shown as Fig. 14–5a is as follows:

(a) The start button is pressed, thereby energizing coil M.

(b) Contacts M_1 and M_2 close. The closing of contact M_1 causes the controlled circuit to be energized; the closing of contact M_2 causes the coil M to remain energized after the pressure on the start button is released.

(c) When the operation of the controlled circuit is to be terminated,

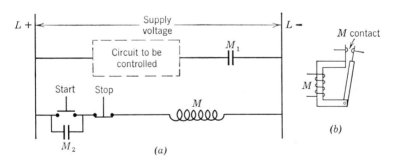

Fig. 14–5. Basic elements of a control circuit.

pressure on the stop button opens the circuit to coil M; thus the coil is de-energized, and contacts M_1 and M_2 open.

Control diagrams are usually drawn for the condition when the circuit is de-energized, as indicated in Fig. 14–5a. Contacts M_1 and M_2 are called **normally open contacts,** the term *normally* being applied to the de-energized condition. **Normally closed contacts** are indicated by the symbol ⊣⊬. Normally open or normally closed contacts that are coil-operated usually carry the same letter designation as the coil that operates them. Examples are contacts M_1 and M_2, which are operated by coil M in Fig. 14–5a. The subscripts for the contact letter designations are employed here for identification purposes; usually, contacts that close at the same time carry the same letter designation, subscripts being reserved for contacts that close in a definite sequence.

Other contacts that may appear in a control circuit are **limit switches** and **thermal overload contacts.** Figure 14–5a is redrawn as Fig. 14–6 to show the inclusion of these contacts.

The limit switch is a normally closed contact that is opened by physical contact with a moving part. Thus, if the travel of a certain motor-operated device is to be limited to a given distance, a limit switch can be placed at the desired limit; when the device reaches the limit switch and opens it, the control circuit is de-energized and the motor stops.

The thermal overload contact is a normally closed contact that is opened by the heat produced in a resistor element in the controlled circuit. When the current and, therefore, the temperature of the resistor element exceed a given value, the thermal overload contact opens and de-energizes the control circuit.

Because the operation of the thermal overload contact depends on temperature, the controlled circuit may have its rated current exceeded

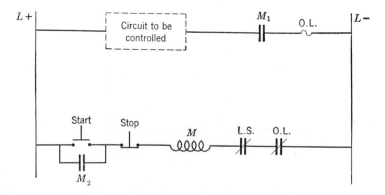

Fig. 14–6. Control circuit with limit switch and thermal overload contact.

for a short time, but an excessive current over a long period causes the contact to open. Thus, when applied to a motor starter, the thermal overload contact does not operate when the large starting current flows, but it does protect the motor against sustained overloads.

The design of thermal overloads varies widely, and no effort is made here to describe them except to point out that a bimetallic strip heated by a coil is the basis for many types.

In summary, the circuits used for control purposes may contain normally open and normally closed contacts that are (a) manually operated, (b) coil operated, (c) operated by impact, and (d) thermally operated. All these contact types may be found in d-c motor starters and in circuits for the control of other operations.

14–6. DEFINITE-TIME AUTOMATIC STARTERS

A **definite-time automatic starter** is one in which the starting resistors are removed from the armature circuit in a definite time, regardless of the motor reaction to its connected load. The complete circuit for the connection of such a starter to a shunt motor is shown as Fig. 14–7a.

The control circuit of Fig. 14–7a is similar to the one previously considered except for an additional contact and resistor in the circuit that bypasses the start button. The contact M_1' closes as soon as coil M is energized but opens again when the starting operation is completed; however, at the instant that contact M_1' opens, M_3 closes and keeps the coil circuit energized. The purpose of the resistor inserted when M_3 closes is to decrease the coil current after the starting operation is completed; thus, the power loss and, therefore, the temperature rise of the coil are kept to the minimum. (Less current is needed in the coil after the starting operation is completed than during the operation.)

The contacts that short-circuit the starting resistors are designated M_{TD1}, M_{TD2}, and M_{TD3}. The letter M signifies that these contacts are closed by action of coil M; the subscript TD, that there is a time delay in the closing of the contacts; and the numerical subscript, the sequence in which the contacts close. The time interval between the closing of contacts is fixed for a particular operation of the starter but may be changed for subsequent operations by adjustment of the mechanism that causes the time delay.

The mechanisms that produce the time delay in definite-time starters are of the dash-pot and mechanical-escapement types. The operation of the dash pot is based upon the fact that the speed of a piston moving against a fluid in a cylinder depends upon the orifice that allows the

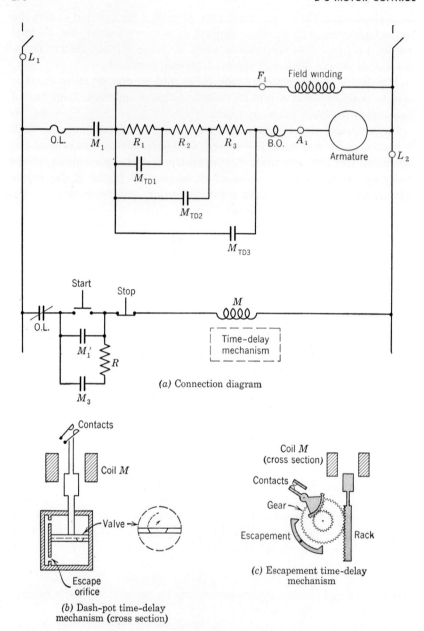

Fig. 14-7. Time-delay automatic starter details.

fluid to escape. The incorporation of a dash pot in the starter is shown schematically in Fig. 14–7b. When coil M is energized, the shaft to which the piston is attached is pulled upward at a rate determined by the setting of the escape orifice. As the shaft moves upward, it closes the M contacts in the desired sequence (only one of the contacts is shown in Fig. 14–7b). When coil M is de-energized, the piston quickly drops to the bottom of the cylinder because of the valve that opens and allows the fluid to flow unimpeded.

One type of mechanical-escapement method of time delay is shown schematically as Fig. 14–7c. When coil M is energized, the shaft and rack are pulled upward at a speed determined by the escapement; thus, the necessary time delay is produced for the closing of the contacts that short-circuit the starting resistors.

The operation of the definite-time automatic starter shown in Fig. 14–7a is as follows:

(a) The start button is pressed, and coil M becomes energized; thus contacts M_1 and M_1' close immediately, and the field and armature circuits are connected to the line.

(b) Contacts M_{TD1}, M_{TD2}, and M_{TD3} close in definite time intervals and thus short-circuit the starting resistors.

(c) At the same instant that M_{TD3} closes, M_1' opens and M_3 closes.

The coil labeled B.O. on the diagram is a magnetic blow-out coil incorporated in the starter to extinguish the arcs formed between the starter contacts when they open.

One disadvantage of a definite-time starter for variable motor loads is that the starting resistors are always removed from the circuit at the same rate. Thus, as an extreme example, if the motor is loaded so that it cannot turn, the starter continues to operate, and, when the last contact closes, the armature is directly across the line. The excessive current that results causes the protective device to operate, but it may damage the motor before the circuit is opened. This is an extreme example, but it does illustrate the disadvantage of this type of starter. The starter is widely employed in applications where the load on the motor remains relatively constant.

14–7. CURRENT-LIMIT AUTOMATIC STARTERS

The operation of the **current-limit** type of automatic starter is based on the fact that the armature current of a motor decreases as the motor accelerates. In the current-limit type of starter, a starting resistor is not short-circuited until the current reaches a predetermined minimum value. The time interval between the closing of contacts therefore depends upon the load connected to the motor shaft. Two types of current-limit starters are the **counter-emf starter** and the **series-lockout starter**.

The connection diagram of a counter-emf starter is shown as Fig. 14–8. The coils that control the short-circuiting contacts are connected in parallel with the motor armature and have high resistance so that the coil current is small compared to the normal armature current. The coils and their contacts are so constructed that the contacts do not close until the voltage across the coils reaches a predetermined value.

The operation of the counter-emf starter shown in Fig. 14–8 is as follows:

(*a*) The start button is pressed, energizing coil M; all contacts labeled M close simultaneously.

(*b*) At the instant the armature circuit is energized, the generated voltage is zero, and the armature current reaches the maximum value established for the starting operation. The voltage drop across the armature at this time is only the armature-resistance drop, which is small compared to the applied voltage, so that the major portion of the applied voltage appears across the external resistors.

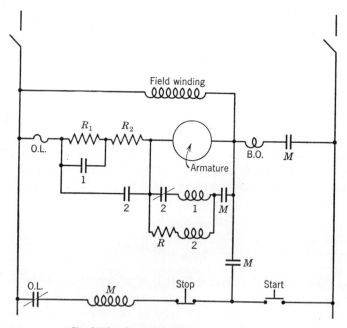

Fig. 14–8. Connection of counter-emf starter.

CURRENT-LIMIT AUTOMATIC STARTERS

(c) As the motor accelerates, the armature current decreases and the voltage drop across the armature increases. When the current decreases to the predetermined minimum value, the voltage across the armature is large enough to cause coil 1 to close contact 1; thus, the first resistor is short-circuited, and the current again rises to the maximum value.

(d) When the armature current again reaches the minimum value, the voltage across the armature is large enough to cause coil 2 to close contact 2; thus, the second starting resistor is short-circuited, and the current again rises to the maximum value. Contact 2 is located so that it short-circuits both R_1 and R_2; the contact is so placed in order that coil 1 may be removed from the circuit after the starting operation is completed. The removal of coil 1 from the circuit is accomplished by connecting the coil in series with a normally closed contact operated by coil 2. Thus, when coil 2 causes the short-circuiting contact to close, it causes the contact in series with coil 1 to open. (The resistor in series with coil 2 is inserted so that identical coils may be employed in the starter. Thus, coils 1 and 2 are identical coils, but the combination of coil 2 and the resistor requires a greater voltage to operate the contact associated with it than coil 1 needs to operate its contact.)

(e) The armature current decreases from the maximum of step (d) to the value determined by the load.

The connection diagram of a series-lockout type of current-limit starter is shown as Fig. 14–9a. Each short-circuiting contact has two coils associated with it; one is labeled T, the other, B. The T and B refer to top and bottom, as indicated in Fig. 14–9b, which shows the schematic arrangement of the coils and the contacts associated with them. The effect of the top coil is to attempt to close the contact; the bottom coil attempts to keep the contact open. Whether the contact closes or remains open depends upon the torques produced on the pivoted member. The contactor is so designed that the magnetic circuit of the top coil saturates but that of the bottom coil does not; therefore, the torque–current characteristics for the two coils are as plotted in Fig. 14–9c. Figure 14–9c indicates that the effect of coil B is greater than that of T at large values of current; therefore, the contact remains open when the starting current is high. When the current falls below a predetermined value, coil T has the greater torque, and the contact closes. As indicated in Fig. 14–9a, this particular starter has two such contactors.

The operation of the series-lockout starter shown in Fig. 14–9a is as follows:

(a) The start button is pressed, energizing coil M; all contacts labeled M close simultaneously.

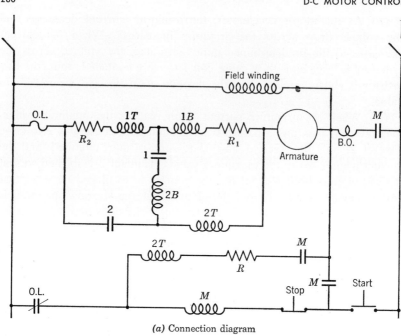

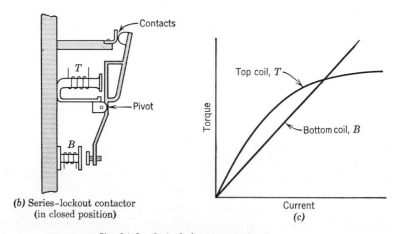

(b) Series-lockout contactor
(in closed position)

(c)

Fig. 14–9. Series-lockout starter details.

(b) At the instant of starting, the current reaches the established maximum. At the maximum current, the effect of coil $1B$ is greater than that of coil $1T$, and contact 1 remains open. When the current decreases to the predetermined minimum, the effect of coil $1T$ becomes greater than that of coil $1B$, and contact 1 closes; therefore, the first starting resistor is short-circuited, and the armature current flows through coils $2B$ and $2T$.

(c) Because the current again increases to the maximum value, the operation of coils $2B$ and $2T$ is similar to that of coils $1B$ and $1T$. When the current again reaches the minimum value, contact 2 closes, and the second starting resistor is short-circuited.

At light loads, coil $2T$ might not be able to hold contact 2 closed, because of vibration in the starter; therefore, an auxiliary top coil, $2T'$, is provided to insure that the last contact remains closed when the starting operation is completed.

14-8. SPEED CONTROL

For the development of the speed–load characteristics of the various types of motors, the voltage equation applying to a motor is written in the form

$$n = \frac{E_T - I_A R_A}{K\phi} \qquad (14\text{--}1)$$

This equation applies equally well in the consideration of those factors that are important in the control of motor speed.

According to Eq. 14–1, the methods that may be applied for the speed control of a given motor are (a) field control, (b) armature-resistance control, and (c) armature-voltage control.

Field control of motor speed is accomplished by the variation of the field-circuit resistance so as to vary the field flux. This method of speed control is probably the most widely used of the three.

Armature-resistance control of motor speed is accomplished by a change in the armature-circuit resistance. Thus, if a motor is driving a given load, an increase in the armature-circuit resistance results in a decrease in motor speed. For the motor to increase in speed, the armature circuit must have some resistance that is removable; this is not always possible. The armature-resistance method of speed control is seldom applied because of the large amount of power wasted in the external resistor.

Armature-voltage control requires a change in the voltage across the armature terminals; the field circuit is energized from a fixed supply. The armature-resistance method of control is really an example of armature-voltage control, since a portion of the supply voltage is taken up by the external resistor, but the term armature-voltage control is not usually applied to that method.

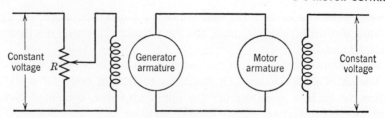

Fig. 14–10. Ward–Leonard system of speed control.

The armature-voltage method of speed control is best illustrated by the Ward–Leonard system, which has the basic components shown in Fig. 14–10.

In the Ward–Leonard system, the motor being controlled has its field winding constantly excited and its armature circuit energized from a separately excited generator. As the voltage of the generator is increased, the motor speed increases; similarly, a decrease in generator voltage results in a decrease in motor speed. If the generator voltage is decreased to zero, the motor stops. The Ward–Leonard system finds wide application where uniform motor acceleration and a wide range of motor speeds are required.

14–9. MOTOR REVERSAL

The direction of the developed torque in a motor depends upon the relative directions of the current and the field flux. Thus, the reversal of the developed torque and, therefore, of the rotation of the motor may be accomplished by a reversal of the connections to either the field winding or the armature circuit. Reversal of the connections to both the field winding and the armature circuit does not change the direction of rotation.

Reversal of the motor in a Ward–Leonard system is accomplished by the reversal of the field current of the generator. The resulting change in the polarity of the generator causes the current in the motor armature to reverse; therefore, the motor rotation reverses.

14–10. MOTOR BRAKING

The method employed to stop a motor after it has been de-energized depends upon the application of the motor. In some applications, it is

MOTOR BRAKING

satisfactory to allow the motor to "coast" and eventually to stop because of its own bearing friction; in others, the motor must be stopped almost instantaneously. Some applications require braking times in between these two extremes.

One device for stopping a motor quickly is the friction brake, shown schematically in Fig. 14–11. The brake shoes are held away from the motor pulley by the action of the energized coil, which exerts a force on the mass of magnetic material attached to the shaft passing through the coil. When the coil is de-energized, the springs cause the brake shoes to contact the pulley and thus provide braking action. The voltage supply to the coil is so arranged that the coil is energized and de-energized at the same time as the motor.

Another method of braking is known as **dynamic braking.** In this method, only the armature is de-energized when the motor is to be stopped; the field winding is so connected that it remains energized. After the motor is de-energized, it continues to rotate because of its inertia; therefore, since the field remains energized, the machine becomes a generator. Dynamic braking consists in the connection of a resistor to the terminals of the de-energized armature so that the resulting generator action provides braking torque. The magnitude of the braking torque depends upon the flux and the armature current and decreases as the motor rotation decreases, because of the decrease in generated

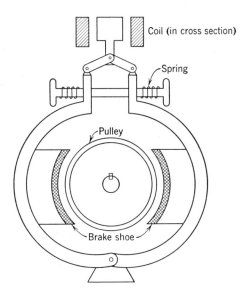

Fig. 14–11. Friction brake.

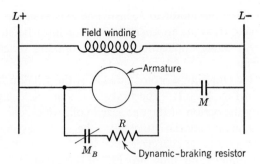

Fig. 14–12. Application of dynamic braking.

voltage. However, if the value of resistance connected to the armature terminals is decreased as the motor speed decreases, the braking torque may be maintained at a high value. At a low value of speed, the braking torque becomes small, and the final stopping of the motor is due to friction.

The application of dynamic braking to a motor having an automatic starter is shown, in part, as Fig. 14–12. The contact M_B is a normally closed contact which opens when the coil in the control circuit is energized and the motor is operating but which closes when the control circuit is de-energized.

PROBLEMS

14–1. The motor of Prob. 13–1 is to be started by connecting it as shown in Fig. 14–1. If the upper and lower limits of the starting current are set at 150% of rated and rated, respectively, how many starting resistors are necessary, and what is the resistance rating of each? (It is assumed that the short-circuiting switches close automatically when the lower limit is reached.)
Ans. 6 resistors.

14–2. If the range of starting current for the starter in Prob. 14–1 is between 200% and 80% of rated, instead of between 150% of rated and rated, what is the final motor speed if full-load torque is connected to the shaft of the motor upon starting? *Ans.* 626 rpm.

14–3. The motor of Prob. 13–1 is to have an automatic counter-emf starter that has starting-current limits of 200% and 65% of rated armature current.

(*a*) How many resistors should the starter have, and what should be the resistance of each? *Ans.* 2; 0.578 ohm and 0.178 ohm.

(*b*) What should be the voltage ratings of the coils that cause the resistors to be short-circuited? *Ans.* 85.6 volts and 111.9 volts.

PROBLEMS

14–4. What is the disadvantage of applying a 5-hp 120-volt starter to a 10-hp 120-volt motor?

14–5. What changes, if any, have to be made in a 5-hp 120-volt automatic definite-time starter to make it suitable for a 10-hp 240-volt d-c shunt motor?

14–6. Repeat Prob. 14–5 for a 5-hp 120-volt automatic counter-emf starter.

14–7. Repeat Prob. 14–5 for a 5-hp 120-volt automatic series-lockout starter.

14–8. A 10-hp 110-volt 900-rpm d-c shunt motor is operating at rated full-load conditions driving a fan (torque proportional to square of speed). Under these conditions, the armature current is 75 amperes and the field current is 2 amperes. The resistance of the armature circuit is 0.08 ohm. The speed is then reduced to 450 rpm by the insertion of resistance in the armature circuit. Determine the new motor output, the new armature current, the resistance added, and the new over-all efficiency.

Ans. 1.25 hp; 18.75 amperes; 3.01 ohms; 40.5%.

14–9. If it is desired to operate a d-c shunt motor at constant speed (the full-load value) from no load to full load, explain briefly and qualitatively three methods that might be applied.

14–10. Draw a diagram showing how two slide-wire rheostats might be connected so that the current in the field winding of the main generator of a Ward–Leonard motor-control system can be adjusted to any value between positive maximum and negative maximum without opening the field circuit.

14–11. When the motor of Prob. 13–1 is operating at rated conditions, the armature is disconnected from the supply and connected to a dynamic braking resistor of 1.0 ohm. What is the initial braking torque in terms of rated torque? *Ans.* $T_B = 1.47 T_{FL}$.

14–12. Draw a circuit diagram for the automatic (push-botton) control of a series motor, incorporating all the following, properly labeled:

(*a*) A two-step counter-emf starter.

(*b*) Reversing controls (there should be a total of only three push buttons: 1 forward, 1 reverse, and 1 stop).

(*c*) Dynamic braking.

14–13. Draw a diagram showing how two widely separated push-button-control stations (each station consisting of one stop button and one start button) may be connected to provide independent start–stop control of a d-c motor having a definite-time starter.

14–14. Why are thermal overload devices, rather than fuses, used in automatic starters?

14–15. Explain precisely why a 1-sec power interruption has little effect on the operation of a d-c motor connected to a supply through a counter-emf automatic starter.

■ 15 *Transformers*

15-1. DEFINITIONS

One reason for the prevalance of alternating current in preference to direct current is the fact that an alternating voltage can be conveniently transformed in magnitude by means of a **transformer.** In its simplest form, a transformer consists of two separate windings with a common magnetic circuit. Although the material of the magnetic circuit may be either magnetic or nonmagnetic, or a combination of the two, the discussion in this chapter is limited to transformers having a magnetic circuit consisting entirely of a magnetic material.

The basic construction of a transformer is shown schematically as Fig. 15-1.

The windings of a transformer are identified according to the direction of power flow. The **primary winding** is connected to the source of

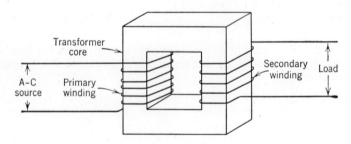

Fig. 15-1. Basic transformer construction.

NO-LOAD CONDITIONS—IDEAL TRANSFORMER

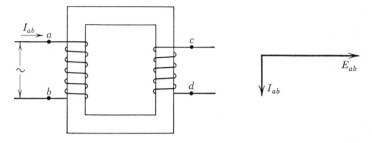

Fig. 15-2. Ideal transformer at no load.

electric energy; the **secondary winding** is connected to the load. Energy is transferred from the primary winding to the secondary winding by electromagnetic induction. Thus, the alternating flux produced by the current in the primary winding induces an alternating voltage of the same frequency in the secondary winding, and this voltage causes current to flow in the load.

The analysis of transformer operation is simplified if it is first made for an **ideal transformer** and is then extended to a consideration of the actual transformer. For purposes of this analysis, an ideal transformer is defined as a transformer that has (a) no winding resistance, (b) no leakage flux, and (c) no core losses (hysteresis and eddy-current losses).

15-2. NO-LOAD CONDITIONS—IDEAL TRANSFORMER

A transformer operates at no load when the load impedance is removed, that is, when the secondary circuit is open. Under no-load conditions, the magnitude of the current taken by the primary winding is determined by the magnitude of the supply voltage and the primary impedance. The primary winding of an ideal transformer is a pure inductive reactance; therefore, the primary current at no load is given by the relation

$$I_{ab} = \frac{E_{ab}}{X} = \frac{E_{ab}}{2\pi f L} \qquad (15-1)$$

The subscripts in Eq. 15-1 refer to the transformer shown in Fig. 15-2a. The vector diagram for the no-load condition is shown as Fig. 15-2b.

One difficulty encountered in the application of Eq. 15-1 to an ideal transformer is the fact that, because the magnetization curve for the magnetic circuit is nonlinear, the primary inductance is not a constant.

Thus, Eq. 15–1 might lead to the belief that the primary current is doubled in value if the frequency is decreased to one half while the primary voltage is held constant. Actually, such a variation in frequency might cause the current to reach a value that is as much as one hundred times the original value, because of the decrease in inductance as the current increases. In its present form, therefore, Eq. 15–1 might lead to erroneous conclusions.

Another approach to the analysis of the no-load conditions in an ideal transformer is to consider that the supply voltage is opposed by an induced voltage that is at all times equal to the supply voltage (since the resistance of the primary winding has been neglected). This induced voltage is produced in the primary winding by the alternating flux in the core of the transformer; the alternating flux, in turn, is produced by the current in the primary winding.

The magnitude of the primary induced voltage at any time is given by the general expression

$$e_i = N_1 \frac{d\phi}{dt} \times 10^{-8} \qquad (15\text{--}2)$$

where N_1 represents the number of turns in the primary winding. If the supply voltage is given by the relation

$$e = E_m \sin(\omega t + \theta) \qquad (15\text{--}3)$$

then, since the supply voltage and the induced voltage are equal,

$$E_m \sin(\omega t + \theta) = N_1 \frac{d\phi_1}{dt} \times 10^{-8} \qquad (15\text{--}4)$$

The solution of Eq. 15–4 for the flux at any time gives the expression*

$$\phi = -\frac{E_m}{N_1 \omega \times 10^{-8}} \cos(\omega t + \theta) \qquad (15\text{--}5)$$

which may be expressed in the general form

$$\phi = -\phi_m \cos(\omega t + \theta) \qquad (15\text{--}6)$$

where ϕ_m is the maximum value of flux. Therefore, from Eqs. 15–5

---

* The constant of integration has been omitted from Eq. 15–5 and subsequent expressions similar to it. The magnitude of the constant of integration is governed by the angle θ, that is, by the magnitude of the supply voltage at the instant the primary circuit is energized, and by the magnitude of the residual flux within the core. In an actual transformer, these two variables determine the magnitude of the transient current. This transient is damped in a few cycles so that, in effect, the constant of integration becomes zero.

NO-LOAD CONDITIONS—IDEAL TRANSFORMER

and 15-6,

$$\frac{E_m}{N_1\omega \times 10^{-8}} = \phi_m \tag{15-7}$$

or

$$E_m = 2\pi f N_1 \phi_m \times 10^{-8} \tag{15-8}$$

If Eq. 15-8 is rewritten in terms of the effective value of the supply voltage,

$$E_{ab} = \frac{2\pi f N_1 \phi_m \times 10^{-8}}{\sqrt{2}} = 4.44 f N_1 \phi_m \times 10^{-8} \tag{15-9}$$

or

$$E_{ab} = 4.44 f N_1 A B_m \times 10^{-8} \tag{15-10}$$

where A is the cross-sectional area of the core, and B_m is the maximum flux density.

According to Eq. 15-10, if the frequency is decreased to one half of its normal value while the supply voltage is held constant, the flux density must double, since the other factors in the equation are constant for a given transformer. Transformers are usually designed to operate, magnetically, at the knee of the magnetization curve when the voltage and frequency are normal. Thus, as indicated in Fig. 15-3, the magnitude of the current necessary to produce double the rated flux density is far greater than twice the normal value, and, as previously indicated, may reach a value many times the original value.

The alternating flux produced by the current in the primary winding also links the turns of the secondary winding and, therefore, causes a

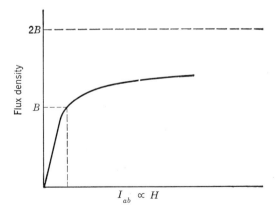

Fig. 15-3. Transformer magnetization.

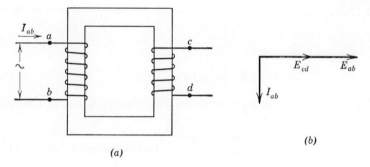

Fig. 15–4. Ideal transformer at no load.

voltage to be induced in the secondary winding. Since the primary and secondary windings differ only in the number of turns, the effective value of the voltage induced in the secondary winding can be expressed by the relation

$$E_{cd} = 4.44 f A N_2 B_m \times 10^{-8} \qquad (15\text{--}11)$$

where N_2 represents the number of turns in the secondary winding.

The ratio of primary voltage to secondary voltage, in the ideal transformer, is found by dividing Eq. 15–10 by Eq. 15–11, with the result

$$\frac{E_{ab}}{E_{cd}} = \frac{N_1}{N_2} \qquad (15\text{--}12)$$

Thus, the ratio of primary voltage to secondary voltage is equal to the ratio of primary turns to secondary turns.

The polarity of the secondary voltage for the transformer shown in Fig. 15–2 (redrawn as Fig. 15–4a) is such that terminal c has the same relative polarity as terminal a of the primary winding. Thus, the complete vector diagram for the ideal transformer at no load is as shown in Fig. 15–4b. (It is assumed that N_2 is less than N_1.)

15–3. LOAD CONDITIONS—IDEAL TRANSFORMER

If, as shown in Fig. 15–5, a load is connected to the secondary winding of an ideal transformer, the magnitude of the secondary current is determined by the magnitudes of the secondary voltage and the load impedance. The phase position of the secondary current with respect to the secondary voltage is determined by the nature of the load impedance (that is, whether the load impedance is resistive, inductive, or capacitive).

The secondary current produces an mmf that tends to change the

LOAD CONDITIONS—IDEAL TRANSFORMER

mutual flux (that which links both the primary and secondary turns). A change in the mutual flux, however, causes a change in the primary induced voltage; therefore, since the induced voltage is originally equal to the applied voltage, the primary current must change. The change in primary current changes the mutual flux and tends to restore it to its original value. The mutual flux must be restored to its original value so that the applied voltage and the primary induced voltage are again equal. Therefore, for the two voltages again to be equal, the change in primary current must be such as to produce an mmf equal to, and opposed to, the mmf produced by the secondary current. Thus,

$$I_{ab}'N_1 = I_{cd}N_2 \qquad (15\text{--}13)$$

where I_{ab}' represents the change in primary current.

Since the mutual flux is restored to its original value, the mmf required to produce the mutual flux is the same as it was originally. Therefore, the total current in the primary may be considered to consist of two components; one produces the mmf required for the mutual flux, the other, the mmf required to balance the secondary mmf. The actual primary current is therefore the vector sum of the two components, as shown in the complete vector diagram shown as Fig. 15–6b.

In Fig. 15–6b, the vector representing the increase in primary current I_{ab}' is shown in phase with the vector representing the secondary current I_{cd}, in agreement with Eq. 15–13, since turns have no vector significance. (The fact that the two currents produce opposing mmf's is shown by the subscripts; that is, when current flows from a to b through the primary winding, current flows from c to d through the load. These currents produce opposing mmf's, as is evident in Fig. 15–6a.)

The magnitude of I_{ab}' is determined by Eq. 15–13. The component of current necessary to produce the mutual flux is labeled $I_{0\text{-}ab}$ to indicate that it is the current that flows in the primary winding at no load. As

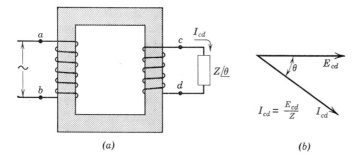

Fig. 15–5. Vector diagram for loaded transformer secondary.

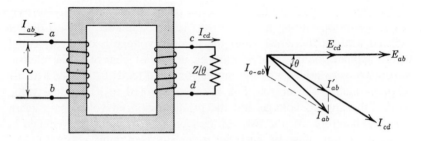

Fig. 15-6. Load conditions for an ideal transformer.

indicated in Fig. 15-6, the actual primary current is the vector sum of the two primary components.

In summary, the connection of a load to the secondary of an ideal transformer results in an increase in the primary current of such a magnitude that the mutual flux remains constant. The vector diagram of Fig. 15-6b shows that the power input to the transformer is equal to the power output.

15-4. THE ACTUAL TRANSFORMER

Thus far, the discussion of the transformer has been limited to the ideal transformer. An examination of the actual conditions in a transformer shows that the assumption of an ideal transformer is satisfactory for most considerations. A knowledge of the differences between the ideal transformer and the actual transformer is required, however, in order that the reasons for considering a transformer from an ideal standpoint may be understood.

Whereas the windings of the ideal transformer are assumed to have no resistance, the windings of the actual transformer do have resistance. The effect of resistance in the transformer windings is to introduce voltage drops that affect the magnitude of the induced voltages and the secondary terminal voltage. Thus, in the primary circuit, the induced voltage cannot be vectorially equal to, and opposed to, the applied voltage but must be of such magnitude that the vector equation

$$\overline{E_{ab}} = \overline{E_{i1}} + \overline{I_{ab}R_1} \qquad (15\text{-}14)$$

is satisfied. In Eq. 15-14, E_{i1} represents the primary induced voltage, and R_1, the primary resistance. Similarly, in the secondary circuit, the terminal voltage must be such that the vector equation

$$\overline{E_{cd}} = \overline{E_{i2}} - \overline{I_{cd}R_2} \qquad (15\text{-}15)$$

THE ACTUAL TRANSFORMER

is satisfied. In Eq. 15–15, E_{i2} represents the secondary induced voltage, and R_2 the secondary resistance.

In addition to the resistance, each winding in an actual transformer has an inductive reactance due to leakage flux, which is neglected in the consideration of the ideal transformer. The leakage fluxes produced by the primary and secondary currents at a given instant are shown schematically in Fig. 15–7.

The amount of leakage flux for a particular winding depends upon the mmf of the winding, and, since the path for leakage flux is principally air, the leakage flux can be assumed to vary directly with this mmf. For a given transformer, therefore, the leakage flux varies directly as the current in the coil, and the inductance due to leakage flux is a constant. The voltage induced in the coil because of the leakage flux can therefore be represented as an inductive reactance drop.

The inclusion of leakage flux in the discussion requires that Eq. 15–14 be changed to

$$\overline{E_{ab}} = \overline{E_{i1}} + \overline{I_{ab}R_1} + \overline{I_{ab}X_i} \tag{15-16}$$

or

$$\overline{E_{ab}} = \overline{E_{i1}} + \overline{I_{ab}Z_1} \tag{15-17}$$

where X_1 is the primary reactance due to leakage flux, and Z_1 is the entire primary impedance. Similarly, Eq. 15–15 must be changed to

$$\overline{E_{cd}} = \overline{E_{i2}} - \overline{I_{cd}R_2} - \overline{I_{cd}X_2} \tag{15-18}$$

or

$$\overline{E_{cd}} = \overline{E_{i2}} - \overline{I_{cd}Z_2} \tag{15-19}$$

where X_2 is the secondary reactance due to leakage flux, and Z_2 is the secondary impedance.

Equations 15–16 and 15–17 indicate that the primary induced voltage cannot remain constant when a load is connected to the secondary, since the primary current and, therefore, the primary impedance drop increase.

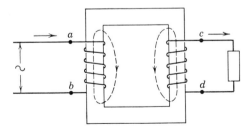

Fig. 15–7. Primary and secondary leakage fluxes.

For a given transformer, the only factor that can be changed in the mathematical expression for primary induced voltage (Eq. 15–10) is the mutual flux, which must therefore decrease. Thus, in an actual transformer, the connection of a load impedance to the secondary winding results in a decrease in the maximum value of the mutual flux. However, the magnitude of the primary and secondary impedance drops under rated conditions is small compared to the normal applied voltage (being of the order of 2 to 5 per cent of rated voltage when the transformer delivers rated current). Also, the primary impedance drop is vectorially added to the induced voltage. Therefore, the mutual flux need only decrease a small amount in order that rated current flow in the primary winding.

In view of the small relative magnitude of the primary and secondary impedance drops and the subsequent smaller effect upon the mutual flux, it is satisfactory to consider, in most practical applications, that the transformer is an ideal transformer and that the mutual flux and secondary voltage remain constant. Of course, when a variation in secondary voltage due to a change in load current is of importance, the assumption of an ideal transformer cannot be made.

In an actual transformer, the primary current at no load (called the **exciting current**) is not in lagging quadrature with the applied voltage, as in the ideal transformer, but lags by some angle smaller than 90 degrees. The actual angle of lag is determined by the losses, principally hysteresis and eddy-current losses, that must be supplied at no load. The magnitude of the exciting current under normal conditions of voltage and frequency is of the order of 3 to 5 per cent of the normal full-load current.

Because the exciting current is so small compared to the normal full-load current, and because it is vectorially added to the change in primary current, one further simplification in the analysis of practical transformer operation is to neglect the exciting current. Thus, the primary current is considered to be that found by multiplying the secondary current by the inverse ratio of turns,

$$I_{ab} = \frac{N_2}{N_1} I_{cd} \qquad (15\text{–}20)$$

Equation 15–20 cannot be applied, of course, when either the frequency or voltage are of such magnitudes that the flux density is greater than normal. Under these conditions, as indicated in Art. 15–2, the exciting current may increase tremendously and may be the major factor in the operation of the transformer.

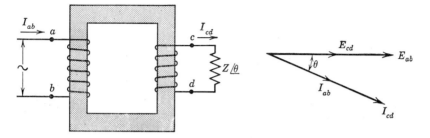

Fig. 15–8. Ideal transformer, no-load current neglected.

If the simplifying assumptions as to voltage and current are combined, the volt-ampere input to the transformer is equal to the volt-ampere output. Thus,

$$\frac{E_1}{E_2} = \frac{N_1}{N_2} = \frac{I_2}{I_1} \qquad (15\text{–}21)$$

or

$$E_1 I_1 = E_2 I_2 \qquad (15\text{–}22)$$

Further, since the exciting current is neglected, the power factor of the primary current is the same as that of the secondary current; therefore, the power input to the transformer is equal to the power output. This conclusion implies a perfectly efficient device, which the transformer obviously cannot be because of the core losses and copper losses already mentioned. However, such an assumption is not too impractical in many applications, because transformers that are 97 to 98 per cent efficient at full load are not uncommon.

If the assumptions that the transformer is ideal and that the exciting current is neglected are incorporated, the vector diagram that applies to a transformer delivering load is as shown in Fig. 15–8.

Unless otherwise specified, subsequent discussions of transformers in this book are based upon the assumptions that the transformer is ideal and that the exciting current is neglected.

15–5. TRANSFORMER RATING

Transformers are rated in terms of the volt-ampere output under specified conditions of voltage and frequency. As in the rating of all electric equipment, the operating temperature is the determining factor in the transformer rating.

The nameplate of a transformer provides the following information:

(*a*) Voltage rating. This is given for both the high-voltage winding and the low-voltage winding. For example, if the transformer has a high-voltage winding rated at 2300 volts and a low-voltage winding rated at 230 volts, the nameplate gives the data as 2300/230 volts. (Note that reference is made to the high- and low-voltage windings, not to primary and secondary windings. Either the high- or the low-voltage winding may be used as the primary.)

(*b*) Kilovolt-ampere rating.

(*c*) Frequency rating.

(*d*) Temperature rise.

The rated currents of the transformer windings are not usually given on the nameplate but can be calculated from the kilovolt-ampere rating of the transformer and the voltage ratings of the windings.

15–6. TRANSFORMER LOSSES

Although transformers are treated as ideal transformers, some consideration should be given to the nature of the power losses in an actual transformer, since these losses may have considerable bearing upon the selection of a transformer for a given application.

The transformer losses are copper losses, which occur in both the primary and the secondary windings, and core losses, which are the hysteresis and eddy-current losses produced by the alternating flux in the transformer core.

The copper losses in the primary and secondary windings are functions of the square of the current in these windings. Since the primary current has a fixed numerical relationship to the secondary current, and since the secondary current is a measure of the delivered load, the copper losses in a given transformer may be assumed to vary as the square of the load.

In a given transformer, the hysteresis losses and eddy-current losses are functions of the flux density in the transformer core and the frequency of the supply voltage. Because the flux density remains practically constant from no load to full load, the core losses for a given transformer connected to a constant-frequency constant-voltage supply are assumed to remain constant from no load to full load.

With reference to the effect of load upon the losses, the copper losses and core losses are sometimes classified as the variable losses and constant losses, respectively.

15-7. SINGLE-PHASE TRANSFORMER CONNECTIONS

In order that various transformation ratios and voltages may be obtained from one transformer, transformers often have high-voltage windings and low-voltage windings consisting of more than one coil, each coil having its own external terminals. As an example, the rating of a transformer may be specified as 100 kilovolt-amperes, 4600–2300/230–115 volts. This rating means that both the high-voltage winding and the low-voltage winding have two coils, which may be connected either in series or in parallel. If the transformer is to be used as a step-down transformer, the primary has a rating of 4600 volts if the two high-voltage coils are connected in series, and a rating of 2300 volts if they are connected in parallel. For either of the primary connections, the secondary voltage is 115 volts if the low-voltage coils are connected in parallel, 230 volts if connected in series. The four possible connections are shown in Fig. 15–9. The rating of the transformer for each of the connections shown is 100 kilovolt-amperes.

The secondary connection shown in Figs. 15–9a and 15–9b is em-

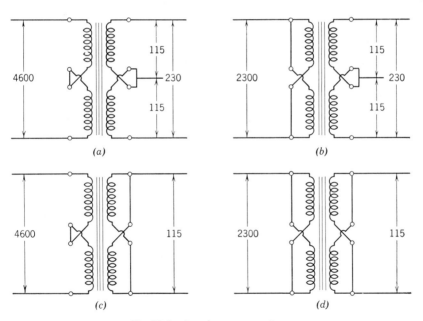

Fig. 15–9. Transformer connections.

ployed by most electric-power companies to supply residential customers. As indicated, three wires are connected to the transformer secondary such that the voltage between either outside wire and the common, or neutral, wire is 115 volts; the voltage between outside wires is 230 volts.

The transformer connections shown in Fig. 15–9 must be made correctly to prevent damage to the transformer, or otherwise faulty operation. As an example, a transformer having one primary coil and two identical secondary coils, as shown schematically in Fig. 15–10a, is to have its secondary coils connected in series. When the transformer is energized, terminals a and c of the secondary have the same relative polarity at a given instant. To indicate this fact, these terminals are marked with the symbol $\pm$.* A series connection in which b and c are joined results in a total secondary voltage that is equal to twice the voltage of one coil. If the series connection is made by joining a and c, however, the voltage output is zero, since the voltages are in opposition. Thus, a random series connection of the secondary coils does not harm the transformer, but it may produce zero voltage output.

If the secondary coils of the transformer shown in Fig. 15–10 are paralleled by the connections of a to c and b to d, as shown in Fig. 15–11a, terminals of like polarity are tied together, and the voltage output of the transformer is equal to the voltage of one coil. If, however, the parallel connection is made by connecting a to d and c to b,

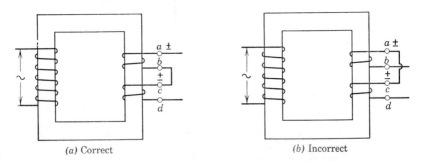

(a) Correct (b) Incorrect

Fig. 15–10. Series connection of secondary coils.

* Standard practice with regard to transformer terminal markings is to designate the high-voltage terminals as H_1, H_2, H_3, etc., and the low-voltage terminals as X_1, X_2, X_3, etc. Thus, for the example shown in Fig. 15–10 (if it is assumed that the high-voltage coil is the primary), the high-voltage terminals would be designated as H_1 and H_2, the low-voltage terminals as X_1, X_2, X_3, and X_4. With this method of designating terminals, the terminals with the odd subscripts (H_1, X_1, X_3) are terminals that have the same relative polarity at a given instant.

AUTOTRANSFORMERS

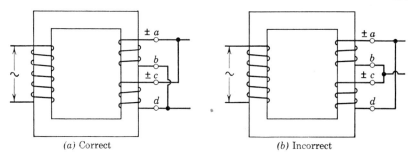

(a) Correct *(b)* Incorrect

Fig. 15–11. Parallel connection of secondary coils.

as shown in Fig. 15–11b, the windings are, in effect, connected properly in series and then short-circuited. This connection would damage the transformer, unless it is protected by fuses or circuit breakers.

If a transformer has two or more primary coils, all of the same voltage rating, the series or parallel connection of these coils must be so made that the mmf's produced by the individual coils are additive; the mutual flux then has the magnitude to produce the proper induced voltage in each coil. If, on the other hand, the series or parallel connection is so made that the mmf's are subtractive, there would be no mutual flux, and therefore no induced voltage. The only impedance to current flow in each coil under these conditions is provided by the primary resistance and leakage reactance. Since the magnitudes of the primary resistance and leakage reactance are relatively small, the current in each winding would be excessive and the transformer would be damaged, unless protected by fuses or circuit breakers.

With respect to terminal polarity designations, the correct connection of primary coils is the same as the correct connection of secondary coils; such a connection produces mmf's that are additive.

15–8. AUTOTRANSFORMERS

Because of possible savings in cost and weight, transformers are often built with only one winding instead of two. A transformer having only one winding is called an **autotransformer**. Connections of an autotransformer for step-up and step-down operation are shown in Fig. 15–12.

In the step-up connection shown in Fig. 15–12a, the voltage induced in the portion of the winding from c to a is produced by the same flux that causes the induced voltage in the portion of the winding from a to b. Therefore,

$$\frac{E_{ca}}{E_{ab}} = \frac{N_2 - N_1}{N_1} \tag{15-23}$$

The secondary voltage E_2 is the voltage from c to b and is given by the relation

$$E_{cb} = E_{ca} + E_{ab} \tag{15-24}$$

or

$$E_{cb} = \frac{N_2 - N_1}{N_1} E_{ab} + E_{ab} = \frac{N_2}{N_1} E_{ab} \tag{15-25}$$

In more general terms, Eq. 15–25 may be written

$$\frac{E_1}{E_2} = \frac{N_1}{N_2} \tag{15-26}$$

Thus, as in a two-winding transformer, the ratio of primary voltage to secondary voltage in the autotransformer is equal to the ratio of the turns employed in the primary to the turns employed in the secondary. Although Eq. 15–26 is derived for the step-up connection, a similar analysis of the step-down connection provides the same result.

Regardless of the autotransformer connection, step-up or step-down, the current in the portion of the winding that is common to both the primary and the secondary is the difference between the input and output currents. The current directions at a given instant for the two types of connection are shown in Fig. 15–13. The relative direction of the current through the common portion of the winding depends upon the connection of the transformer, since the type of connection determines whether the input current or the output current is the larger.

In an ideal autotransformer in which the exciting current is neglected,

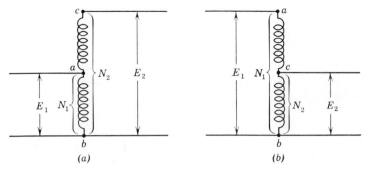

Fig. 15–12. Autotransformer connections.

AUTOTRANSFORMERS

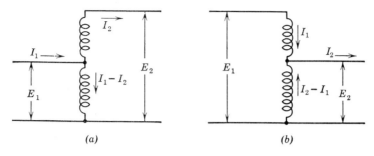

Fig. 15–13. Autotransformer current relations.

the value of current in the common portion of the winding approaches zero as the ratio of primary to secondary voltage approaches unity. In an actual autotransformer, however, the value of current in the common portion of the winding approaches the value of exciting current necessary to produce the mutual flux. For transformation ratios near unity, therefore, the common portion of the winding can be wound with wire of smaller cross-sectional area; thus, savings in construction are effected.

One type of autotransformer is so constructed that the secondary voltage can be easily varied from zero to approximately 130 per cent of the applied voltage. Such a variation is obtained by a change in the position of one secondary connection, which can be moved along the winding in the same manner as the slider on a wire-wound rheostat. The schematic diagram for such an autotransformer is shown in Fig. 15–14.

A transformer having more than one winding can also be connected as an autotransformer and therefore have the number of possible transformation ratios and voltage ratings extended. Thus, the 2300/230-volt two-winding transformer shown in Fig. 15–15a can be connected as an

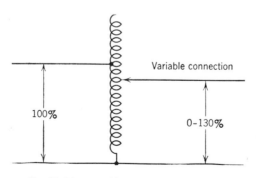

Fig. 15–14. Variable-output autotransformer.

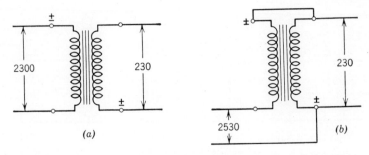

Fig. 15–15. Connection of a two-winding transformer as an autotransformer.

autotransformer by having its coils connected in series, as shown in Fig. 15–15b. The voltage rating of the transformer connected as an autotransformer is 2530/230 volts.

The disadvantage of an autotransformer is that the primary and secondary circuits are not electrically isolated from each other as they are in a two-winding transformer. Thus, if the primary circuit of an autotransformer should come in contact with a high-voltage circuit, the secondary circuit is similarly affected, and damage to equipment or injury to personnel may result. Similarly, if the ratio of primary to secondary voltage is appreciably greater than unity, an accidental opening of the common portion of the winding introduces the higher primary voltage into the secondary circuit, with the same chances for damage or injury.

15–9. POLYPHASE CONNECTION OF TRANSFORMERS

The type of transformer connection for transforming the voltages of a polyphase system depends upon the requirements of the circuit that is to be connected to the secondary.

Since the primary winding of a transformer is a receiver of electric energy, the connection of the primary windings of several transformers of proper rating into a polyphase system can cause no harm to the transformer or the system. An improper connection of the secondary windings of these same transformers, however, can damage the transformers and system, or it can result in a secondary system that does not exhibit the polyphase characteristics of the primary. The discussions to follow, therefore, are mainly concerned with the connection of the secondary windings.

Wye–Wye Connection. Figure 15–16a shows the proper wye–wye

POLYPHASE CONNECTION OF TRANSFORMERS

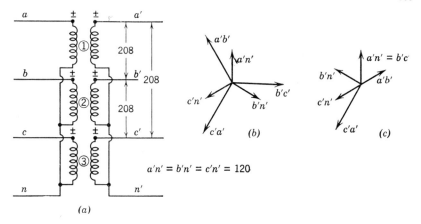

Fig. 15–16. Wye-wye connection of single-phase transformers.

connection of three single-phase transformers into a three-phase four-wire system. The primary and secondary windings of the same transformer are labeled with the same number. The $\pm$ symbols associated with the primary and secondary windings of a given transformer are the terminals that have the same relative polarity. If the phase sequence of the primary is assumed to be a–b–c, the vector diagram applying to the secondary voltages is as indicated in Fig. 15–16b.

If the secondary winding of one of the transformers is reversed, for example, number 2, the vector diagram of secondary voltage is as shown in Fig. 15–16c. As indicated, the resultant system is not a balanced three-phase system.

The type of connection shown in Fig. 15–16a is used where lighting and power loads are to be supplied from the same group, or bank, of transformers. Thus, the 120-volt lighting load is connected between line

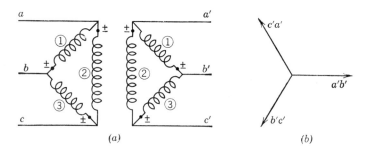

Fig. 15–17. Delta-delta connection of single-phase transformers.

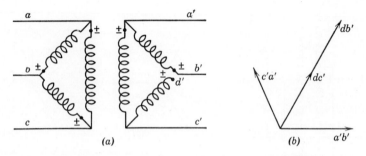

Fig. 15-18. Incorrect delta connection of secondary windings.

and neutral of the secondary; the 208-volt power load (motors) is connected between lines.

Delta–Delta Connection. Figure 15–17a shows the proper delta–delta connection of three transformers into a three-phase system. The vector diagram applying to the secondary voltages is shown in Fig. 15–17b. (Phase sequence is assumed to be a–b–c.)

If, in the connection of the secondary windings, one is reversed, the voltage between the last two terminals to be connected is not zero, as it should be, but is considerably larger. If the connection is made under these conditions, the transformers would be damaged unless suitably protected by fuses or circuit breakers. The improper delta connection of the secondary windings is shown in Fig. 15–18a with the delta unclosed. The voltage between the open terminals, which would normally be joined to complete the delta connection, is indicated in the vector diagram of Fig. 15–18b to be twice the voltage across a single transformer secondary. Connection of the terminals would be equivalent to placing a short circuit across a source having this value of voltage.

Because of the possibility of damage to the transformer, it is com-

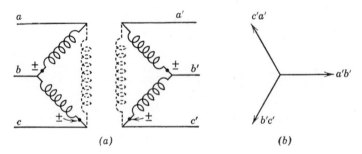

Fig. 15-19. Open-delta connection of transformers.

POLYPHASE CONNECTION OF TRANSFORMERS

mon practice to measure the voltage between the last two terminals to be joined to make certain that the connection is a proper one.

Open-Delta Connection. One advantage in using the delta–delta connection of transformers is the possibility of quick and convenient conversion to the open-delta connection, should one of the transformers be damaged and have to be removed from service. The open-delta connection allows the three-phase system to continue to operate, although at a reduced power rating. The proper open-delta connection is shown as Fig. 15–19a, with the prior connection of the transformer that has been removed shown in dotted form.

The basis of operation of the open-delta connection is the fact that the vector sum of any two of the line voltages in a balanced three-phase system is equal to the third line voltage. Thus, even though a transformer has been removed, the voltage between the terminals to which it was connected remains unchanged. The vector diagram applying to the secondary voltages of the open-delta connection is shown as Fig. 15–19b.

Six-Phase Connection. A transformer connection applied in rectifier

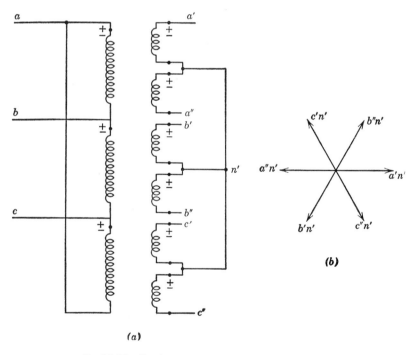

Fig. 15–20. Six-phase connection of secondary windings.

circuits consists of the secondary windings of three transformers in a three-phase system that are so connected as to produce a six-phase secondary system. Each of the transformers for this connection has one primary winding and two secondary windings that are connected as shown in Fig. 15-20a. The vector diagram applying to the secondary voltages is shown as Fig. 15-20b.

15-10. INSTRUMENT TRANSFORMERS

The usual practice in the measurement of alternating voltages and currents that exceed 300 volts and 25 amperes, respectively, is to reduce the quantities by means of instrument transformers so as to measure them with low-range instruments. For voltage measurements, and some current measurements, this procedure is also necessary as a safety precaution to isolate the instrument, and therefore the operator, from a high-voltage circuit.

A typical connection of instrument transformers in a circuit in which it is desired to measure current, voltage, and power is shown schematically in Fig. 15-21.

The theory of operation and the connection of the potential transformer are the same as for the voltage transformer previously discussed. The magnitudes of the current and power involved, however, are small, since voltmeters and voltage elements are the only loads on these transformers. As indicated in Fig. 15-21, the transformer terminals having the same relative polarity are labeled, to facilitate the correct connection of the wattmeter voltage element to the transformer secondary. (Standard practice is to use the H_1, H_2, X_1, X_2 notation previously mentioned.)

When a potential transformer is connected into a circuit, the indication of the voltmeter multiplied by the ratio of primary to secondary turns of the potential transformer is the voltage of the circuit. The ratio of primary to secondary turns is usually marked on the nameplate of the transformer as a multiplier. For the circuit of Fig. 15-21, the multiplier must also be applied to the indication of the wattmeter.

As indicated in Fig. 15-21, the primary of a current transformer is connected in series with the load. The impedance of the transformer primary is low, and its addition to the circuit has negligible effect upon the remainder of the circuit. The current in the primary of the current transformer depends therefore only upon the load impedance and the supply voltage, and it is constant if the supply voltage and the load impedance are constant.

The secondary winding of the current transformer has more turns

INSTRUMENT TRANSFORMERS

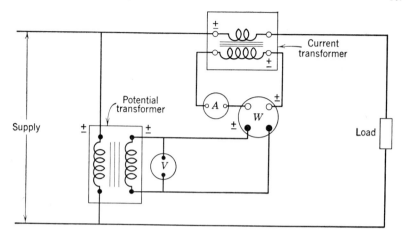

Fig. 15-21. Connection of instrument transformers.

than the primary winding so that, in effect, the voltage drop across the primary winding is stepped-up in the secondary. The current in the secondary winding produces an mmf opposing the primary mmf, and, since the primary mmf is constant, the value of mutual flux decreases to a low value. (As previously mentioned, the primary current of a current transformer is determined only by the supply voltage and the load impedance of the circuit in which the transformer is connected.) Very little mutual flux is needed, however, to produce the voltage necessary to cause rated current to flow in the ammeter and the wattmeter current element connected in the transformer secondary. The exciting-current component of the primary current is therefore very small compared to the total primary current, and, for all practical purposes, the primary mmf equals the secondary mmf. Thus,

$$I_1 N_1 = I_2 N_2 \qquad (15\text{-}27)$$

or

$$I_2 = \frac{N_1}{N_2} I_1 \qquad (15\text{-}28)$$

Since there are more secondary turns than primary turns, the current in the load circuit (the primary current of the transformer) is reduced for measurement purposes in the secondary of the transformer.

Current transformers are usually rated at 5 amperes in the secondary. Thus, a current transformer having an indicated ratio of 10-1 is rated at 50 to 5 amperes. Portable current transformers are usually provided with several primary taps and, therefore, have several ratios, so that a wide range of currents can be measured with a 5-ampere ammeter connected to the secondary of the transformer. The multiplier applied to the ammeter indication for the circuit of Fig. 15–21 must also be applied to the wattmeter.

The secondary circuit of the current transformer must not be opened when the primary winding carries current. Opening of the secondary under these conditions removes the opposing mmf produced in the secondary and allows the mutual flux to reach a value determined only by the primary mmf. This value of mutual flux is far greater than the original value and may result in saturation of the transformer core or an excessive value of voltage across the secondary winding. The effect of saturation may be to leave the transformer with a large value of residual flux that would seriously impair the accuracy of the transformer; the effect of the excessive voltage may be to puncture the insulation of the transformer or to provide an unpleasant shock to the operator opening the circuit. To prevent these occurrences, portable current transformers are usually provided with short-circuiting switches so that the secondary winding may be short-circuited and the circuit maintained before an instrument is removed, or before some other change is made that would ordinarily require that the circuit be opened. If such a switch is not provided with the transformer, one should be placed in the circuit if any change in the secondary circuit is anticipated.

PROBLEMS

15–1. A transformer has an approximately sinusoidal flux with a maximum value of 1,100,000 lines and alternating at a frequency of 60 cycles. There are 800 turns in the primary winding and 80 turns in the secondary.

(a) What is the effective value of the voltage induced in the primary winding? *Ans.* 2340 volts.

(b) What is the effective value of the voltage induced in the secondary winding? *Ans.* 234 volts.

15–2. Given: two 13,200-volt transformers, one designed for 60 cycles, and the other for 25 cycles. The flux density, number of turns, and length of magnetic circuits are to be the same. How do the weights of the cores compare? *Ans.* $W_{(60)} = 25/60\ W_{(25)}$.

15–3. Given: two 13,200-volt transformers, one designed for 60 cycles, the other for 25 cycles. The cores and flux densities are identical. What is the difference in the design of the two transformers? How do the exciting currents compare? *Ans.* $N_{(60)} = 25/60\ N_{(25)}$; $I_{(60)} = 60/25\ I_{(25)}$.

15–4. A 60-cycle transformer is to be operated on a 30-cycle system in the

PROBLEMS

future. What change, if any, must be made in its voltage, its current, and its kilovolt-ampere ratings, and why?

Ans. $E_{(30)} = \frac{1}{2} E_{(60)}$; $I_{(30)} = I_{(60)}$; $kva_{(30)} = \frac{1}{2} kva_{(60)}$.

15–5. Plot a curve of voltage versus exciting current for an ideal 60-cycle transformer operated at constant frequency up to 120% of rated voltage. Explain the shape of the curve.

15–6. A 60-cycle voltage having a magnitude between 110 and 120 volts is needed. Available are a 440-volt 60-cycle supply and the following transformers:

(1) 10 kva, 450/115 volts, 90 cycles.
(2) 10 kva, 450/115 volts, 30 cycles.

If one or both of these transformers can provide the desired voltage, determine how much unity power-factor load can be supplied, and determine what the secondary voltage would be. If one or both of these transformers cannot be applied, briefly explain why not.

15–7. Given: a 15-kva 460/115-volt 60-cycle transformer that has one high-voltage winding and one low-voltage winding. Upon testing for continuity, you find that one of the windings is open-circuited internally. Is it possible to determine, by means of a nondestructive test, whether the high-voltage winding or the low-voltage winding is open-circuited? Explain. (Assume that the leads to the terminals are not visible and that any a-c equipment you might need is available.)

15–8. A 5-kva 440/110-volt 60-cycle transformer draws a no-load current of 0.6 ampere when energized from a 440-volt 60-cycle supply. What current is drawn from the supply if a capacitor of 58 μf is connected to the 110-volt secondary?

15–9. (*a*) A 1-ohm resistor is normally connected to the low-voltage side of a 440/110-volt transformer. If a single resistor is to replace the transformer and the 1-ohm resistor so as to be equivalent, what should be the complete rating of the replacement resistor?   *Ans.* 16 ohms, 12.1 kw.

(*b*) Compare the resistance ratio of the replacement resistor found in (*a*) to the 1-ohm resistor with the turns ratio of the transformer.

15–10. The low-voltage winding of a transformer has 125 turns and is used as the secondary. What will be the effect on the operation of the transformer if one of these turns is perfectly short-circuited?

15–11. (*a*) A 10-kva 2300/230-volt 60-cycle transformer that has been damaged has its high-voltage winding rewound with 20% fewer turns than originally. What effect will this shortage have on the operation of the transformer if the transformer is normally energized from a 2300 volt 60-cycle supply? Explain.

(*b*) Repeat question (*a*) except that the low-voltage winding is rewound with 20% fewer turns than originally.

15–12. Explain the effect, if any, upon the rating of a 10-kva 2300/230-volt 60-cycle transformer of cutting a small air gap in its iron core. (The transformer is to be energized from a 2300-volt 60-cycle supply.)

15–13. The low-voltage winding of a 100-kva 2500/250-volt 60-cycle transformer is connected to a 250-volt 60-cycle supply, and its high-voltage winding is open-circuited. The following data are obtained by measurement: 250 volts; 20 amperes; 500 watts. What values of current and power are measured if the high-voltage winding is connected to a 2500-volt 60-cycle supply and the low-voltage winding is open-circuited? Justify your answer.

Ans. 2 amperes; 500 watts.

15–14. The schematic diagram of a 230-kva 4600–2300/230–115-volt 60-cycle transformer with its four coils unconnected is shown as Fig. 15–22. The voltage ratings of the coils are tabulated.

Coil	Voltage Rating
a–b	2300
c–d	2300
e–f	115
g–h	115

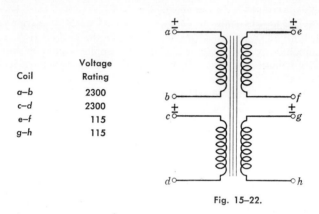

Fig. 15–22.

(*a*) When coil *a–b* is connected to a 2300-volt 60-cycle supply and the other windings are open-circuited, the current to the transformer is 5 amperes. What current will flow if *b* is connected to *d* and a 2300-volt 60-cycle supply is connected to *a* and *c*? Explain briefly.

(*b*) What current will flow if *b* is connected to *c* and a 2300-volt 30-cycle supply is connected between *a* and *d*? Explain briefly.

15–15. Prove that the efficiency of a transformer is maximum when the load is such that the copper losses in the windings are equal to the core losses.

15–16. A 5-kva 440/110-volt 60-cycle transformer is to be connected as an autotransformer with the ratio of 550/440 to a 550-volt supply. What maximum safe kilovolt-amperes can the transformer deliver? *Ans.* 25 kva.

15–17. A certain transformer is rated at 10 kva, 460–230/230–115 volts, 60 cycles.

(*a*) What is the maximum voltage that this transformer can deliver when connected as an autotransformer to a 115-volt 60-cycle supply? What maximum safe kilovolt-amperes can be delivered under these conditions?

Ans. 690 volts; 6 kva.

(*b*) It is desired to have a voltage between 100 and 130 volts when the only available supply is 460 volts, 90 cycles. Discuss the possible application of the given transformer to this problem; include a consideration of the connection, the kilovolt-ampere rating, and the operation of the transformer.

15–18. A balanced three-phase load of 30 kva at a power factor of 0.866

lagging is connected to two transformers connected in open-delta to give a 230-volt three-phase system. Compute the power delivered by each transformer. *Ans.* 8.66 kw; 17.32 kw.

15–19. A three-phase supply having a line voltage between 200 and 250 volts is desired. Available are three 25-kva 7600–3800/230–115-volt 60-cycle transformers and a three-wire 60-cycle 13,800-volt three-phase supply. Is it possible to connect the transformers so as to satisfy the requirements without exceeding the voltage rating of the coils by more than 10%? Explain.

■ 16 The Three-Phase Induction Motor

16–1. DEFINITIONS

The operation of an electric motor depends upon the fact that a current-carrying conductor in a transverse magnetic field has a force exerted upon it. In a d-c motor, the field winding produces the magnetic field, and the armature supports the current-carrying conductors upon which the force is exerted. Both the field winding and the armature winding of the d-c motor are connected to a voltage source. An **induction motor**, on the other hand, has only a field-producing winding connected to a voltage source; the winding containing the current-carrying conductors receives its energy from the energized winding by electromagnetic induction. The induction motor may be considered to be a transformer with a rotating secondary, and it may therefore be described as a "transformer-type" a-c machine in which electric energy is converted to mechanical energy.

The field-producing winding, corresponding to a transformer primary winding, is mounted on the stationary frame, or **stator,** of the induction motor and is called the **stator winding.** The winding containing the current-carrying conductors is mounted on the rotating member, or **rotor,** and is called the **rotor winding.**

Induction motors are commercially available in both single-phase and polyphase types. In the latter classification, the three-phase induction motor is the most common and is the type considered in this chapter. The principles to be considered, however, are applicable

THE STATOR

to all types of polyphase induction motors. The single-phase induction motor is discussed in Chapter 18.

16–2. THE STATOR

The stator of an induction motor is built up of thin laminations of sheet steel similar in shape to the two types shown in Fig. 16–1. The tooth-and-slot construction on the inner periphery of the laminations is similar to that on the outer surface of a d-c machine armature.

The stator slots are filled with insulated conductors connected to form a balanced three-phase wye or delta circuit. Thus, the circuit diagram for the stator winding appears as shown in Fig. 16–2a or 16–2b. Each coil represents the connection of one third of the total number of conductors in the stator slots.

Consideration of the details of the stator winding is limited in this book to a simplified machine having six equally spaced stator slots and conductors, as shown in Fig. 16–2c. Each conductor of the six-slot machine represents a group of conductors in a machine having many slots, and, therefore, the discussion of the simplified machine also applies to machines available commercially. The six conductors are arranged in three coils by the connection of conductors that are 180 degrees apart. The conductors are joined on the side of the stator away from the reader, and the letter designations represent the ends of the conductors toward the reader, these ends being the open ends of the coils, as shown by the side view in Fig. 16–2d. The proper connection of the three coils into wye and delta circuits is as shown in Figs. 16–2a and 16–2b.

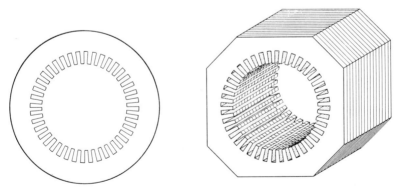

Fig. 16–1. Stator laminations.

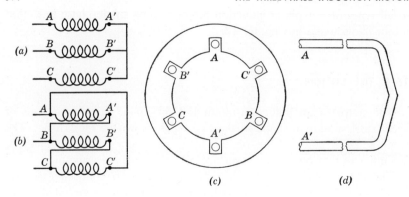

Fig. 16–2. Stator circuit details.

16–3. THE ROTOR

The rotor of an induction motor is also built up of thin laminations of sheet steel. The laminations are circular in shape and have a tooth-and-slot construction on their outer periphery. Except for the shape of the slots, the rotor laminations are similar to those used for d-c machine armatures. The shape of the slot of an induction motor rotor depends upon the type of winding inserted and upon the machine characteristics desired.

Classified as to winding, there are two types of induction motor rotors: **squirrel-cage rotor** and **wound rotor**. The squirrel-cage rotor has slots filled with uninsulated conductors tied together at each end by a connecting ring made of the same material. Thus, the configuration formed by the conducting material resembles a squirrel cage open at both ends. The conductors, usually copper or aluminum, are uninsulated because the rotor voltage is small when the machine is operating

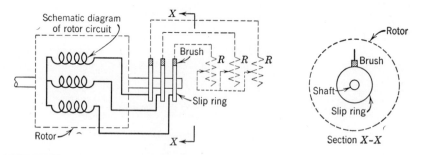

Fig. 16–3. Wound rotor details.

and because the steel paths in parallel with the conductors are of very high resistance owing to the varnished surfaces of the laminations. If the operation of an induction motor is compared to that of a transformer, the squirrel-cage type of rotor winding corresponds to a short-circuited secondary winding.

The wound rotor has slots filled with insulated conductors connected in series by groups to form a balanced wye-connected circuit. The open ends of the wye circuit are brought out to insulated slip rings on the rotor shaft so that external resistance can be added to each phase of the rotor circuit, as shown schematically in Fig. 16–3. The effect of the addition of the external resistor upon the operating characteristics of the motor is discussed later in this chapter.

When the external resistance is decreased to zero, the wound rotor is short-circuited and has the same characteristics as a squirrel-cage rotor. In appearance, the wound rotor resembles a d-c machine armature, except that the wound rotor has slip rings instead of a commutator.

16–4. THE TWO-POLE ROTATING FIELD

The operation of a three-phase induction motor depends upon the magnetic field produced by the stator winding when it is properly energized. This field is such that its poles do not remain in a fixed position on the stator but revolve around the stator; for this reason, it is called a **rotating field.**

For the consideration of the rotating field, it is assumed that the stator winding of the simplified machine of Fig. 16–2c is wye-connected and is so energized that the phase currents vary with time as shown in Fig. 16–4. (The double-subscript notation is the same as that used previously in this book: that is, when the plot of current versus time is

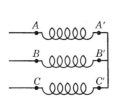
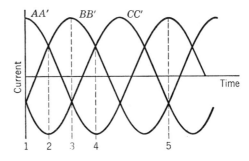

Fig. 16–4. Current versus time for stator circuit.

positive, or above axis, the current flows in the direction indicated by the subscripts—A to A', B to B', or C to C'; when the plot is negative, or below axis, the current direction is reversed.)

At the instant of time indicated as 1 on the current–time diagram, the current directions in the stator conductors are as shown in Fig. 16–5a. The direction of the resultant magnetic field produced by the current in the stator conductors is from right to left across the air gap; therefore, the polarity of each magnetic pole is as indicated on the diagram. The stator mmf's are so combined that only two magnetic poles are produced. At some later instant of time the currents in the stator conductors are in such a direction as to produce the resultant magnetic field shown in Fig. 16–5b, which indicates that the location of the poles on the stator has shifted in a clockwise direction. The strength of the field shown in Fig. 16–5b is the same as that shown in Fig. 16–5a, however, since the net stator mmf is the same in each case.

Figures 16–5c and 16–5d show the direction of the field at time instants 3 and 4, respectively. These figures show that the field con-

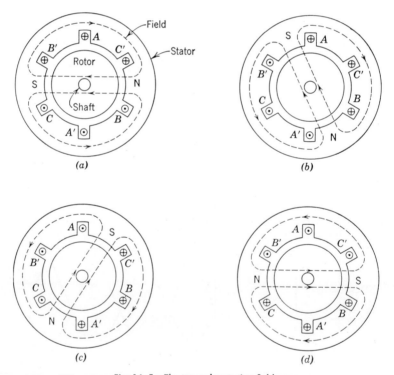

Fig. 16–5. The two-pole rotating field.

tinues to rotate in a clockwise direction and that it is constant in magnitude. It is to be noted that the magnetic effect revolves around the stator but that there is no physical movement of the stator itself; the stator and stator conductors remain fixed, and the poles of the magnetic field change position.

The time instant 4 represents the completion of one half of a cycle of the alternating current from time instant 1. During this one-half cycle, the field has rotated 180 degrees. At a time instant represented by 5 on the diagram, or one complete cycle from the origin, the field has completed one revolution. Thus, if the stator is connected to a 60-cycle supply, the field rotates at a rate of 60 revolutions per second or 3600 revolutions per minute.

The phase sequence of the three-phase voltage applied to the stator winding in the preceding analysis is $A–B–C$. If this sequence is changed to $A–C–B$ and a similar analysis performed, it is observed that the only change in the conclusions from the preceding analysis concerns the direction of rotation of the field—the field rotates counterclockwise instead of clockwise; the number of poles and the speed at which they revolve are unchanged. Thus, it is necessary only to change the phase sequence to change the direction of rotation of the field. This fact is of practical importance because the direction of rotation of the rotor is the same as that of the rotating field.

16-5. THE FOUR-POLE ROTATING FIELD

In order to produce a four-pole rotating field, the simplified machine must have four conductors in each stator phase, placed as shown in Fig. 16–6a. As in the previous analysis, each conductor represents a group of conductors, and the letter on each conductor designates the end of the conductor toward the reader.

The four conductors in each phase are connected in series in the following manner:

(a) The conductor end opposite to A is connected to the conductor end opposite to A'.

(b) A' and A'' are connected.

(c) The conductor end opposite to A'' is connected to the conductor end opposite to A'''.

The other phases are connected in a similar manner. Therefore, if it is assumed that the conductors of which A and A' are a part form one coil, and that those of which A'' and A''' are a part form another coil,

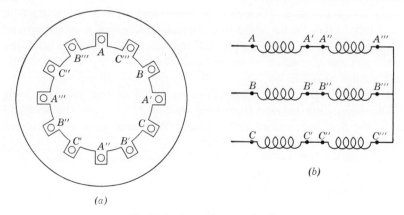

Fig. 16–6. Four-pole stator details.

the correct wye connection of the four-pole stator winding is as shown in Fig. 16–6b.

The stator winding of this machine is assumed to be so connected to a three-phase supply that the phase currents vary with time as shown in Fig. 16–7a. The analysis of the magnetic field produced by the currents in the conductors at the instants of time indicated as 1, 2, 3, and 4 in Fig. 16–7 is then shown diagrammatically in Figs. 16–7b through 16–7e. This analysis reveals (a) that the magnetic field has four poles and (b) that it rotates one fourth of a revolution in a time interval corresponding to one half of a cycle of the supply frequency. Thus, the speed of rotation of the field is one half of the supply frequency.

A similar analysis of a simplified machine in which there are six equally spaced conductors in each phase would reveal that the rotating field has six poles and that it rotates at a speed corresponding to one-third of the supply frequency.

16–6. SYNCHRONOUS SPEED

The speed at which the field of an induction motor rotates is called the **synchronous speed.** As pointed out in the preceding analyses, the synchronous speed varies directly as the frequency of the supply voltage and inversely as the number of poles. The synchronous speed for a two-pole machine in revolutions per second is equal to the supply frequency in cycles per second, and the synchronous speed for a four-pole machine is equal to one-half the supply frequency; therefore, the

SYNCHRONOUS SPEED 319

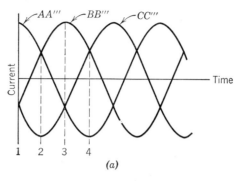

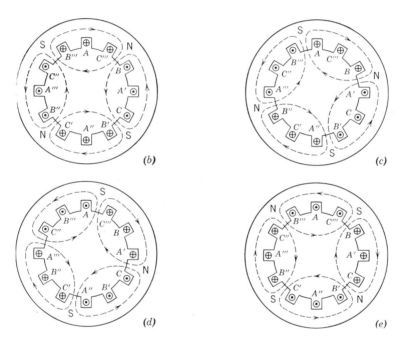

Fig. 16-7. The four-pole rotating field.

number of poles, the frequency, and the speed are related by the expression

$$n = \frac{60f}{P/2} = \frac{120f}{P} \qquad (16\text{-}1)$$

where n is the synchronous speed in revolutions per minute, f is the supply frequency in cycles per second, and P is the number of poles.

16–7. INDUCTION MOTOR TORQUE AT STANDSTILL

If the rotor of an energized two-pole squirrel-cage induction motor is blocked so that it cannot turn, the rotor conductors cut the flux of the rotating field at a rate that is a function of the supply frequency, since the speed of the rotating field is a function of the supply frequency. The alternating voltages induced in the rotor conductors therefore have the same frequency as the supply voltage, and, if the field flux is sinusoidally distributed in the air gap, the induced voltages are sinusoidal in wave form. The polarities of the rotor voltages for a given position and rotation of the field are shown schematically in Fig. 16–8a. (The ⊕ in the conductor cross section indicates that the polarity of the induced voltage is such as to tend to send current into the paper, away from the reader; the ⊙ in the cross section, the reverse.) In Fig. 16–8a, a single line represents the center, or position of maximum flux density, of the rotating field. (The field is assumed to be sinusoidally distributed on each side of this maximum position.) The rotor conductor that has the maximum induced voltage is therefore the conductor that lies on the line representing the maximum field.

The magnitude and phase position of the currents that flow in the rotor conductors depend upon the impedance of the rotor circuit. This impedance is inductive in nature, since the rotor conductors are imbedded in slots surrounded by a high-permeability material, and is small in magnitude. At standstill, the reactive component of this impedance is larger than the resistive component, being of the order of 4 to 5 times as great in a typical machine. The maximum rotor current therefore lags the maximum rotor voltage by a large power-factor angle.

The rotor power-factor angle is shown as θ_2 in Fig. 16–8b, which shows the current directions in the rotor conductors relative to the position of the rotating field. In Fig. 16–8b, the line I–I represents the position of the conductors carrying maximum current and is drawn θ_2 degrees behind the line representing the center of the rotating field and therefore the position of the conductor having maximum induced voltage. (Since this is a two-pole machine and the field completes one revolution in one cycle of the supply frequency, each degree in space is equal to a degree in time. A similar drawing for a four-pole machine would have the lines separated, in degrees, by one half the power-factor angle.)

Figure 16–8b also shows the directions of the torques developed in the rotor because the rotor conductors carry current while in a magnetic field. As indicated, there are more conductors that have a

INDUCTION MOTOR TORQUE AT STANDSTILL

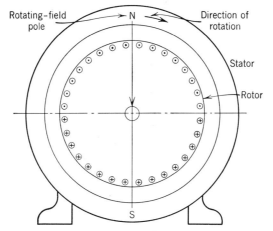

(a) Polarity of induced voltages in rotor

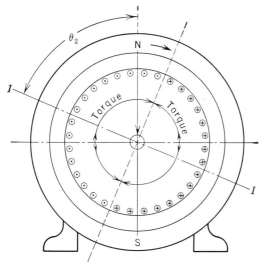

(b) Current and torque directions

Fig. 16–8. Effect of rotating field upon blocked rotor.

developed force in the direction of the rotating field than there are acting in the opposite direction. The net torque on the rotor is in the direction of the rotating field, and this is the direction in which the rotor rotates when unblocked.

The instantaneous torque on a rotor conductor depends upon the current in the conductor and upon the flux surrounding the conductor at the same instant. Since the flux and the rotor current are sinusoidal quantities differing in phase by the angle θ_2, the average torque developed by a rotor conductor may be written

$$T = K\phi I_2 \cos \theta_2 \qquad (16\text{--}2)$$

where ϕ is the air-gap flux, and I_2 is the effective value of current in a rotor conductor. Since the total average torque developed on the rotor is equal to the product of the number of conductors on the rotor and the average torque per conductor, Eq. 16–2 may be applied as the total average torque by assuming that the number of rotor conductors is included in the constant K.

Since

$$\cos \theta_2 = \frac{R_2}{Z_2} = \frac{R_2}{\sqrt{R_2^2 + X_2^2}} \qquad (16\text{--}3)$$

and

$$I_2 = \frac{E_2}{Z_2} = \frac{E_2}{\sqrt{R_2^2 + X_2^2}} \qquad (16\text{--}4)$$

Eq. 16–2 may be written

$$T = \frac{K\phi E_2 R_2}{R_2^2 + X_2^2} \qquad (16\text{--}5)$$

where E_2 is the effective value of voltage induced in a rotor conductor at standstill, R_2 is the resistance of each rotor conductor, and X_2 is the inductive reactance of each rotor conductor at the supply frequency, which is also the frequency of the rotor voltage and current at standstill. For a wound-rotor motor, Eq. 16–5 is equally applicable if E_2, R_2, and X_2 are taken to be per-phase values and if the constant K is considered to be appropriately modified.

16–8. SLIP

As previously indicated, the net developed torque on an induction-motor rotor at standstill acts in the direction of the rotating field, and, if the rotor is free to turn, the rotor rotates in that direction. As the rotor accelerates, however, the rotor conductors cut the field flux at a

EFFECTS OF SLIP

decreasing rate, which results in a reduction in both the magnitude and frequency of the rotor voltage.

If the rotor were to accelerate until it was rotating at synchronous speed, the rotor conductors would cut no flux and, therefore, would have no voltage induced in them, since the rotor and the rotating field would be rotating at the same speed. Under these conditions, there would be no rotor current and, therefore, no developed torque. Since a developed torque is necessary to turn the rotor, even without a load connected to the shaft, the rotor cannot operate at synchronous speed but must reach equilibrium at some speed less than synchronous. The actual speed of the rotor depends upon the torque necessary to turn the rotor and its connected load.

The difference between the synchronous speed of an induction motor and the actual speed of the rotor is called the *slip*. The slip may be expressed in revolutions per minute, as a percentage of the synchronous speed, or as a fraction of the synchronous speed. Thus, in revolutions per minute,

$$\text{Slip} = \text{Synchronous speed} - \text{Actual speed} \tag{16-6}$$

or, expressed as a percentage,

$$\% \text{ slip} = \frac{\text{Synchronous} - \text{Actual}}{\text{Synchronous}} \times 100 \tag{16-7}$$

The slip at full-load torque for typical squirrel-cage induction motors and wound-rotor motors with no external rotor resistance is in the range from 3 to 6 per cent.

16-9. EFFECTS OF SLIP

The effective value of voltage induced in the rotor conductors is directly proportional to the relative velocity of the rotating field with respect to the rotor; therefore, the rotor voltage is directly proportional to the slip. Thus, when the rotor is in a standstill position, the voltage induced in the rotor conductors is a maximum; when the rotor rotates at synchronous speed, the voltage induced in the rotor conductors is zero. The effective value of voltage induced in the rotor conductors at any time is therefore given by the expression

$$E_2' = sE_2 \tag{16-8}$$

where E_2' is the effective value of voltage induced in the rotor at any time, E_2 is the effective value of voltage induced in the rotor at standstill, and s is the slip, expressed as a fraction.

A similar analysis of the effect of slip upon rotor frequency reveals that the frequency of the rotor voltage at any time is given by the expression

$$f_2' = sf_2 = sf_1 \tag{16-9}$$

where f_2' is the frequency of the rotor voltage at any time, f_2 is the frequency of the rotor voltage at standstill (and is the same as the supply frequency f_1), and s is the slip, expressed as a fraction.

The rotor reactance at any time is given by the expression

$$X_2' = 2\pi f_2' L_2 = 2\pi s f_2 L_2 = sX_2 \tag{16-10}$$

where X_2' is the rotor reactance at any time, L_2 is the rotor inductance (assumed constant), and X_2 is the rotor reactance at standstill.

From Eq. 16–2, the average developed torque in the induction motor at any time is given by the expression

$$T' = K\phi I_2' \cos \theta_2' \tag{16-11}$$

where I_2' is the rotor current at any time and θ_2' is the power-factor angle of the rotor at the same time. Equation 16–11 may be written

$$T' = \frac{K\phi E_2' R_2}{R_2^2 + X_2'^2} \tag{16-12}$$

By the substitution of Eqs. 16–8 and 16–10, Eq. 16–12 is converted to

$$T' = \frac{K\phi s E_2 R_2}{R_2^2 + s^2 X_2^2} \tag{16-13}$$

16–10. THE TORQUE–SLIP CURVE

As the load torque connected to an induction motor increases, the developed torque must increase, or the motor stalls. According to Eq. 16–13, if the supply voltage is constant, the only factor that can change to cause the motor to develop an increased torque is the slip. Thus, the ability of a given induction motor to handle a given load torque can be determined from a curve that shows the variation in developed torque with slip, or a torque–slip curve.

The torque–slip curve for a typical squirrel-cage induction motor is shown as Fig. 16–9a. Although the speed characteristics of motors are usually plotted in terms of speed versus load or speed versus torque, as shown in Fig. 16–9b, the torque–slip curve is useful in analyzing the

INDUCTION-MOTOR OPERATION

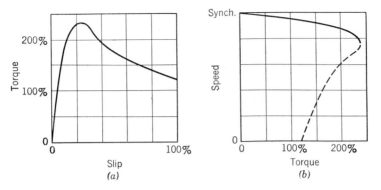

Fig. 16-9. Torque–slip and speed–torque characteristics.

performance of an induction motor because the curve is a plot of Eq. 16-13. (It is to be noted that the abscissa of the torque–slip curve is scaled from 0 to 100 per cent in terms of slip and not in terms of synchronous speed.)

As shown in Fig. 16-9a, the torque–slip curve is essentially linear from 0 slip to a slip slightly beyond that corresponding to full-load torque. The explanation for this linearity lies in the fact that, for small values of slip, the $s^2 X_2^2$ factor in Eq. 16-13 is negligible compared to the R_2^2 term. Thus, for small values of slip, Eq. 16-13 reduces to

$$T' = \frac{K\phi s E_2 R_2}{R_2^2} = \frac{K\phi s E_2}{R_2}$$

or

$$T' = K's \qquad (16\text{-}14)$$

At slips beyond the full-load value, the $s^2 X_2^2$ factor increases rapidly, which accounts for the shape of the curve beyond the full-load point.

The maximum torque developed in an induction motor is called the **pull-out torque** or **breakdown torque** and, when the motor is operated at rated voltage and frequency, is at least twice the full-load value. The maximum torque is developed when the rotor resistance is equal to the rotor reactance, or when R_2 is equal to sX_2.*

16-11. INDUCTION-MOTOR OPERATION

For purposes of discussion, it is assumed that the induction motor has the torque–slip characteristic shown as Fig. 16-9a and that a constant

---

* The proof of this statement is left as a problem at the end of the chapter.

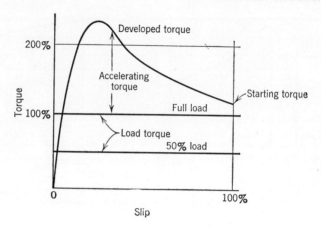

Fig. 16–10. Torque–slip characteristics.

rated-full-load torque is connected to the shaft. The torque–slip characteristics of the motor and the load are shown as Fig. 16–10.

At the instant of starting, the developed torque of the motor is greater than the resisting torque; therefore, the motor accelerates. As the slip decreases, the torque available for acceleration increases until the slip that corresponds to maximum developed torque is reached. Then, the accelerating torque decreases, but the motor continues to accelerate until the developed torque is equal to the resisting torque. At torque equilibrium, represented by the intersection of the motor torque–slip curve and the load torque–slip curve, the motor rotates at constant speed.

If the load on the motor is decreased, the motor accelerates until the developed torque decreases to that necessary to drive the load at constant speed, as shown in Fig. 16–10.

If the load on the motor is increased, the motor must slow down until it again develops a torque equal to the resisting torque. If the increase in load torque is such that the load torque exceeds the maximum developed torque, the motor slows down and stops.

16–12. INDUCTION-MOTOR RATING

The nameplate of a three-phase induction motor provides the following information: (*a*) horsepower, (*b*) line voltage, (*c*) line current, (*d*) speed, (*e*) frequency, and (*f*) temperature rise. The horsepower rating is the mechanical output of the motor when it is operated at rated line voltage, rated frequency, and rated speed. Under these conditions, the line current is that specified on the nameplate, and the temperature rise does not exceed that specified.

The speed given on the nameplate is the actual speed of the motor at rated full load; it is not the synchronous speed. Thus, the nameplate speed of a given 60-cycle motor might be listed as 1710 rpm. This speed rating indicates that the synchronous speed is 1800 rpm and that the motor has four poles, since the rated full-load slip for an induction motor is of the order of 3 to 6 per cent.

16–13. INDUCTION-MOTOR CURRENT

Thus far, no reference has been made to the effect upon the stator circuit of an increased shaft load, although, certainly, an increased mechanical output must be accompanied by an increased electric input. The manner in which the electric input increases is similar to the action of a transformer under increased load conditions. Thus, upon an increase in shaft load, the rotor current, which corresponds to the secondary current of a transformer, increases because of the increased slip. The increased rotor mmf, which opposes the stator mmf, causes the induced voltage of the stator to decrease. The stator current then increases until the rotating field, which corresponds to the mutual flux of a transformer, is nearly equal to its original value.

If the rotor of an induction motor is rotated at synchronous speed by some form of prime mover, there is no voltage induced in the rotor and, therefore, no rotor current to react with the revolving field. This condition compares with the operation of a transformer with its secondary winding open-circuited; the only current flowing in the stator winding is that necessary to produce the flux of the rotating field and to supply the stator core losses and copper losses. Because of the air gap in the magnetic circuit, this exciting current is larger than that of a transformer of the same voltage and current rating.

The current taken by the stator winding of a motor at no load may be considered to consist of two components—one, the exciting current discussed in the preceding paragraph, and the other, a component necessary to provide the power to keep the rotor turning when the prime mover considered in the preceding paragraph is removed. The power provided by the second component of current therefore represents the friction losses of the rotor and the copper losses in stator and rotor due to this component of current. The vector sum of these two components of current represents the no-load current of the motor and may range from

20 to 30 per cent of rated current. The no-load power factor is of the order of 10 to 20 per cent.

The stator current at any load other than no load may be considered to be the vector sum of a component representing the no-load current and a component representing the increase in rotor current because of the increased load.

16–14. STARTING CURRENT AND TORQUE

Upon starting, the voltage induced in an induction-motor rotor is a maximum, and, since the rotor impedance is low, the rotor current may be excessively large. This large current is reflected in the stator, because of the transformer action, and results in a stator current of the order of 4 to 10 times the normal full-load current. Because of the short duration if the motor accelerates normally, this value of current does not harm the motor, but it may cause an objectionable drop in the supply voltage at the motor terminals because of a large impedance drop in the supply lines. In many installations, it is necessary to reduce the magnitude of the current upon starting, and several methods are applicable.

If it is assumed that the rotor impedance at standstill is constant, the rotor current is proportional to the rotor voltage. The rotor voltage, in turn, is proportional to the stator voltage. Since the stator current is directly related to the rotor current, the stator current upon starting is therefore directly proportional to the stator voltage. Thus, reducing the stator voltage upon starting reduces the starting current in the same proportion.

A reduction in stator voltage, however, also affects the starting torque. The torque at starting is normally in the range from 1.1 to 1.5 times the full-load torque and is given by the expression

$$T = \frac{K\phi E_2 R_2}{R_2^2 + X_2^2} \qquad (16\text{--}15)$$

The rotor voltage E_2 as previously indicated, is proportional to the stator voltage; the flux ϕ of the rotating field is also proportional to the stator voltage. The starting torque is therefore proportional to the square of the stator voltage. Thus, any plan to reduce the stator voltage in order to reduce the starting current must also take into account the effect on the starting torque.

The starting current of a wound-rotor motor can also be reduced by adding resistance to the rotor circuit, thus increasing the impedance of the rotor circuit. This method is not applicable to squirrel-cage motors

STARTING CURRENT AND TORQUE 329

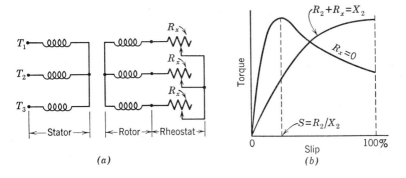

Fig. 16-11. Wound-rotor–motor torque characteristics.

because of the manner in which the squirrel-cage motor is constructed.

The effect on starting torque of adding resistance to the rotor circuit depends on the amount of resistance added. Since the maximum torque of an induction motor occurs when the rotor resistance is equal to the rotor reactance, or when R_2 is equal to sX_2, adding enough external resistance to make the total rotor resistance equal to the rotor reactance at standstill causes maximum torque to be developed upon starting. The effect on the torque–slip curve of the motor of adding rotor resistance is shown in Fig. 16-11b. As indicated in Fig. 16-11b, the addition of enough resistance to make the rotor resistance equal to the rotor reactance at standstill has shifted the torque–slip curve so that the maximum value of torque occurs at standstill.

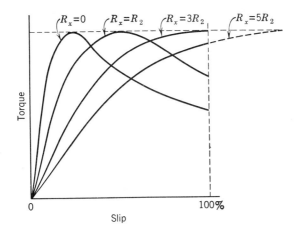

Fig. 16-12. Effect of various values of external rotor resistance upon torque–slip characteristics.

The addition of resistance to the rotor circuit does not change the maximum value of torque but does change the value of slip at which the maximum torque occurs. Figure 16–12 shows the effect on the normal torque–slip curve of a wound-rotor motor of adding three different values of external resistance. The curves indicate that the maximum torque is unchanged and that the starting torque depends upon the amount of resistance added. (The machine for which the curves of Fig. 16–12 are drawn has a rotor-winding reactance that is equal to four times the rotor resistance at standstill.)

16–15. STARTING METHODS

The method to be employed in starting a given induction motor depends upon the size of the motor, the type of motor, and the nature of the voltage supply to which the motor is to be connected. The voltage supply must be considered because of the large value of current taken by the induction motor upon starting; if the source supplies other loads as well as the motor, and if the other loads might be affected by the drop in line voltage caused by the large current flowing in the line conductors, an effort must be made to reduce the starting current taken from the line. If the reduction in supply voltage because of line drop is not a serious problem, the motor may be connected directly to the line.

The common methods used to start induction motors are: (*a*) across-the-line, (*b*) starting compensator, and (*c*) rotor rheostat.

Methods (*a*) and (*b*) are applicable to both squirrel-cage and wound-rotor motors; method (*c*) applies only to the wound-rotor motor.

Across-the-Line Starting. The across-the-line method of starting is just what the name implies—the motor is started by connecting it directly to the line. The starting current is then 4 to 10 times the full-load current, and the starting torque is 1.1 to 1.5 times the full-load torque. The devices used to connect the motor to the line range from a three-pole knife switch, manually operated, to an automatic push-button-controlled contactor.

Starting Compensator. The starting compensator is a device that connects the motor to a reduced-voltage supply upon starting, and then connects it to the full-voltage supply after it is running. The purpose of a starting compensator is to reduce the starting current taken by the motor by reducing the supply voltage at the motor terminals. As previously indicated, the amount of voltage reduction permissible depends upon the starting-torque requirements as well as the desired starting current.

STARTING METHODS

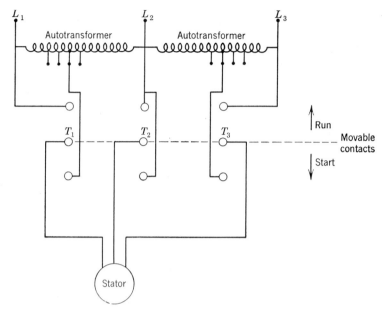

Fig. 16–13. Connection of starting compensator.

A schematic diagram for the connection of an induction motor to a starting compensator is shown as Fig. 16–13. The starting compensator consists of two step-down autotransformers, open-delta-connected, and a control device that connects the motor either to the output of the transformers or directly to the line.

When the motor is to be started, the movable contacts, to which the motor leads are connected, are moved in the *start* direction. This operation connects the motor to the reduced-voltage supply. After the motor has started, the movable contacts are moved in the *run* direction, connecting the motor to the line. The compensator is provided with a holding-coil circuit that maintains the movable contacts in the run position until the holding coil is de-energized.

Rotor Rheostat Starting. The rotor rheostat method of starting consists in connecting resistance in each phase of the rotor circuit of a wound-rotor motor so as to reduce the starting current. As previously indicated, proper selection of the external resistance results in maximum torque at starting. After the motor has started, the resistance is reduced until the motor is operating with no external resistance in its rotor circuit. The connection of a rotor rheostat to a wound-rotor motor is shown schematically in Fig. 16–11a.

16-16. SPEED CONTROL

Induction motors fail to meet the requirements of many motor applications because the speed of an induction motor cannot be conveniently adjusted or controlled. Factors that can be changed to adjust the speed of the motor are the frequency, the number of poles, and, in the case of the wound-rotor motor, the rotor resistance.

Changing the frequency is not a convenient or practical method of adjusting the speed since, in most applications, the supply frequency is fixed and a variable-frequency source that can be applied to the motor is not available. If a variable-frequency source is available, it must be one that maintains the ratio of voltage to frequency constant at frequencies below the rated frequency; if this ratio increases, the exciting current taken by the motor may become excessive.

Induction motors that have their stators so wound that the number of poles can be changed are available commercially. Such motors usually have but two combinations of poles, the number of poles for the two combinations having the ratio of 2 to 1. Thus, a motor may have its stator so wound that by changing connections it has either 4 or 8 poles, 6 or 12 poles, etc. Such an arrangement does not provide for fine speed adjustment or control, but it serves many applications where a coarse adjustment is all that is necessary.

As previously indicated, the addition of external resistance to the rotor circuit of a wound-rotor motor shifts the torque–slip curve so that the slip is greater for a given torque. The addition of rotor resistance, therefore, serves not only to reduce the starting current but also to adjust the speed of the motor. This method has the disadvantage, however, of introducing an additional, and in some cases considerable, power loss in the external resistor.

For a given torque, the change in slip that occurs when resistance is added to the rotor circuit of a wound-rotor motor is such that the ratio of slips is equal to the ratio of resistances: that is, if the total rotor resistance is doubled, the slip is doubled; if the resistance is tripled, the slip is tripled, etc. This fact may be shown by substituting the new value of resistance in the torque equation and solving for the new slip. Thus, if the rotor resistance is doubled and the resisting torque remains constant,

$$T = \frac{K\phi s E_2 R_2}{R_2^2 + s^2 X_2^2} = \frac{K\phi s' E_2 (2R_2)}{4R_2^2 + s'^2 X_2^2} \qquad (16\text{--}16)$$

SPEED CONTROL

Solving Eq. 16–16 for the new slip s' gives

$$s' = 2s$$

or the slip is doubled when the rotor resistance is doubled.

PROBLEMS

16–1. Prove that two equal alternating magnetic fields that are 90 degrees apart in space and 90 degrees apart in time form a constant rotating field having a magnitude equal to one of the original fields.

16–2. What is the maximum speed at which the magnetic field of a three-phase 60-cycle induction motor can rotate? *Ans.* 3600 rpm.

16–3. A 40-hp 220-volt 60-cycle three-phase squirrel-cage induction motor has a full-load speed of 1710 rpm. (In the following calculations, neglect the no-load slip, and consider that the developed torque is equal to the output torque.)

(*a*) What are the slip and the torque at full load? *Ans.* 5%; 123 lb-ft.

(*b*) At what speed does this motor operate to develop one-half full-load torque? *Ans.* 1755 rpm.

(*c*) At approximately what speed does this motor operate to deliver 10 hp? *Ans.* 1778 rpm.

(*d*) At what speed does this motor operate if it is connected to a constant torque load of 100 lb-ft? *Ans.* 1727 rpm.

16–4. Prove that the maximum torque is developed in an induction motor when R_2 is equal to sX_2.

16–5. Explain completely the effect of driving an energized induction motor beyond synchronous speed.

16–6. (*a*) What is the significance of a negative slip?

(*b*) What is the significance of a slip greater than 100%?

16–7. Is it possible for a squirrel-cage induction motor to have a steady-state operating point to the right of the maximum torque point on the torque–slip curve? Explain.

16–8. The nameplate data for a wound-rotor induction motor are as follows: 25 hp, 220 volts, 66 amperes, three-phase, 60 cycles, 855 rpm. The resistance measured between two terminals of the wye-connected rotor is 0.20 ohm. A similar measurement between two terminals of the wye-connected stator indicates 0.28 ohm. A load test of the motor indicates that the breakdown torque is reached at a speed of 675 rpm.

(*a*) What torque, in pound-feet, is developed in this motor at starting? *Ans.* 188 lb-ft.

(*b*) What is the full-load efficiency of this motor (assuming a full-load power factor of 85%)? *Ans.* 87%.

16–9. A 15-hp 440-volt three-phase induction motor has a starting torque equal to 1.5 times the full-load torque when it is started with rated voltage applied to the stator. Under the same conditions, the starting current is 4.5 times the full-load current.

(a) Calculate the starting torque and the starting current, in terms of full-load values, if the motor is started at 80% of rated voltage.

Ans. 96% T_{FL}; 360% I_{FL}.

(b) What starting voltage must be applied to make the starting torque equal to the full-load torque? What is the starting current when this voltage is applied? *Ans.* 359 volts; 367% I_{FL}.

(c) What starting voltage must be applied to make the starting current equal to the full-load current? What starting torque is developed at this voltage? *Ans.* 97.8 volts; 7.4% T_{FL}.

16–10. A three-phase wound-rotor induction motor is to be used to drive a fan. The nameplate data of the motor are as follows: 10 hp, 220 volts, 60 cycles, 1712 rpm. The resistance per phase of the rotor circuit is found to be 0.22 ohm. The motor and fan combination rotate at 1723 rpm when the rotor terminals of the motor are short-circuited, but a lower speed is desirable. If a speed of 1400 rpm is satisfactory, how much resistance should be added to each phase of the rotor circuit to produce this speed? (Torque of the fan is proportional to the square of the speed.) *Ans.* 1.51 ohms.

16–11. Assume that a squirrel-cage induction motor is operating at full-load torque driving a constant-torque load, and the applied voltage is reduced to one half. Completely explain the effect on the operation of the motor.

16–12. A certain squirrel-cage induction motor is designed to operate with its stator coils delta-connected, but provision is made to switch to a wye connection.

(a) How will the starting torque when wye-connected compare with that when delta-connected, if the line voltage is the same for each case?

Ans. $T_y = \frac{1}{3}T_D$.

(b) How is the line current at starting affected?

Ans. $I_Y = \frac{1}{3}I_D$.

16–13. Why is it more desirable to use a starting compensator than inductors in series to reduce the applied voltage when starting an induction motor?

16–14. Briefly, explain how it is possible to determine the speed of a wound-rotor induction motor by measuring the voltage at the rotor terminals with a d-c voltmeter.

16–15. A wound-rotor motor has a rotor reactance at standstill equal to 5 times its rotor resistance. If an external resistor, equal in magnitude to the resistance of the rotor, is added to each phase of the rotor circuit, how are the starting torque and starting current affected, quantitatively?

Ans. $T_{S2} = 1.8T_{S1}$; $I_{S2} = 0.94I_{S1}$.

16–16. The nameplate data for a wound-rotor induction motor are as follows: 25 hp, 220 volts, 60 amperes, three-phase, 60 cycles, 855 rpm. The resistance measured between two terminals of the wye-connected rotor is 0.20 ohm. It is proposed that the speed of the load to which this motor is connected be maintained constant at the full-load value from one-fourth full load to full load by varying the resistance in the rotor circuit. What should be the resistance of each phase of a wye-connected rotor rheostat to satisfy the requirements? *Ans.* 0.30 ohm.

16–17. A 40-hp 220-volt 60-cycle three-phase wound-rotor induction motor has a full-load speed of 1710 (with the wye-connected rotor terminals

PROBLEMS

short-circuited). The rotor reactance per phase at standstill is equal to four times the rotor resistance per phase. If an external resistor with a resistance value three times that of the rotor is connected to each phase of the rotor, how will each of the following be affected, quantitatively? (Sufficient work should be shown in each part to justify the answer given.)

(a) Starting torque (in terms of normal starting).
(b) Starting current (in terms of normal starting).
(c) Stator current at full-load torque (in terms of normal).
(d) Speed at full-load torque (in revolutions per minute).
(e) Frequency of rotor current (in cycles per second).

16–18. The solution of Eq. 16–16 for s' involves a quadratic equation in which one of the roots is found to be $2s$. What is the significance of the second root?

16–19. As the chief engineer of a certain company, you find on your desk a wiring diagram submitted by a subordinate. The diagram shows a wound-rotor induction motor with external capacitance in each phase of the rotor circuit. No explanation is submitted with the diagram. What is evidently the purpose of the connection? Comment on the proposal.

16–20. Draw a connection diagram showing a push-button control circuit applied to an automatic across-the-line induction-motor starter.

16–21. In terms of the normal starting current, what current is taken from the line when an induction motor is started by means of a starting compensator that reduces the motor voltage to 85% of the line value?

Ans. 72%.

16–22. A given wound-rotor induction motor operates at one-half load driving a constant-torque load. What is the effect upon the motor current of inserting a resistor in each phase of the rotor circuit so that the resistance per phase is doubled? Explain.

16–23. What effect does undervoltage operation have upon the temperature of a motor at full-torque load? At no load? Explain.

17 Three-Phase Synchronous Machines

17-1. INTRODUCTION

A **synchronous machine** is an a-c machine that normally operates at synchronous speed. Alternating-current machines that do not normally operate at synchronous speed are called **asynchronous machines.**

A synchronous generator, or an **alternator,** is a synchronous machine that transforms mechanical energy into electric energy. A synchronous motor is a synchronous machine that transforms electric energy into mechanical energy.

The synchronous machine is comparable to the d-c machine in that both have an armature winding and a field winding. In a synchronous machine, the armature winding is an a-c winding; that is, it is connected to or produces an a-c supply. The field winding is connected to a d-c supply.

Although the construction of a synchronous machine may be similar to that of a d-c machine, the synchronous machine usually has the armature winding on the stator and the field winding on the rotor. Reasons for this type of construction are:

(a) A stationary armature is more easily insulated for the high voltage for which the synchronous machine is usually designed.

(b) Only two, relatively small slip rings are required as connections to the d-c rotor circuit.

The fixed-armature rotating-field-pole type of construction is the only one considered in this book. The principles to be discussed, however, apply also to the rotating-armature fixed-pole type of construction.

STATOR CONSTRUCTION

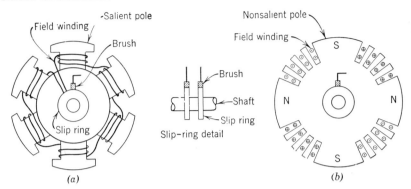

Fig. 17-1. Salient-pole and nonsalient-pole rotors.

17-2. ROTOR CONSTRUCTION

The field-pole construction of the rotor of a synchronous machine may be one of two types—**salient** or **nonsalient**. The salient-pole type of construction is common to synchronous machines having six or more poles; the nonsalient construction, to alternators having two or four poles.

The salient-pole rotor has field poles that are bolted to a cylinder on the shaft, as shown schematically in Fig. 17-1a. The individual field-pole windings are connected in series to form a single winding that is energized by the d-c source through brushes on the slip rings that serve as the terminals for the winding.

The nonsalient-pole rotor is so constructed that it forms a smooth cylinder, as shown schematically in Fig. 17-1b. The field windings are embedded in slots milled into the solid cylinder, and, as in the salient-pole rotor, they are connected in series to the slip rings through which they are energized.

17-3. STATOR CONSTRUCTION

The stator and stator winding of a synchronous machine are identical with the stator and stator winding of a three-phase induction motor. Thus, the stator is built up of sheet-steel laminations that have slots on their inner periphery. These slots are filled with insulated conductors connected in groups to form a balanced three-phase wye or delta circuit.

The stator winding of a synchronous machine is so wound that the number of rotating-field poles that the winding produces when ener-

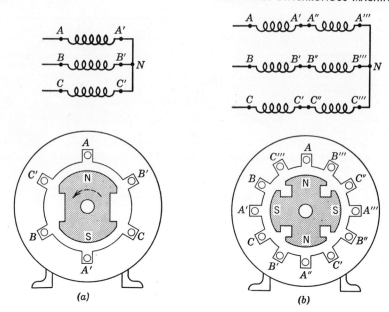

Fig. 17-2. Simplified two-pole and four-pole synchronous machines.

gized by the proper a-c supply is equal to the number of poles on the rotor. Stator windings for simplified two-pole and four-pole wye-connected synchronous machines are shown in Fig. 17-2. Each stator conductor of the simplified machines represents a group of conductors, and the stator conductors are connected into coils as shown schematically in the connection diagrams of Fig. 17-2.

17-4. THE SYNCHRONOUS GENERATOR

If the rotor winding of the two-pole machine shown in Fig. 17-2a is energized from a d-c supply, and the rotor is then rotated, alternating voltages are induced in the stator conductors. For a sinusoidal flux distribution, the voltages induced in the stator conductors are sinusoidal. The phase relationship among the voltages induced in the three stator coils for counterclockwise rotation of the rotor is shown in Fig. 17-3. (It is assumed that the stator circuit is open: that is, that the terminals of the stator winding are not connected to a source or to a load.)

Since the alternating voltage of a given stator conductor completes one cycle for one revolution of the rotor, the frequency of the voltage induced in cycles per second is equal to the speed of rotation of the

SYNCHRONOUS-GENERATOR OPERATION

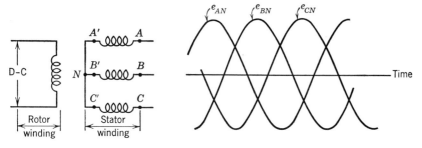

Fig. 17-3. Phase relationships among stator voltages.

rotor in revolutions per second. An analysis of the stator voltages induced in the four-pole machine of Fig. 17-2b reveals that one cycle of the alternating voltage corresponds to one half of a revolution of the rotor, or that the frequency is twice the speed of rotor rotation.

The expression relating frequency, poles, and speed of rotation is

$$f = \frac{P}{2} \times \frac{n}{60} = \frac{Pn}{120} \qquad (17\text{-}1)$$

where f is the frequency in cycles per second, P is the number of poles, and n is the speed of the rotor in revolutions per minute. (Equation 17-1 is identical with Eq. 16-1, which gives the speed of the rotating field of an induction motor in terms of the number of stator poles and the supply frequency.)

The magnitude of the voltage induced in each stator phase depends upon the rotor field flux, the number and position of the conductors in the phase, and the speed of the rotor. For a given machine, the effective value of the induced, or generated, voltage is given by the expression

$$E_g = K\phi f \qquad (17\text{-}2)$$

or

$$E_g = K\phi n \qquad (17\text{-}3)$$

where E_g is the generated voltage in a stator phase, K is a constant for the given machine, ϕ is the rotor field flux, f is the frequency of the stator voltage, and n is the speed of the rotor.

17-5. SYNCHRONOUS-GENERATOR OPERATION

When the rotor is rotating and energized, a voltage is generated in each stator phase, and the circuit diagram for the generator with its stator circuit open may be represented as shown in Fig. 17-4a. The

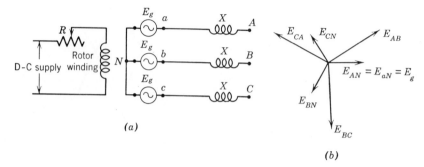

Fig. 17–4. Synchronous generator at no load.

rheostat in the rotor circuit is for changing the field current and, therefore, the generated voltage. Each phase of the stator circuit is represented by an alternating-voltage source in series with an inductor, which represents the stator reactance.* (The resistance of the stator is usually so small compared to the stator reactance that it is reasonable to neglect the effect of stator resistance and to consider only the stator reactance.)

With the stator circuit open, the magnitude of the voltage across each stator phase is equal to the generated voltage of that phase. Thus, for the circuit of Fig. 17–4a,

$$|E_{AN}| = |E_{BN}| = |E_{CN}| = |E_g| \qquad (17\text{–}4)$$

The magnitude of the voltage between terminals of the stator is equal to $\sqrt{3}$ times the phase voltage. Thus,

$$|E_{AB}| = |E_{BC}| = |E_{CA}| = |\sqrt{3}\, E_g| \qquad (17\text{–}5)$$

The vector diagram showing the phase relationship of the phase voltages and line voltages of the generator is drawn as Fig. 17–4b.

If a three-phase load is connected to the stator terminals, as shown in Fig. 17–5, the magnitude of the voltage across each phase is no longer equal to the generated voltage but differs from it by the drop across the reactance. Thus, for the A phase,

$$\overline{E_{AN}} = \overline{E_{aN}} - \overline{I_{NaA}X} \qquad (17\text{–}6)$$

and is therefore dependent upon the magnitude and phase angle of the

---

* The value of stator reactance referred to here is larger than the value of stator leakage reactance in order to take into consideration the effects of armature reaction; that is, the effects of the mmf produced by the armature current. In practice, this larger value of reactance is referred to as synchronous reactance. Since the discussion of synchronous machines in this book is to be qualitative, a detailed analysis of the determination of the synchronous reactance is ommitted.

SYNCHRONOUS-GENERATOR OPERATION

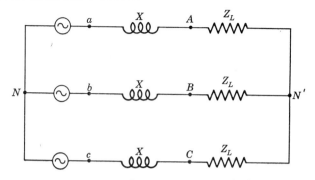

Fig. 17–5. Synchronous generator with load.

stator current. For a balanced load, the stator current is given by the relation

$$I_{NaA} = \frac{E_{aN}}{jX + (R_L \pm jX_L)} \tag{17-7}$$

where $(R_L \pm jX_L)$ represents the impedance of the load. Rearranging Eq. 17–7 results in Eq. 17–6 when it is recognized that

$$I_{NaA}(R_L \pm jX_L) = E_{AN'} = E_{AN} \tag{17-8}$$

The vector diagrams of Fig. 17–6 are drawn for loads having unity power factor, lagging power factor, and leading power factor to show the effect of load power factor on the phase voltage. For these diagrams, the generated voltage is assumed to be constant, and the magnitude of the load impedance is so modified that the stator current is the same for each. (The vector diagrams are drawn for only one stator phase, because the load is assumed to be balanced, and therefore the vector diagrams for the other two phases are similar to the one indicated. The angle θ in Figs. 17–6b and 17–6c is the power-factor angle of the load impedance.) The diagrams reveal that, for a given stator current, the more lagging the power factor, the smaller the phase voltage. A leading

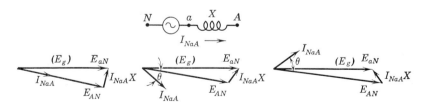

Fig. 17–6. Vector diagrams for unity, lagging, and leading power-factor loads.

power factor, on the other hand, causes the voltage across the phase to increase.

Application of the double-subscript notation for power derived in Art. 7–2 to the three vector diagrams of Fig. 17–6 reveals that, since the product of $E_{AN'}$, I_{NaA}, and the cosine of the angle between them is negative for each example, the stator phase is a source of electric energy in each example. Since the other two phases have similar vector diagrams, the machine acts as a synchronous generator. (The diagrams of Fig. 17–6 are drawn for the machine acting as a generator; the application of the double-subscript power convention is to illustrate the convention.)

17–6. GENERATOR-TO-MOTOR CONVERSION

The complete circuit diagram and the vector diagram for a synchronous generator supplying a balanced three-phase resistance load are shown as Fig. 17–7.

If each phase of the load shown in Fig. 17–7a is paralleled by an a-c source that has a terminal voltage of the same magnitude, phase position, and frequency as the voltage across the particular phase, no change occurs in the operation of the generator; that is, the current delivered by each phase of the generator has the same magnitude as before the individual sources are connected, and this current still flows to the resistors. Further, if the magnitude, frequency, and phase position of the voltages of the individual sources are assumed to be constant and unaffected by current flow, and the sources are assumed to be capable of receiving power, the resistors may be removed without affecting the

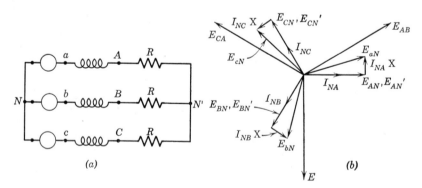

Fig. 17–7. Synchronous generator with a balanced load.

GENERATOR-TO-MOTOR CONVERSION

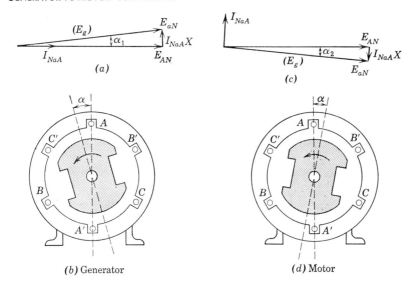

Fig. 17–8. Rotor positions when terminal voltage is maximum.

operation of the generator. With the resistors removed, the power output of the generator is taken by the individual sources, but the current and power of the generator remain the same as they were when the resistors were connected, and therefore the vector diagram of Fig. 17–7b is still applicable. Also, with the resistors removed, the individual sources constitute a three-phase wye-connected source connected to the terminals of the original generator.

Because the voltages of the three-phase source that replaces the load are constant in magnitude and frequency, the vectors representing these voltages are constant in length and constant in rotational velocity. The vectors representing the generated voltages, however, have lengths dependent upon the rotor field flux and the speed of the rotor, and rotational velocities dependent upon the speed of the rotor. Because of these facts concerning the sets of vectors, the vector can be used to indicate what is happening physically to the rotor. Thus, the vector diagram for the A phase of Fig. 17–7 (redrawn as Fig. 17–8a) shows that the generated voltage vector E_{aN} reaches its maximum α degrees before the phase voltage, E_{AN}. Therefore, at the instant that the phase voltage is a maximum, the rotor pole has gone α degrees beyond the plane of the coil representing the phase, as shown in Fig. 17–8b, because the generated voltage is maximum for that phase when the center line of the rotor coincides with the plane of the coil. Thus, the position of the

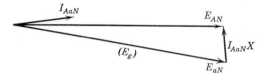

Fig. 17-9. Synchronous–motor vector diagram.

rotor with respect to the coil at the instant the phase voltage is a maximum corresponds to the position of the generated-voltage vector with respect to the phase-voltage vector. This fact is helpful in analyzing synchronous-machine operations.

If the power input to the prime mover driving the synchronous generator is decreased, the rotor decelerates. Vectorially this has the effect of causing the generated-voltage vector in Fig. 17–8a to drop back and, in a short time if the power is sufficiently decreased, to fall behind the phase-voltage vector E_{AN}, which is supplied by the external source and is constant in frequency. The vector diagram that applies under these conditions is as drawn in Fig. 17–8c; the position of the rotor when the phase voltage is maximum is then as shown in Fig. 17–8d. The vector diagram of Fig. 17–8c is drawn for the same equation as that in Fig. 17–8a; thus, the current subscripts are the same, but the position of the current vector indicates that at a given instant in time the current direction is reversed from that shown in Fig. 17–8a.

Application of the double-subscript power convention to the vector diagram of Fig. 17–8c reveals that the average power for the phase is positive; therefore, the machine is a receiver of electric energy and is a synchronous motor. The power received by the motor is converted to mechanical power to maintain the rotation of the rotor; thus, the vector representing the generated voltage in Fig. 17–8c lags the vector representing the applied voltage by an angle dependent upon the power necessary to maintain the rotation at synchronous speed.

The vector diagram of Fig. 17–8c is redrawn as Fig. 17–9 with both the position and the order of subscripts of the current vector reversed. The corresponding alteration of the voltage equation represented by the vector diagram is

$$E_{AN} = E_{aN} + I_{AaN}X \tag{17-9}$$

17-7. SYNCHRONOUS-MACHINE TORQUE

When the synchronous machine is a generator, the torque developed within the machine opposes the rotation; when the machine is a motor,

SYNCHRONOUS-MACHINE TORQUE

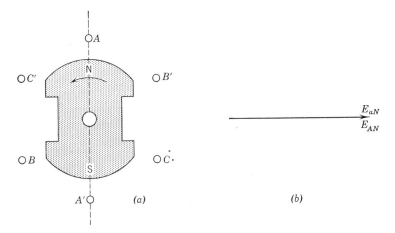

Fig. 17–10. Rotor position when generated voltage and terminal voltage are in phase.

the developed torque produces the rotation. The magnitude of the developed torque is a function of the current in the stator and the flux produced by the rotor.

If a synchronous generator is so connected to a voltage source that the generated voltage is equal to and in phase opposition to the applied voltage, no current flows in the stator winding. The position of the rotor corresponding to the maximum value of the generated and applied voltages is as shown in Fig. 17–10a; the vector diagram is shown in Fig. 17–10b.

An increase in the power input to the prime mover tends to increase the speed of rotation of the rotor; this is equivalent to the generated-voltage vector's moving ahead of the applied-voltage vector (since the applied voltage is assumed to have a constant frequency). When the generated-voltage vector moves ahead, current flows in the stator winding, since there is now a difference between the generated voltage and the applied voltage, and since any difference in voltage must be applied to the reactance of the winding. The vector diagram applying to the machine after the increase in power input to the prime mover is therefore as shown in Fig. 17–11a; the position of the rotor corresponding to the maximum value of the applied voltage is shown in Fig. 17–11b (the axis of the rotor field is beyond the plane of the coil at the instant that $E_{AN'}$ is a maximum).

The force developed on the rotor (because the stator conductor, which is fixed, carries current while in the magnetic field produced by the rotor) is in a direction to oppose the rotation of the rotor, as indicated

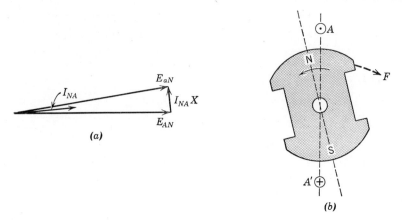

Fig. 17–11. Electromagnetic force developed on rotor of synchronous generator.

in Fig. 17–11b. The torque developed by the force depends upon the position of the rotor with respect to the stator conductors, as well as the stator current and the rotor flux. Therefore, when the power input to the prime mover is increased, the vector corresponding to rotor position moves ahead of the applied-voltage vector until the torque developed in the machine equals the increased torque applied to the shaft of the machine. When the generated-voltage vector reaches the torque equilibrium position, the rotor continues to rotate at synchronous speed.

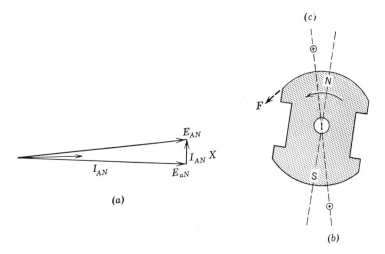

Fig. 17–12. Electromagnetic force developed on rotorosynchronous motor.

A MECHANICAL ANALOGY

If the prime mover is removed, the rotor decelerates. The generated-voltage vector therefore drops behind the applied-voltage vector, as shown in Fig. 17–12a; the position of the rotor corresponding to the maximum value of the applied voltage is shown in Fig. 17–12b.

In Fig. 17–12b, the force developed by the current in the stator conductors and the flux of the rotor produces a torque in the direction of the original rotation. Therefore, the vector corresponding to rotor position drops back of the applied-voltage vector until the torque developed is sufficient to supply the torque requirements of the machine (in this example, the friction losses). The rotor then continues to rotate at synchronous speed.

17–8. A MECHANICAL ANALOGY

A mechanical analogy is applicable, and helpful, in an analysis of the operation of a synchronous machine. The basis for this analogy is the fact that a rotating field is produced in the stator if only the stator winding is energized. If it is assumed that the external supply produces such a rotating field when the rotor is energized and rotating, then the poles of the rotating field of the stator so react with the poles of the rotor that the rotor operates at the same speed as the rotating field, as shown schematically in Fig. 17–13a.

A mechanical analog for the synchronous machine is shown schematically as Fig. 17–13b. The components of this analog are: the member A, which is assumed to be driven by a constant-speed prime

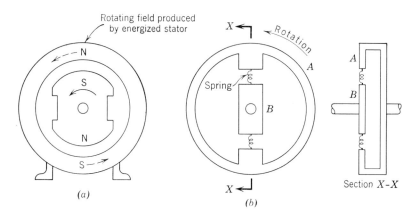

Fig. 17–13. Mechanical analogy for a synchronous motor.

mover; the member B, which is connected to a shaft to which a prime mover or a mechanical load can be connected; and the springs that connect member A to member B. Member A corresponds to the rotating field produced by the constant-frequency stator supply, member B corresponds to the d-c excited rotor, and the tension in the springs corresponds to the attractive force between the rotating-field poles and the poles of the rotor.

When member B is driven by a prime mover at the same speed as member A, the central axes of the two members coincide, as shown in Fig. 17–13b. For this condition, there is tension in the springs, but, since the force is radial, no torque is exerted on either member.

If the power input to the prime mover of B is increased, B tends to rotate faster than A. However, at the instant that B moves ahead of A, the tension in the springs increases and acts to develop a restraining torque on B (member A rotates at constant speed). Therefore, B moves ahead until the restraining torque developed by the springs equals the torque developed by the increased power input to the prime mover; B then rotates at the same speed as A but has its central axis ahead of that of A. Thus, the action of the mechanical device is analogous to the operation of a synchronous generator when the power input to the prime mover is increased.

If the prime mover of B is disconnected, B tends to decelerate. However, when the central axis of B falls behind that of A, the tension in the springs increases and produces a torque on B to keep it rotating. Therefore, B drops behind A until the torque developed by the springs is sufficient to drive B at the same speed as A. This action is comparable to the operation of a synchronous motor.

If a mechanical load is placed on its shaft, B tends to slow down. In doing so, its central axis lags farther behind that of A, and the tension in the springs increases. The increased angle between the central axes and the increased tension in the springs cause the torque developed on B to increase. When the developed torque is equal to the retarding torque caused by the mechanical load, B maintains its position with respect to the central axis of A and continues to rotate at the same speed as A. Since the mechanical output of B is increased, the power input to the prime mover of A has to be increased to maintain the speed of A constant.

A further increase in the mechanical load on the shaft of B causes the angle between the two axes to increase further. If the angle between the two axes becomes too great, however, the springs may be stretched beyond their elastic limit and may break, causing B to slow down and stop.

17–9. SYNCHRONOUS-MOTOR OPERATION

The operation of a synchronous motor is similar to that of the mechanical analog. Because of the attractive force between poles of opposite magnetic polarity, the synchronous motor rotates at the same speed as the rotating field produced by the stator. If the rotor poles coincide with the stator poles, no torque is developed on the rotor. Therefore, in order for torque to be developed, the rotor poles must lag behind the stator poles.

At no load, the only torque necessary to turn the rotor is that to supply the rotational losses; therefore, the rotor poles lag only slightly behind the stator poles at no load. An increase in the mechanical load on the rotor shaft causes the rotor to change position with respect to the rotating field until the developed torque equals the retarding torque caused by the load.

If the retarding torque placed on the rotor shaft is greater than the torque that can be developed, the rotor slows down and stops. The load torque that causes the motor to stop is called the **pull-out, or breakdown, torque.**

The effect of mechanical load upon the electric input to a synchronous motor can be determined by an examination of the vector diagram applying to the motor, if it is assumed that the vector for the supply voltage represents the position of the rotating field, and that the vector for generated voltage represents the position of the rotor. The vector diagram that applies to a synchronous motor at no load is shown in Fig. 17–14a.

An increase in the mechanical load causes the rotor and, therefore, the generated-voltage vector to drop back, increasing the angle between $E_{AN'}$ and $E_{aN'}$, as shown in Fig. 17–14b. Since the vectorial difference between $E_{AN'}$ and $E_{aN'}$ is the stator reactance drop, and since this difference increases when the angle between $E_{AN'}$ and $E_{aN'}$ increases, the stator current increases. The angle between $E_{AN'}$ and $E_{aN'}$ increases

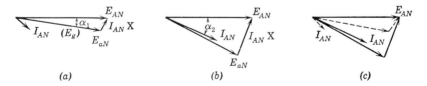

Fig. 17–14. Vector diagrams for a synchronous motor at different loads.

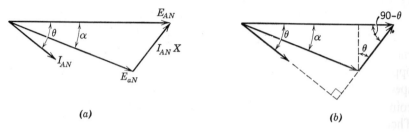

Fig. 17-15. Synchronous-machine vector relations.

until the increased electric-power input is equal to that required because of the increased mechanical output. To show that the electric-power input is increased when load is placed on the rotor shaft, the vector diagrams of Fig. 17–14a and 17–14b are combined as Fig. 17–14c. Figure 17–14c reveals that the product, $E_{AN'}I_{AN'} \cos \theta$, is greater for the diagram of Figure 17–14b than for the diagram of Fig. 17–14a, thus indicating that the power input is greater when the mechanical load is added.

A mathematical relation can be developed to show that the angle α between the generated voltage and phase voltage must be increased when the motor develops an increased torque. Thus, for the synchronous-motor vector diagram shown in Fig. 17–15a, the power input to the motor is given by the relation

$$P = E_{AN}I_{AN} \cos \theta \qquad (17\text{--}10)$$

a line drawn from the intersection of the E_{aN} and $I_{AN}X$ vectors normal to the E_{AN} vector, as shown in Fig. 17–15b, has a magnitude given by both $E_{aN} \sin \alpha$ and $I_{AN}X \cos \theta$. Therefore, from this identity,

$$I_{aN} \cos \theta = \frac{E_{aN}}{X} \sin \alpha \qquad (17\text{--}11)$$

Substituting for $I_{AN} \cos \theta$ in Eq. 17–10 the identity given in Eq. 17–11 gives

$$P = \frac{E_{AN}E_{aN}}{X} \sin \alpha \qquad (17\text{--}12)$$

Because the speed of a synchronous motor is constant, the torque developed is proportional to the power. Therefore, Eq. 17–12 can be modified to

$$T = K \frac{E_{AN}E_{aN}}{X} \sin \alpha \qquad (17\text{--}13)$$

Thus, as indicated in Eq. 17–13, an increase in load torque for constant generated and phase voltages must be accompanied by an increase in the angle α. For this reason this angle is referred to as the **torque angle**.

17-10. THE SYNCHRONOUS CONDENSER

The preceding article considers the effect of an increased mechanical load upon a synchronous motor. The discussion is based upon a constant generated voltage, since the rotor field current is unchanged, and a constant supply voltage. This article considers the effect of changing the rotor field current while the mechanical load is maintained constant.

The vector diagram of Fig. 17–16a represents the stator conditions for a synchronous motor driving a constant load. Increasing the rotor field current increases the rotor flux and, therefore, the generated voltage. The increased strength of the rotor field poles tends to strengthen the attractive force between the rotor field poles and the poles of the rotating field, with the result that the vector representing the generated voltage not only increases in length but also draws closer to the vector representing the supply voltage, as shown in Fig. 17–16b. As indicated in Fig. 17–16b, the current is increased in magnitude and now leads the applied-voltage vector.

Since the mechanical load is assumed to be unchanged, the stator current of the motor must have a magnitude and phase angle such that the product, $E_{AN}I_{AN'}\cos\theta$, is the same after the rotor field strength is increased as before it is increased; this is so indicated in Fig. 17–16c, the combination of Figs. 17–16a and 17–16b.

The effect of an increased field current, therefore, is to cause the motor to operate at a more leading power factor. For this reason, synchronous motors are often installed in industrial plants solely for the purpose of correcting the total power factor of the plant. When the synchronous motor is applied in this manner, it is operated at no load;

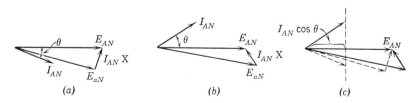

Fig. 17–16. Vector diagrams illustrating the effect of a change in field current upon synchronous-motor operation.

therefore, the only power input is that necessary to supply the motor losses. If the rotor field current is greater than that which produces unity power factor, the motor is said to be overexcited and operates at a leading power factor (thus acting as a capacitor). A further increase in the field current increases the capacitance effect. A synchronous motor used solely for the purpose of power-factor correction is called a **synchronous condenser.**

The power-factor-correction possibilities of a synchronous motor are not limited to no-load operation. If the motor is used to drive a load, overexciting the motor causes it to operate with a leading power factor. The amount of power-factor correction is limited, under any conditions, by the rated current of the stator winding; therefore, the more mechanical load a synchronous motor drives, the more its power-factor-correction ability is reduced.

17-11. STARTING SYNCHRONOUS MOTORS

Energizing the stator winding of a synchronous motor when the rotor is energized but at a standstill results in a rotating field that produces no average torque on the motionless rotor and, therefore, no motion of the rotor. This condition is shown schematically by Figs. 17–17a and 17–17b.

In Fig. 17–17a, the pole of the rotating field is shown to be approaching that of the rotor and to be exerting an attractive force on the rotor pole; therefore, the rotor tends to rotate in a direction opposite to that of the rotating field. Figure 17–17b represents the conditions an instant later; it indicates that the attractive force tends to rotate

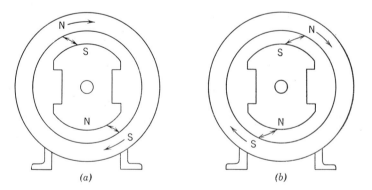

Fig. 17–17. Forces on synchronous-motor rotor at standstill when stator winding is energized.

the rotor in the direction of the rotating field. The net effect of the forward and backward torques is zero, and the rotor remains motionless. A synchronous motor is therefore inherently not self-starting, and some auxiliary means of starting must be provided.

Two methods of starting synchronous motors might be called **induction-motor starting** and **prime-mover starting.** These names are applied here for convenience in identifying the methods.

Induction-Motor Starting. In order to start a synchronous motor by the induction-motor-starting method, it is necessary that the rotor have a squirrel-cage winding, known as an **amortisseur**, or **damping, winding.** The damping winding consists of bars inserted in the pole faces parallel to the shaft and connected at the ends to form, in effect, a partial squirrel-cage winding. A salient-pole rotor with a damping winding is shown schematically as Fig. 17–18.

With the induction-motor method of starting, the rotor winding is left unenergized when the stator is energized. The motor starts as an induction motor and reaches a speed slightly less than synchronous, the actual value of slip being dependent upon the amount of load connected to the motor. If the rotor winding is then energized, the attractive force between the poles of the stator and rotor causes the rotor to pull into step with the rotating field of the stator, and the rotor operates at synchronous speed. The maximum torque that the motor can develop to cause the rotor to pull in is called the pull-in torque. In addition to the slip, the stator field, and the rotor field, the pull-in torque developed in a synchronous motor depends upon the rotational inertia of the rotor and its connected load.

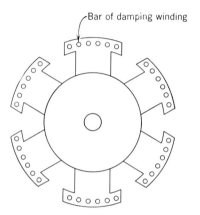

Fig. 17–18. Salient-pole rotor with damping winding.

If the adjacent stator and rotor poles are of the same magnetic polarity when the rotor winding is energized, the force between the poles is one of repulsion rather than of attraction. Under these conditions, the rotor slips a pole, or drops back so that poles of opposite magnetic polarity are adjacent. Depending on the load, the rotor may slip several times and create a noticeable disturbance before it finally falls into step, or it may not pull in at all.

If the load on a motor with salient poles is light, and the rotational inertia is small, the rotor may pull in without having its winding energized. The torque in this case is that developed on the magnetic material of the field poles by the rotating field.

Although the rotor winding is unenergized upon starting, it is not allowed to remain open, because of the large voltage that might be induced by the rotating field linking the large number of turns of the winding. This voltage might be large enough to cause breakdown of the winding insulation. Therefore, when the motor is started, the rotor winding is often short-circuited by all or part of the rheostat that is normally connected in series with the winding. In some applications, a field-discharge resistor is connected to the field winding when the switch to the d-c supply is open; it is usually removed when the d-c supply switch is closed.

Starting a synchronous motor by the induction-motor method may produce the same objectionable starting current as that produced in starting a normal induction motor. If so, the synchronous motor is started at reduced voltage by means of a starting compensator similar to that described for starting induction motors. When the motor is started in this manner, the rotor winding may be energized while the motor is operated at the reduced voltage, but only after the rotor has reached a steady speed.

The damping winding serves to provide starting torque and to reduce hunting. When the rotor is turning at synchronous speed, the damping winding is inoperative because no voltage is induced in it. If the load on the motor is suddenly increased, however, the rotor tends to drop back farther than it should. If the damping winding is not present, the rotor tends to oscillate about its final position with respect to the rotating field produced by the stator winding. Such an oscillation is called **hunting**. Hunting is materially reduced in a machine having a damping winding, because of the induction-motor torque developed by the damping winding when the rotor momentarily operates below synchronous speed. Similarly, if the load on the motor is suddenly reduced, the rotor tends to run faster than synchronous speed in order to reach a new position in relation to the rotating field. In this instance, the torque developed by the damping winding is in a direction to oppose rotation, and thus the oscillation of the rotor is damped.

Prime-Mover Starting. The prime-mover method, applied to motors to be started at no load, consists in (*a*) driving the synchronous motor as a synchronous generator by means of a prime mover, (*b*) connecting the generator in parallel with the voltage supply, and (*c*) removing the prime mover so that the generator becomes a motor as described in a preceding article.

The prime mover to drive the synchronous motor as a generator may be the d-c machine that is often connected to the motor shaft to provide the rotor-winding excitation, or it may be a prime mover that can be uncoupled after the synchronous machine is connected to the supply and is operating as a motor.

When the motor is to be started, the switch to the stator supply is open, and the speed of the prime mover is increased until the synchronous machine is operating at a speed corresponding approximately to the frequency of the supply; the rotor winding is then energized. The rotor field current is adjusted until the voltage of the generator is equal to that of the supply. If the generator has the same phase sequence as the supply, as it should, the next step is to close the stator switch at the instant that the terminals that are to be connected together have the same relative polarity; at this instant, the generator is said to be in step with the supply. (This procedure of connecting the generator to the supply is called **synchronizing**.)

To close the switch at any time other than when the generator is in step would mean that one voltage source would send current to the other voltage source, the amount of current being dependent upon the relative positions of the vectors corresponding to the voltages of the generator and of the supply, and upon the impedance of the two sources. Correct closing of the switch corresponds to paralleling two equal-voltage batteries by connecting terminals of like polarity together; for this connection, no current flows between the batteries. Incorrect closing of the stator switch corresponds to paralleling the batteries by connecting terminals of unlike polarity; for this connection, the current that flows is limited only by the internal resistance of the batteries.

After the machine is synchronized to the line and the line switch is closed, the prime mover is removed, and the machine then operates as a synchronous motor.

17-12. SYNCHRONIZING METHODS

The synchronizing process described for the starting of synchronous motors by the prime-mover method is also the process by which synchronous generators are placed in parallel. The difference between the two operations is that the prime mover is not removed when the synchronous machine is to operate as a generator in parallel with other generators.

The time when the generator is in synchronism with the supply can be determined by means of lamps or a synchroscope.

Two lamp methods of synchronizing are in common use—the dark-lamp method and the two-bright–one-dark method. The connections of the lamps for these methods are shown in Fig. 17–19.

In the dark-lamp method of Fig. 17–19a, the three lamps are dark when the terminals to which they are connected have the same relative polarity, and the switch should be closed at this instant. At any other time the lamps glow with the same brilliance, reaching a maximum in brilliance when the voltage between the terminals to which the lamps are connected is a maximum. Thus, if there is a slight difference between the frequency of the generator and the frequency of the supply, the lamps flicker together at a rate corresponding to the difference in frequencies.

In the two-bright–one-dark method of Fig. 17–19b, the lamp connected between the terminals that are to be joined together is dark, and the other two are of equal brilliance, at the instant the switch should be closed.

The two-bright–one-dark method is considered to be superior to the dark-lamp method because it gives a more positive indication of the instant when the relative polarities of the terminals are the same. Because of the relatively large voltage needed to make the lamps glow visibly, there may be a voltage between the terminals to be connected even though the lamps in the dark-lamp method are dark.

The synchroscope is an instrument that indicates, by means of a revolving pointer, the instant for closing the stator switch. The dial and terminals of a synchroscope are shown schematically in Fig. 17–20. As indicated in the diagram, provision is made for connecting single-phase voltages from the supply and from the generator.

The switch that connects the generator to the supply should be closed at the instant that the pointer of the synchroscope is pointing vertically upward. At this instant, the generator is in step. If the frequencies

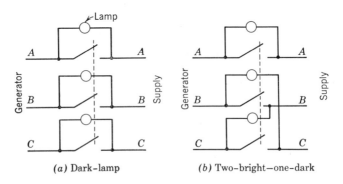

Fig. 17–19. Lamp methods of synchronizing.

SYNCHRONIZING METHODS

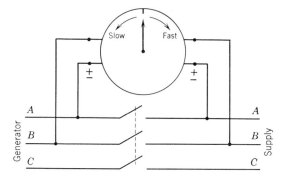

Fig. 17-20. Synchroscope connections.

of the generator and the supply are exactly equal but the generator is not in step, the pointer of the synchroscope assumes a position corresponding to the angular amount by which the generator is out of step. If the frequencies are different, the pointer rotates in the direction corresponding to the frequency of the generator with respect to the frequency of the supply—fast if the generator frequency is higher, slow if the generator frequency is lower.

The synchroscope method is superior to either of the lamp methods because it not only gives a positive indication of the time to close the switch but also indicates the adjustment to be made should there be a difference between the frequencies of the generator and the supply.

PROBLEMS

17-1. While on an inspection trip through the synchronous-machine division of an electrical manufacturing company, you are shown three synchronous-machine rotors. The first rotor has two poles, the second has six poles, and the third has forty poles. All are designed for 60-cycle machines. Which rotor is probably used in the machine to be driven by a Diesel engine? by a waterwheel? by a steam turbine?

17-2. A synchronous generator is synchronized to a constant-voltage constant-frequency three-phase supply in such a manner that no current flows between the generator and the supply; the prime mover is left connected to the generator. What is the effect, upon the operation of the generator, of increasing the rotor field current? Explain.

17-3. Instead of the rotor field current of the generator of Prob. 17-2 being changed, the fuel input to the prime mover is increased. Explain the effect upon the operation of the generator.

17–4. A three-phase synchronous generator is operating in parallel with a constant-voltage constant-frequency three-phase supply. The generator is delivering rated kilovolt-amperes at 0.8 lagging power factor. Using carefully drawn vector diagrams to substantiate the comparison, discuss the effect of a complete failure in the fuel supply to the prime mover of the generator upon the power, the current, the power factor, and the general operation of the generator.

17–5. Plot and explain the curve of stator current versus rotor current for a three-phase synchronous motor operating at full-load torque. Completely label the plot.

17–6. A three-phase synchronous generator is operating in parallel with a constant-voltage constant-frequency supply and is delivering rated kilovolt-amperes at 80% leading power factor. Would it be desirable or undesirable from the standpoint of the operation of the generator to increase the rotor field current? Explain.

17–7. Explain completely the effect of an increase in the supply voltage upon the operation of a synchronous motor that is operating at normal voltage at one-half rated torque and a power factor of 0.707 lagging.

17–8. A three-phase synchronous motor is operating at unity power factor driving a constant-torque load. Explain completely the effect on the operation of the motor of connecting an inductor in each line between the stator terminals and the supply switch, at which the voltage is maintained constant in magnitude and frequency. (The inductor has a reactance of the same order of magnitude as the reactance of the motor.)

17–9. A three-phase synchronous machine is synchronized to a three-phase 60-cycle constant-voltage supply in such a manner that no current flows between the machine and the supply; the prime mover is left connected to the machine.

(*a*) What would be the final effect on the operation of the machine if the supply frequency were to increase slowly to 62 cycles while the supply voltage remained constant? Explain.

(*b*) Starting from the original conditions, adjustments are made so that the machine acts as a synchronous generator delivering power to the line at a 0.707 lagging power factor. Explain the effect of the operation of the machine of connecting an inductive reactor in each line from the generator to the supply.

17–10. If the phase sequence of a machine being synchronized to a three-phase supply is different from that of the supply, what is the reaction of lamps connected for the "dark lamp" method of synchronizing? Explain.

■ 18 Single-Phase Motors

18–1. INTRODUCTION

The most common type of electric motor is the single-phase type, which finds wide industrial and domestic applications. No attempt is made here to list all the applications of this type of motor, but some examples are refrigerators, vacuum cleaners, mixers, fans, and other domestic appliances.

Single-phase motors may be classified generally as being of the induction type or of the commutator type, although one motor considered in this chapter is a combination of the two types.

Induction-Type Single-Phase Motors

18–2. THEORY OF OPERATION

The basic components of an induction-type single-phase motor are a single-phase stator winding and a squirrel-cage rotor. These components are shown schematically in Fig. 18–1a.

When the stator winding is energized from an a-c supply, the current in the winding produces a magnetic field that alternates along the axis A–A shown in Fig. 18–1a. This oscillating field causes voltages to be so induced in the rotor conductors that, at the instant the flux is increasing

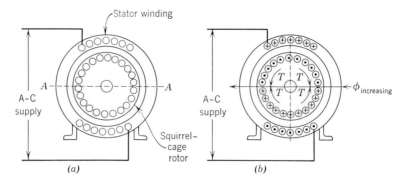

Fig. 18–1. Induction-type single-phase motor.

in a right-to-left direction, current flows as indicated in Fig. 18–1b.

Because the individual rotor conductors carry current in a transverse magnetic field, each rotor conductor has a force exerted upon it. However, since as many conductors contribute to a torque acting in a clockwise direction as contribute to a torque in a counterclockwise direction, the net torque on the rotor is zero, and the rotor remains stationary. The induction type of single-phase motor is inherently not self-starting, and some auxiliary means must be provided for producing a starting torque. If the rotor of the motor shown in Fig. 18–1a is started by an auxiliary means, and the starting device is then removed, the rotor continues to rotate in the direction in which it is started.

One theory for the operation of a single-phase motor once it is started

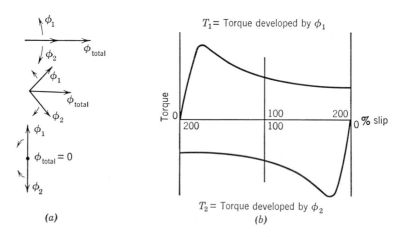

Fig. 18–2. Double-rotating-field details.

THEORY OF OPERATION

is known as the **double-rotating-field theory.** This theory is based on the fact that the alternating field produced by the stator winding can be represented as the sum of two rotating fields, each equal to one-half the maximum value of the oscillating field, and rotating in opposite directions, each rotating at the frequency of the oscillating field. If these fields are represented by vectors that rotate in opposite directions, as shown in Fig. 18–2a, the summation of the vectors is a stationary vector that oscillates in length along the horizontal axis. When the rotating vectors are in phase, the resultant is the maximum; when they are out of phase by 180 degrees, the resultant is zero.

The torque–slip curves corresponding to each of the rotating fields considered independently are shown in Fig. 18–2b. As indicated, the torques developed by the two rotating fields are in opposite directions; each field develops a torque that tends to drive the rotor in the direction in which the field rotates. Thus, the point of zero slip for one field corresponds to 200 per cent slip for the other. The value of 100 per cent slip, or standstill, is the same for both fields.

The average torque developed on the rotor is the algebraic summation of the torques produced by the two rotating fields. The resultant torque–slip curve is therefore as shown in Fig. 18–3. This curve shows that the average torque at standstill is zero and, as previously stated, the motor is not self-starting. When the rotor is started in either direction, however, the average torque developed causes the rotor to continue to rotate in the direction in which it is started.

The resultant torque–slip curve also shows that the net torque is also zero at some value of slip near zero, and indicates that a single-phase

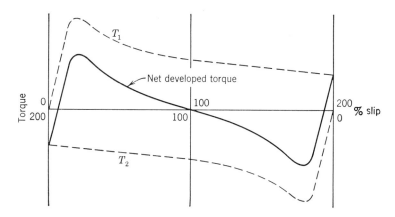

Fig. 18–3. Torque–slip characteristic of a single-phase induction motor.

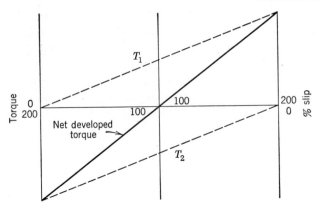

Fig. 18-4. Torque–slip characteristic for high-resistance rotor.

motor operates with a greater percentage slip at full load than a corresponding three-phase induction motor.

If the rotor of the single-phase motor has a relatively high resistance so that the ratio of resistance to reactance is much larger than normal, the torque–slip curves for the individual rotating fields and the resultant torque–slip curve are as shown in Fig. 18-4. As indicated, the large resistance-to-reactance ratio moves the maximum of the normal torque–slip curve to a greater slip, and, for this example, results in approximately linear curves for the individual rotating fields. The net developed torque is such that, if the motor is started in one direction, the developed torque opposes the rotation. Thus, when the torque causing rotation is removed, the motor stops; it cannot operate as a single-phase induction motor. This characteristic is important in control applications where a deviation in the quantity being controlled causes a driving torque to be applied to the motor until the controlled quantity is again at the proper level. In such applications the rotation of the motor is such as to affect the controlled quantity.

18-3. STARTING

Induction-type single-phase motors are normally classified according to the auxiliary means used to provide the necessary starting torque. The classification of single-phase induction motors is as follows: (*a*) **split-phase**, (*b*) **resistance-start**, (*c*) **capacitor-start**, (*d*) **capacitor**, and (*e*) **shaded-pole**.

All starting methods depend upon two alternating fields displaced in

STARTING

space and phase; two fields so displaced produce a rotating field which, reacting with the squirrel-cage rotor, provides the starting torque. One field is produced by the **main winding;** the other by an **auxiliary, or starting, winding.**

If the main winding and the auxiliary winding are so displaced that they produce fields 90 degrees apart in space, as shown in Fig. 18–5a, and carry currents that produce fields equal in magnitude and 90 degrees apart in time, as shown in Fig. 18–5b, the resultant rotating field is constant in magnitude and is equal to the maximum value of one of the fields. This rotating field may be represented by a rotating vector that is constant in magnitude and thus describes a circle in each revolution, as shown in Fig. 18–5c. Each revolution of the vector corresponds to one cycle of the supply frequency.

If the main and auxiliary winding are displaced 90 degrees in space but produce fields that are either not equal or not 90 degrees apart in time, as shown in Fig. 18–6a, the resultant field is a rotating field but is not constant in magnitude. This type of field may be represented by a rotating vector that changes in length as it rotates, the length being dependent upon the position of the vector with respect to the axes that represent the axes of the coils. The figure described by one revolution of the vector depends upon the relative magnitudes and the phase displacement of the component fields and is elliptical in form, as shown in Fig. 18–6b. Similarly, the resultant of two fields displaced both in time and space by some angle other than 90 degrees is a nonuniform rotating field.

One effect of a nonuniform rotating field is the production of a developed torque that is nonuniform and that therefore causes noisy operation of the motor. Starting windings that tend to produce nonuniform rotating fields in conjunction with the main winding are disconnected from the circuit after the motor is started, and the operation of the motor then becomes smoother.

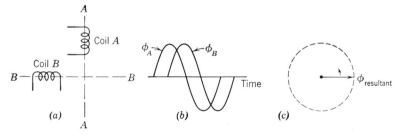

Fig. 18–5. Production of a uniform rotating field.

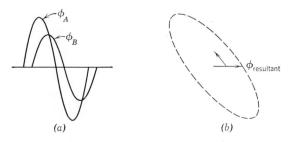

Fig. 18-6. Nonuniform rotating field.

Another, more important aspect of the nonuniform rotating field is its effect upon the starting torque. The minor axis of the ellipse representing the rotating field in Fig. 18-6b becomes smaller as the component fields become more nearly in phase in space or time, and the resultant field approaches an oscillating one. Thus, when the ellipse has a small minor axis, the resultant field can be considered to represent the summation of a small rotating field and a large oscillating field. Since the starting torque depends upon the rotating field, the starting torque under these conditions is small. Thus, in the comparison of two machines of the same rating, the motor having the more uniform rotating field has the larger starting torque.

18-4. SPLIT-PHASE MOTORS

With the exception of the shaded-pole motor, all the single-phase motor classifications in the preceding article are, in reality, split-phase types. The split-phase motor has main and auxiliary windings displaced 90 degrees in space; the time displacement of the fields produced by

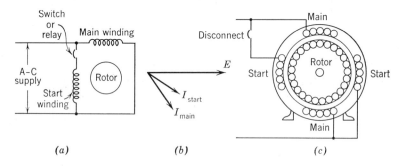

Fig. 18-7. Split-phase motor.

SPLIT-PHASE MOTORS

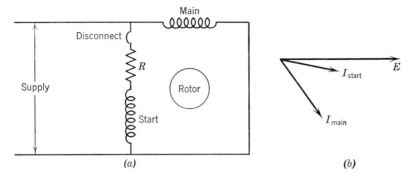

Fig. 18-8. Resistance-start motor diagrams.

the two windings depends upon the specific type of motor: that is, whether it is split-phase, resistance-start, capacitor-start, or capacitor.

The split-phase motor has main and auxiliary windings that so differ in resistance and reactance that the currents in the windings, and therefore the fields produced by the windings, differ in time phase by some angle less than 90 degrees. The connection diagram, vector diagram, and schematic representation of this type of motor are shown in Fig. 18-7. Provision is made for opening the auxiliary winding by means of a centrifugal switch or a relay after the motor has started.

The direction of rotation of the rotating field produced by the combination of the main and auxiliary windings, and therefore the direction of rotation of the motor, depends upon the relative connections of the windings to the a-c supply. These directions can be reversed by a change in the connections of one of the windings.

The resistance-start motor is a split-phase type of motor that has a

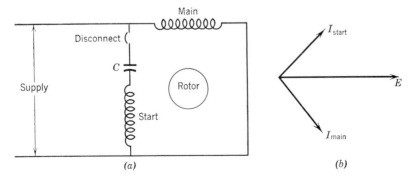

Fig. 18-9. Capacitor-start motor diagrams.

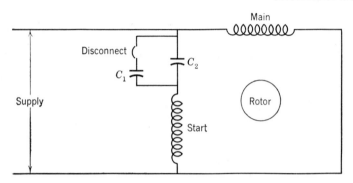

Fig. 18–10. Capacitor-motor connection diagram.

resistor connected in series with the auxiliary winding, as shown in the connection diagram in Fig. 18–8a. The connection of the resistor produces a greater time–phase displacement of the fields than that attained in the split-phase motor; therefore, the resistance-start motor has a greater starting torque than the split-phase motor.

The capacitor-start motor is similar to the resistance-start motor except that capacitance is connected in series with the auxiliary winding. This connection makes possible a 90-degree time–phase displacement and, therefore, a greater starting torque than can be attained in either the split-phase or the resistance-start types. The connection diagram and vector diagram for a capacitor-start motor are shown in Fig. 18–9.

The capacitor motor is similar to the capacitor-start type except that the auxiliary winding remains in the circuit after the motor is started. Such a motor operates at a higher power factor than the split-phase motors previously considered and, in addition, has a larger pull-out torque. The capacitor connection is usually so made that the capacitance is decreased after the motor reaches a speed of the order of 75 per cent of the final speed. This change in capacitance is accomplished by means of a centrifugal switch or relay, as shown in the connection diagram of Fig. 18–10. (For best operation, the capacitance should be changed continuously as the motor accelerates. A single change between two values in the actual motor represents an economic compromise.)

18–5. SHADED-POLE MOTORS

The shaded-pole motor differs from the split-phase type of induction motor in that there is only one winding to be connected to the a-c supply.

SHADED-POLE MOTORS

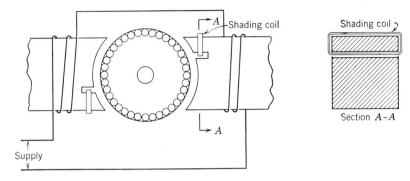

Fig. 18–11. Shaded-pole motor.

This winding is mounted on field poles similar in form to those of the d-c machine, and the starting torque is produced by the action of a one-turn short-circuited coil, known as a **shading coil,** that encircles a portion of each pole. The construction of a shaded-pole motor is shown schematically in Fig. 18–11.

Because the flux in the field poles is alternating, a voltage is induced in each of the shading coils, which act as short-circuited transformer secondary windings. The polarity of the induced voltage is such as to produce an mmf that opposes the change in flux linkage of the coil. Thus, when the flux in the pole is increasing, the shading-coil current produces an mmf opposing the increase; therefore, the flux density of the unshaded portion of the pole is greater than that of the shaded portion. When the pole flux is decreasing, the shading-coil current flows to oppose the decrease; therefore, the flux density of the unshaded portion is less than that of the shaded portion. Thus, the effect of the shading coil is to produce a shift in flux from the unshaded portion to the shaded portion. This shift in flux may be considered to be a partially rotating field and is sufficient to produce a small starting torque. Provided there is little or no load on the motor, the motor starts and runs in the direction of the shift in flux.

Because of the small starting torque developed in the motor and the relatively large power losses in the shading coil, the shaded pole motor is built only in small sizes, the most common application being to fans.

The shaded-pole principle of starting is also applied to self-starting electric-clock motors, although the construction and principle of operation of this type of motor differ from those of the shaded-pole induction motor.

Commutator Single-Phase Motors

18–6. SERIES MOTORS

Because a d-c shunt motor or a d-c series motor continues to operate with the same direction of rotation when the polarity of the line connections is reversed, it follows that either motor is capable of operation from an a-c supply. Although the series motor operates much more satisfactorily from an a-c supply than the shunt motor, both have disadvantages that make their application to an a-c supply unsatisfactory.

For the shunt motor, there is a large time-phase difference between the armature current and the field flux, because of the large reactance of the field winding compared to the armature winding, and only a small torque can be developed. Therefore, the shunt connection is not used in a-c motors.

For the d-c series motor, the armature current and field flux are in phase and produce a reasonable torque, but the large reactance of the field and armature windings, the increased core losses, and the sparking at the brushes make it impractical to operate the motor on an a-c supply. Although similar to the d-c series motor, the a-c series motor must differ in construction to obtain satisfactory operation.

To keep core losses to the minimum, the a-c series motor is constructed of material having a low hysteresis loss, and the entire magnetic circuit is laminated.

The large reactance drop in the armature circuit is due to the armature reaction flux created by the alternating armature current. This reactance drop is eliminated or appreciably reduced in large series motors by the installation of a compensating winding. The compensating winding is connected in series with the armature and field windings but has its conductors embedded in the pole faces parallel to the armature conductors. The winding is so connected that it produces an mmf that effectively neutralizes the mmf produced by the current in the armature conductors. A series motor with a compensating winding is shown schematically in Fig. 18–12.

Since the reactance of the field winding is a function of the square of the number of turns in the winding, the motor is designed to operate at a relatively low flux density and, therefore, requires only a small

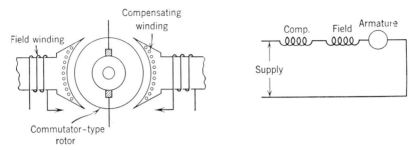

Fig. 18-12. A-c series motor connections.

number of turns in its field winding; thus, the reactance of the field winding is reduced.

The sparking at the brushes of a series motor connected to an a-c supply is caused by the voltage induced by the alternating main field in the coils short-circuited by the brushes during the commutation period. To reduce the sparking in an a-c series motor, the number of turns in each armature coil is kept small, and high-resistance leads connect each coil to the commutator segments. Designed in this manner, the armature winding of an a-c series motor has more coils and, therefore, more commutator segments than a comparable d-c series motor.

The speed–current and torque–current curves for an a-c series motor are similar to those previously developed for a d-c series motor. Since it operates on either an a-c or a d-c supply, the a-c series motor, often called a **universal motor,** finds wide application.

18-7. REPULSION MOTORS

The **repulsion motor** is a commutator-type motor that has its armature winding short-circuited, as shown schematically in Fig. 18-13a. The stator winding is wound like the main winding of a split-phase motor and, when energized, produces an oscillating magnetic field.

When the brushes are in a plane perpendicular to the alternating stator field, as shown in Fig. 18-13a, each armature coil has a voltage induced in it by the oscillating field. However, because of the brush location, the net induced voltage between brushes is zero; therefore, no current flows in the armature conductors, and no torque is developed on the rotor.

When the plane of the brushes is parallel to the field, as shown in

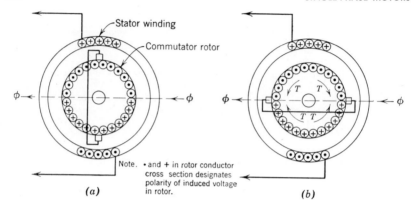

Fig. 18-13. Repulsion-motor details.

Fig. 18–13b, the net induced voltage between brushes is a maximum; therefore, the current in the short-circuited armature is a maximum. Because each conductor carries current while located in a transverse magnetic field, each conductor has force exerted on it; the directions of the forces are such, however, that the net torque on the rotor is zero, as indicated in Fig. 18–13b.

If the plane of the brushes is at some angle other than zero or 90 degrees to the plane of the stator field, a net torque is developed on the rotor, and, if the resisting torque is not greater than the developed torque, the rotor accelerates. The direction of rotation depends upon the direction in which the brushes are shifted from the plane parallel to the field. Thus, in Fig. 18–14, the brushes are shifted clockwise, and the net torque acts in a clockwise direction; shifting the brushes in the

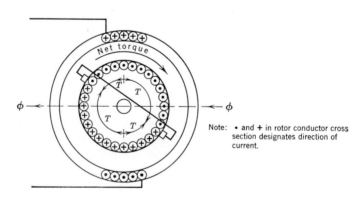

Fig. 18-14. Repulsion-motor torque.

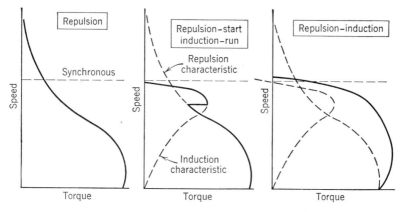

Fig. 18-15. Repulsion-motor characteristics.

opposite direction causes the rotation to reverse. (The axis parallel to the plane of the field is called the neutral axis.)

The repulsion motor has a speed–load characteristic similar to that of the d-c series motor. It has the advantage, in comparison with other single-phase a-c motors, of a high starting torque and a relatively low starting current.

The high starting-torque and low starting-current characteristics of the repulsion motor and the good speed regulation of the induction motor are effectively combined in the **repulsion-start induction-run motor,** which starts as a repulsion motor and, upon reaching a certain speed, has its brushes lifted and its commutator short-circuited by a centrifugal mechanism. The motor then continues to operate as a squirrel-cage induction motor, with the torque–speed characteristic common to that type of motor.

The **repulsion–induction motor** has the armature winding and commutator of the normal repulsion motor and, in addition, the squirrel-cage winding of the induction motor. Both windings are effective during the entire operating cycle of the motor. The motor has a high starting torque and a starting current that is only slightly larger than that of the repulsion motor.

The torque–slip characteristic of the repulsion–induction motor is similar to that of the repulsion-start induction-run motor in that the repulsion characteristic dominates up to about two thirds of normal speed, at which time the characteristic takes the form of that of the induction motor. The transition from the repulsion characteristic to the induction characteristic is smooth, however, without the sharp break that occurs in the repulsion-start induction-run motor when the centrifugal mechanism operates. The repulsion–induction motor operates at a speed slightly above synchronous at light loads because of the repulsion characteristics, but the amount above synchronous is limited by the induction-generator action of the squirrel-cage winding when the motor speed exceeds the synchronous value.

The approximate speed–torque characteristics of the three types of repulsion motor are shown in Fig. 18–15.

PROBLEMS

18–1. What will be the effects of energizing a split-phase motor when the starting winding is open?

18–2. The centrifugal switch of a certain split-phase motor fails to *close* when it should. How does this failure affect the operation of the motor?

18–3. A certain split-phase motor is found to be quieter when its starting winding is open than when a resistor is connected in series with the starting winding. Explain.

18–4. A certain repulsion–induction motor fails to start by itself, but, once started by some external means, it appears to operate normally at no load. Give two possible reasons for this trouble.

18–5. Draw a complete connection diagram showing a push-button control circuit applied to start and stop a capacitor motor.

18–6. A student says to his professor, "My electric shaver works better on 120-volts direct current than on 120-volts alternating current. Why?"

(*a*) Knowing that the student did not make any electrical measurements, what did he probably mean by "works better"?

(*b*) Assuming that the shaver has a universal motor, answer the student's question.

■ 19 Control Equipment

19–1. INTRODUCTION

Many industrial processes require a degree of control that can best be attained by the use of electric equipment that serves not only to indicate that the process has deviated from normal but also to initiate the necessary corrective measures. Systems incorporating such control are called closed-loop systems, and they have a wide field of application.

One application of the closed-loop system is in the field of temperature control. As an example, if the voltage output of a thermocouple circuit is compared with a battery voltage that corresponds to the desired value of temperature, the difference in voltage is a function of the deviation of the temperature from the desired value. The polarity of the voltage difference is dependent upon the sense of the deviation; that is, whether the temperature is above or below the desired value. If this difference in voltage is impressed upon an amplifying device that controls the heating circuit, the control circuit adjusts the heating circuit so as to make the desired correction in temperature and restore the voltage balance.

Many positioning-control systems in which the position of a device is to be controlled from a remote location are closed-loop systems. An example is a method that might be applied to the positioning of a television antenna that is too far away for a mechanical linkage to be practicable. In this application, the operator could change the magnitude, or phase, of a voltage that is normally balanced by a voltage corresponding to the position of the antenna. The difference in voltage could then be applied to an amplifying device that controls the motor to

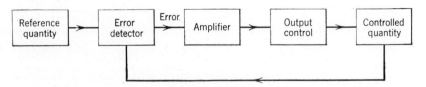

Fig. 19-1. Block diagram of a closed-loop system.

rotate the antenna. When the antenna reaches the position desired by the operator, the voltage corresponding to antenna position is again equal to the voltage at the control of the operator, and the driving motor is de-energized.

Any closed-loop system can be represented by a block diagram as shown in Fig. 19-1.

As indicated in Fig. 19-1, the controlled quantity is compared with the reference quantity by the error-detecting device. Any difference between the reference quantity and the controlled quantity is transmitted through an amplifier to the output control, which changes the controlled quantity so that the difference, or error, no longer exists. In the example for temperature control, the reference quantity is a constant voltage (corresponding to the desired temperature) which is compared with a voltage corresponding to the actual temperature. In the example of television-antenna positioning control, the reference quantity is a variable voltage (corresponding to the desired position of the antenna) which is compared with a voltage corresponding to the actual position of the antenna.

The amplifier is sometimes unnecessary in closed-loop systems, the error being applied directly to the output control. In general, however, the signals representing the reference quantity and the controlled quantity are of such small magnitude that some form of amplification is necessary.

Although control systems of the type described often contain electronic components, particularly in the amplifier section, it is the purpose of this chapter to consider two machine components that appear in many control systems. One machine component is the **synchro,** which is used as an error detector; another is the **rotating amplifier,** which, as the name indicates, is used as an amplifier. The chapter also considers the **Magnetic Amplifier.**

19-2. SYNCHRO UNITS

In addition to closed-loop control systems, the synchro unit finds wide application in remote indicating systems in which the position of a device

SYNCHRO VOLTAGES

—for example, a weathervane or a water-level float—is indicated on a remote indicator by means of an electric connection. If the electric connection between the generator (the synchro unit at the device) and the motor (the synchro unit at the indicator) is broken and is restored at some later time, the rotor of the repeater aligns to give the correct indication of the position of the device. Because the rotors of the synchro units seek the same corresponding positions with respect to their stators, the units are said to be self-synchronous; it is from this term that the name of the unit is derived, synchro being an abbreviation of self-synchronous. (The trade names applied by the manufacturers of synchro units also refer to the self-synchronous characteristic of the units: Selsyn, General Electric; Diehlsyn, Diehl Electric; Autosyn, Bendix.)

The construction of a synchro unit depends upon the function of the unit in the system of which it is to be a part. Classified as to function, the units are (*a*) **synchro generators** or **motors**, (*b*) **differential synchro generators** or **motors**, and (*c*) **synchro control transformers.**

The construction of a synchro generator or motor is similar to that of a synchronous machine in that the unit has a three-phase winding and a single-phase winding, either of which may be on the stator. In this book, the three-phase winding, which is wye-connected, is considered to be the stator winding, and the single-phase winding to be the rotor winding. An important difference between a synchro generator or motor and a synchronous machine is that the single-phase rotor winding of the synchro unit is energized from an a-c, rather than a d-c, supply.

The rotor of a synchro generator or motor has two poles of the salient type. The rotors of the generator and motor differ in that the motor has an oscillation damper, in the form of a lead ring, mounted on its shaft.

The construction of differential synchro units and synchro control transformers differs somewhat from that of the synchro generator and motor; these differences are considered when the operation of the differential units and the control transformer is discussed.

19–3. SYNCHRO VOLTAGES

If the rotor of a synchro unit is energized from a single-phase a-c supply, as shown in Fig. 19–2, alternating voltages are induced in each phase of the stator winding by transformer action, and each phase of the stator may be considered to be the secondary winding of a transformer of which the rotor winding is the primary.

The effective value of voltage induced in a particular stator phase

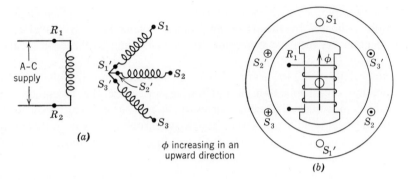

Fig. 19-2. Synchro details.

depends upon the position of the rotor with respect to the phase. With the rotor in the position shown in Fig. 19-2, the voltage induced in the coil representing phase 1 is zero because no flux links the coil. At the same time, the effective values of the voltages induced in coils 2 and 3 are equal, with the primed terminal of one coil having the same relative polarity as the unprimed terminal of the other. The voltage–time diagram for the rotor position of Fig. 19-2 is shown as Fig. 19-3.

The voltages induced in coils 2 and 3 are equal and in phase for the rotor position of Fig. 19-2. The fact that the voltages are in phase is stressed because it is important to realize that, *although the stator winding is wound and connected as a three-phase winding, the three coils are actually the secondary windings of a single-phase transformer; three-phase voltages do not exist in the winding.*

For the rotor position shown in Fig. 19-2, the voltage induced in stator coil 1 is zero. If the rotor position is changed, the effective value of voltages induced in coil 1 is changed, and it reaches a maximum when the axis of the rotor is perpendicular to the axis representing the plane

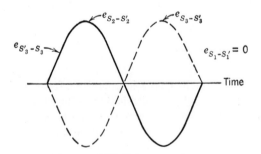

Fig. 19-3. Synchro stator voltage versus time.

SYNCHRO GENERATORS AND MOTORS

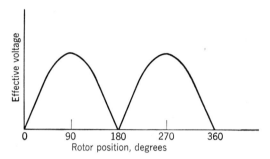

Fig. 19-4. Voltage versus rotor position for a stator coil.

of the coil. A plot of the effective value of voltage induced in coil 1 versus the position of the rotor with respect to coil 1 is shown as Fig. 19-4. (For purposes of the plot, the rotor position shown in Fig. 19-2 is taken as the zero position.)

The effective values of the voltages induced in stator coils 2 and 3 vary with rotor position in the same manner as the voltages induced in coil 1 except that the maximum effective values occur at different rotor positions in each instance. Since the effective voltage measured between any two stator terminals is the summation of two coil voltages that vary sinusoidally with rotor position, the effective voltage measured between any two stator terminals also varies sinusoidally with rotor position.

19-4. SYNCHRO GENERATORS AND MOTORS

The simplest synchro system consists of a synchro generator and motor connected as shown in Fig. 19-5. The units of the synchro system are assumed to be identical; that is, the zero positions of the rotors correspond, and the variations of stator voltage with respect to rotor position are the same.

If the rotors of the units of Fig. 19-5 are in the same relative position with respect to their stator windings when the system is energized, the voltages induced in the stator coils of the generator are equal to and in phase opposition to those of the motor, and no current flows in the stator circuit (the units are said to be in **correspondence**). If the rotor of the generator is then displaced and the rotor of the motor is restrained, the stator voltage balance between the units is altered, and a circulating current flows in the stator windings. This circulating current reacts with the rotor flux of each unit to produce a torque tending to bring

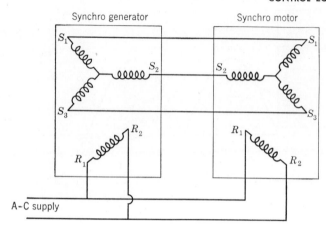

Fig. 19–5. A simple synchro system.

the rotors to a position where the stator voltages are again equal and opposite. Thus, with both rotors unrestrained, any motion given to the rotor of the synchro generator is transmitted to, and duplicated by, the rotor of the synchro motor.

One position of the rotors when the units are in correspondence is shown schematically in Fig. 19–6a (the units are in correspondence whenever the rotors have the same relative position with respect to their stators; therefore, the position shown in Fig. 19–6a is only one of many possible correspondence positions). The polarities of the voltages induced in the stator coils at an instant when the rotor flux is increasing in an upward direction (as shown in Fig. 19–6a) are shown in the circuit diagram of Fig. 19–6b. As previously indicated, no current flows in the stator conductors because the voltages of the stator coils that are connected together are equal and in phase opposition.

If the rotor of the generator is displaced, as shown in Fig. 19–7a, the induced-voltage balance of the stator circuit is altered, as shown in Fig. 19–7b. (The polarities of the induced voltages are indicated for the same conditions of rotor flux as Fig. 19–6: that is, when the rotor flux is increasing in an upward direction.) Therefore, a current flows in the stator windings, and a force is so exerted on the motor rotor because of the reaction between the rotor flux and the current in the stator windings that the rotor tends to move to restore the balance, as indicated in Fig. 19–7b (a similar force in the opposite direction is exerted on the rotor of the generator, but the generator is assumed to be restrained in its new position). The movement of the generator rotor is therefore accompanied by a similar movement of the motor rotor.

SYNCHRO GENERATORS AND MOTORS

The amount of torque developed in the synchro units when the rotors are displaced is a function of the displacement angle (the angle that one unit is displaced from correspondence). The variation of torque with displacement angle for one type of synchro unit is shown in Fig. 19-8, which also shows the variations of stator current and rotor current with displacement angle.

As previously indicated, when there is zero displacement between the rotors of the synchro units, the torque and stator current are zero. As indicated in Fig. 19-8, the torque is also zero at a displacement angle of 180 degrees, but the stator and rotor currents are at their maximum values. The point of 180 degrees displacement is unstable, however, and the units do not settle at this point but always tend to come into correspondence, thus exhibiting their self-synchronous characteristic.

If reasonable accuracy in positioning is desired, the synchro system shown in Fig. 19-5 is useful only when the motor is lightly loaded, since the torque developed in the units at small angles of displacement is low and may not be large enough to overcome the normal friction of the

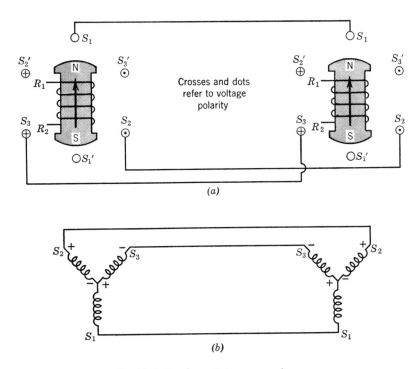

Fig. 19-6. Synchro units in correspondence.

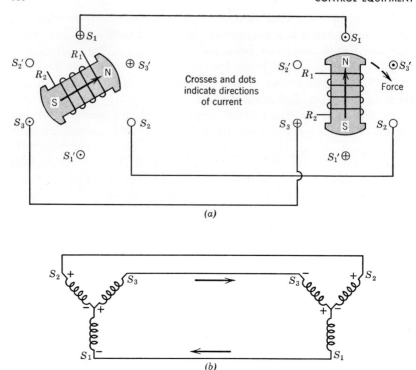

Fig. 19–7. Synchro-system rotors displaced.

system to bring the units into correspondence. Thus, such a system could be applied to indicate the approximate location of a device if only a pointer is attached to the shaft of the synchro motor; the pointer would then indicate a value on a scale calibrated in terms of device position.

If it is desired to have the position of a device indicated at more than one location, additional motors may be connected in parallel with the first, as shown in Fig. 19–9. The operation of the multiple system is, in general, the same as that described for the system containing a single generator and motor. In the multiple system, however, the generator unit must be larger than the individual motor units in order that the normal torque be developed in each motor for a given displacement angle.

19–5. DIFFERENTIAL SYNCHRO UNITS

A differential unit differs in construction from a synchro motor or generator in that both stator and rotor contain a three-phase wye-

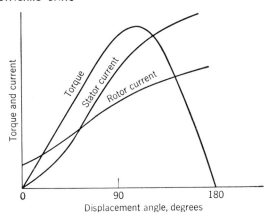

Fig. 19-8. Synchro torque and current characteristics.

connected winding, and the rotor is a cylinder with a slotted periphery. Thus, the construction of a differential unit is similar to that of a three-phase wound-rotor induction motor. The function of a differential unit, however, as of synchro generators and motors, is that of a single-phase transformer. The construction of a differential synchro motor differs from that of a differential synchro generator in that the motor has an oscillation damper mounted on its shaft.

Differential Synchro Generators. The connection of a differential synchro generator into a synchro system is shown in Fig. 19-10.

When the units are connected as shown in Fig. 19-10, the voltage distribution in the rotor coils of the differential unit is the same as that in the stator coils of the synchro generator. If, in the differential unit,

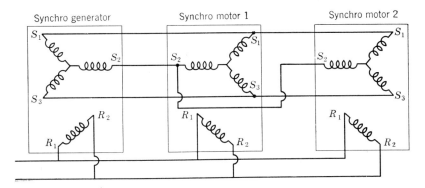

Fig. 19-9. Multiple connection of synchro motors.

the corresponding rotor coils are physically parallel to the stator coils, the voltage distribution in the stator coils is a duplication of that in the rotor coils and thus corresponds to the voltage distribution of the synchro generator. (The voltage ratings of the stator and rotor coils of the differential unit are usually the same; therefore, the differential unit is a one-to-one transformer when the coils are physically parallel.) Therefore, with the rotor coils of the differential unit held in parallel with the stator coils, the system operates in the same manner as the synchro system considered earlier; any movement of the rotor of the synchro generator is duplicated by the rotor of the synchro motor.

If the rotor of the synchro generator is restrained and that of the differential generator is displaced, the voltage distribution in the stator circuit of the differential generator is altered by an amount equal to that produced by a similar displacement of the synchro-generator rotor. The rotor of the synchro motor therefore moves through an angle equal to that by which the rotor of the differential synchro generator is displaced. The connection of the differential generator therefore provides a synchro system in which the synchro motor can be controlled from two separate locations.

If the rotors of the synchro generator and the differential synchro generator are both displaced at the same time, the rotor of the synchro motor rotates through an angle equal to the algebraic sum of the angles by which the rotors of the synchro generator and the differential unit are displaced.

Differential Synchro Motors. The connection of a differential synchro motor into a synchro system is shown in Fig. 19–11.

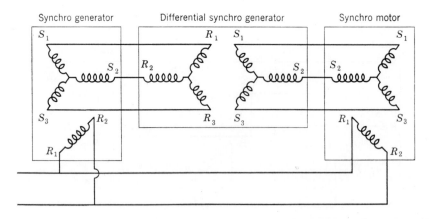

Fig. 19–10. Connection of a differential synchro generator.

SYNCHRO-CONTROL TRANSFORMERS

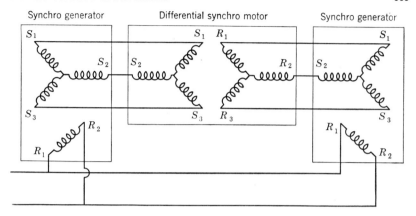

Fig. 19-11. Connection of a differential synchro motor.

If the rotors of the synchro generators are in corresponding positions, the rotor of the differential synchro motor takes a position such that rotor coils of the differential unit physically parallel its stator coils, and the differential unit corresponds to a transformer with a one-to-one ratio.

If the rotor of one of the synchro generators is displaced and the rotor of the other restrained, the voltage balance is altered, and the rotor of the differential unit moves through an angle equal to that by which the generator rotor is displaced. If the generator rotor that is displaced is now restrained, and the rotor of the other generator is displaced by the same amount, and in the same direction, as the first, the rotor of the differential motor moves back to its original position. The displacement of the differential motor is therefore the algebraic difference of the displacements of the two generator rotors.

The differential synchro system is identical with a mechanical differential in that both have three shafts and the relationships among the three shaft rotations are the same.

19-6. SYNCHRO-CONTROL TRANSFORMERS

A synchro-control transformer is similar in construction to a synchro generator and has the function of indicating, by means of an output voltage, the magnitude and direction of the displacement of a generator rotor in the synchro system of which the generator is a part. The connection of a synchro-control transformer into a synchro system is shown as Fig. 19-12.

The rotor circuit of the control transformer is shown to be open in Fig. 19-12. The usual connection of the rotor is to a circuit that has a high impedance so that a small current flows. Under these conditions, little or no torque is developed on the control-transformer rotor.

Because the voltage distribution in the stator coils of the control transformer is the same as that in the generator-stator coils, the resultant field produced by the current in the stator coils of the control transformer corresponds to that produced by the generator rotor, as shown in Fig. 19-13. (Figure 19-13 is drawn for the instant of time when the flux is increasing in an upward direction in the generator rotor as shown.) Thus, when the position of the rotor of the control transformer corresponds to that of the generator rotor, the effective value of the voltage induced in the rotor winding of the control transformer is the maximum.

If the rotor of the generator in Fig. 19-12 is displaced, the effective value of the voltage of the control-transformer rotor winding varies with the angle of displacement and reaches a zero value when the rotors are displaced by 90 degrees. Increasing the displacement angle beyond this point causes the transformer output voltage to increase, and it again reaches the maximum value when the rotors are displaced by 180 degrees from the original position.

The transformer rotor is in the same physical location when it is displaced 180 degrees from the generator rotor as it is when it is considered to correspond with the generator rotor. The difference between the two positions is in the relative polarities of the transformer-rotor terminals with respect to the generator-rotor terminals; the change in relative polarity takes place after the rotor displacement has exceeded 90 degrees.

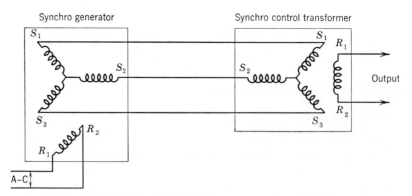

Fig. 19-12. Connection of a synchro-control transformer.

SOME SYNCHRO APPLICATIONS

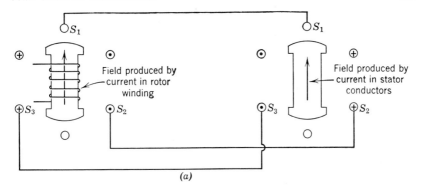

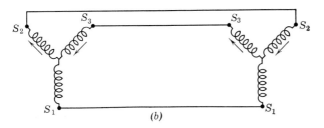

Fig. 19-13. Details for circuit of Fig. 19-12.

The normal position of the transformer rotor when the unit is connected into a synchro system is such that the voltage output is zero. Thus, in the synchro system considered in the preceding discussion, the rotor of the control transformer would normally be displaced 90 degrees from the generator rotor. Any change in the position of the generator rotor then causes a voltage to appear at the transformer-rotor terminals. The magnitude of this voltage is a function of the angle of displacement. The relative polarity of the rotor terminals with respect to the generator-rotor terminals is a function of the direction in which the generator rotor is displaced. When properly applied, the magnitude and polarity information determines how the remainder of the synchro system should operate.

19-7. SOME SYNCHRO APPLICATIONS

The application of synchro units to the problem of antenna positioning control discussed earlier in this chapter is shown schematically in Fig.

19-14. In this application, the voltage output of the control transformer is zero when the antenna is in the position desired. Under these conditions, there is no input to the amplifier and, therefore, no operation of the output control, which consists of a motor having a developed torque dependent upon the magnitude of the control-transformer output voltage and a direction of rotation dependent upon the relative polarity of the transformer output terminals with respect to the synchro-generator-rotor terminals.

If the control dial (and therefore the synchro-generator rotor) is turned to a new position, a voltage is developed at the terminals of the control-transformer rotor. Application of this voltage to the amplifier and, subsequently, to the output control causes the antenna to be rotated. As the antenna rotates, the rotor of the control transformer is rotated a similar amount, because of the geared connection to the antenna shaft. When the antenna (and therefore the control-transformer rotor) has rotated through the same angle as the control dial is originally displaced, the voltage output of the control transformer is again zero, and rotation ceases.

As stated earlier, the rotor circuit of the control transformer is connected to a high-impedance circuit in the amplifier so that negligible current flows, and, therefore, negligible torque is exerted by the control transformer.

If it is desired that the antenna be controlled from either of two locations, the partial schematic circuit shown as Fig. 19–15 might be applied. The output connections are omitted in Fig. 19–15, but they are assumed to be the same as shown in Fig. 19–14.

The operation of the system shown in Fig. 19–15 is, in general, the same as that described for Fig. 19–14 except that the position of the

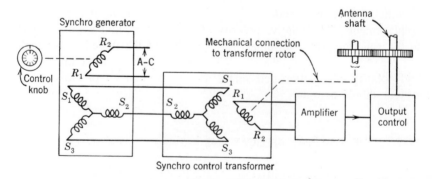

Fig. 19–14. Synchro application to antenna position control.

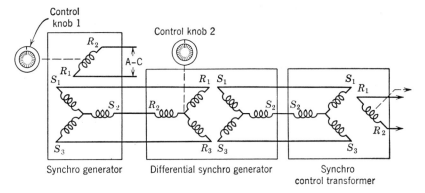

Fig. 19-15. Synchro application.

antenna can be controlled from either of two locations. In this application, however, the control dials do not indicate the true location of the antenna since the antenna may be rotated by turning one control dial without affecting the position of the other control dial. If it is desired that the true position of the antenna be known at either or both of the control locations, the system must be augmented by an energized synchro generator having its rotor connected to the rotor of the control transformer and its stator winding connected to the stator windings of synchro motors at the desired locations. The rotors of the synchro motors would carry pointers that indicate the true position of the antenna.

19-8. ROTATING AMPLIFIERS

Some factors to be considered in applying an amplifier to a control system are (a) the nature of the input signal, (b) the desired power amplification, and (c) the required speed of response. There are, of course, other factors, such as economy and reliability, but in the continuing search for new and improved methods of controlling power the first three factors predominate.

The d-c generator is a form of power amplifier, and its field rheostat and field winding constitute one of the earliest advancements in control. The voltage of a generator can be changed more rapidly and more accurately by adjusting the field rheostat than by changing the speed of the generator, and thus the time of response is decreased. As an example of the d-c generator as an amplifier, a 10-kilowatt generator delivering full load to a constant-resistance load requires a power input to its field winding of approximately 100 watts. Since the power output of the generator is zero when the power input to the field is zero, a change in the field power from zero to 100 watts results in a change in power output of 10 kilowatts. Thus, if amplification is defined as the ratio of the change in controlled power to the change in controlling power, the amplification of the generator is 100.

Consideration of the d-c generator as an amplifier for modern control applications reveals the following disadvantages: (a) because of the large number of turns in the field winding and the resulting high inductance, the time of response is too long for most applications; (b) the controlling power cannot be considered to be a low-power signal, a requirement in most control applications; and (c) an amplification of 100 is seldom large enough.

Although the d-c generator, as such, has disadvantages as an amplifier, it forms the basis of several rotating amplifiers that find wide application in the field of control. These rotating amplifiers are known in industry by the trade names applied by the various manufacturers. Examples are the Amplidyne (General Electric), the Regulex (Allis-Chalmers), and the Rototrol (Westinghouse). Although these amplifiers differ in construction and operating details, they have the same field of application and are similar in that they all produce good amplification of a low-power signal in a short time.

19–9. THE AMPLIDYNE

The 10-kilowatt d-c generator discussed in the preceding article is shown schematically as Fig. 19–16. As indicated in the diagram, the load resistor is assumed to have a resistance of 1 ohm. Additional assumptions are (a) that the field input of 100 watts produces a generated voltage of 100 volts, and (b) that the armature resistance of 0.01 ohm causes negligible voltage drop. Thus, the output current is 100 amperes, and the output power is 10 kilowatts.

The mmf resulting from the armature current produces an armature-reaction field perpendicular to the main field, as shown in Fig. 19–16. The armature-reaction field may be, and is assumed to be, of the same magnitude as the main field, but it does no useful work because the voltages that are generated on account of it do not appear between the brushes as they are now placed. The polarities of the voltage generated in the armature conductors by the armature-reaction field are shown schematically in Fig. 19–17. As indicated, the net voltage generated by the armature-reaction field is zero between the brushes.

THE AMPLIDYNE

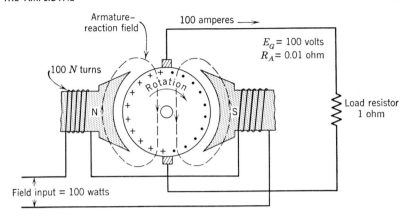

Fig. 19-16. Conventional d-c generator.

If the number of turns on the field winding of the generator is reduced to $\frac{1}{100}$ of the original number, the field voltage required to produce the same field current as before is $\frac{1}{100}$ of the original voltage, and the power input is $\frac{1}{100}$ of the original power, or 1 watt. For a linear magnetization curve, the generated voltage under these conditions is 1 volt (since it is assumed that a 100-watt field produces a generated voltage of 100 volts).

If, with the field input thus reduced to 1 watt, the load resistor is removed, and the armature is then short-circuited, the current flowing in the short circuit is 100 amperes, and the armature-reaction field is therefore the same as under the original conditions. The schematic

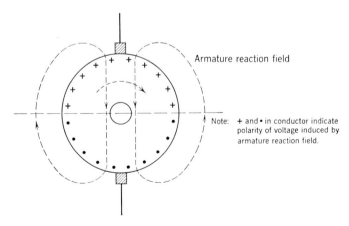

Fig. 19-17. Armature-reaction field.

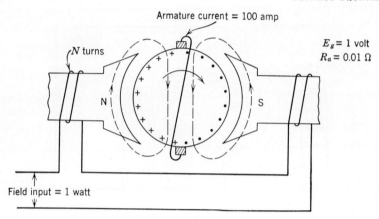

Fig. 19–18. Conventional d-c generator with field input reduced and armature short-circuited.

diagram of the generator operating under the new conditions is shown as Fig. 19–18.

If additional brushes are placed on the quadrature axis, or the axis perpendicular to the axis of the original brushes, the voltage appearing between the additional brushes is that due to the armature-reaction field. Since it is assumed that the armature-reaction field produced by a total armature current of 100 amperes is equal to the original main field produced by a 100-watt input, the value of the voltage between the quadrature brushes is 100 volts. This condition is represented schematically in Fig. 19–19.

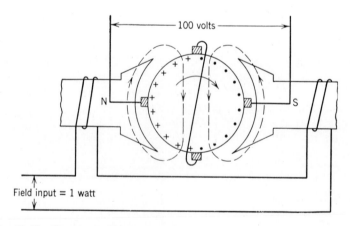

Fig. 19–19. Short-circuited generator with brushes added on the quadrature axis.

THE AMPLIDYNE

If the original 1-ohm load is connected to the quadrature brushes, the current to the load is 100 amperes, and the power output of the machine is 10 kilowatts. Since the power input to the field is 1 watt, the power amplification is 10,000.

The armature-reaction field produced by the 100-ampere load current is in such a direction as to oppose, and thus to reduce, the main field. To prevent this reduction, a compensating winding is connected in series with the load and wound on the main-field poles to produce an mmf that is equal and opposite to that produced by the armature current flowing in the circuit between the quadrature brushes, as shown in Fig. 19–20.

The armature shown schematically in Fig. 19–20 contains two rows of crosses and dots which indicate directions of currents, although there is only one armature conductor at each of the locations. The outer row of symbols represents the directions of currents due to the voltage generated by the main field in the circuit between the short-circuited brushes; the inner row represents the direction of currents due to the voltage generated by the armature-reaction field in the circuit between the quadrature brushes. The current actually flowing in the armature conductors, therefore, is the algebraic sum of the two currents. For the example chosen, conductors in the first and third quadrant carry zero current; those in the second and fourth quadrants carry 100 amperes. Because of this unbalance and the larger armature conductors required, an Amplidyne is physically larger than a conventional d-c generator of the same output rating.

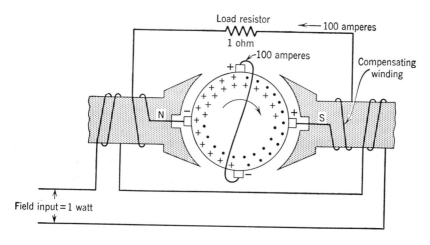

Fig. 19–20. The Amplidyne.

A comparison of the Amplidyne and a conventional d-c generator as amplifiers reveals (*a*) that the speed of response of the Amplidyne is much greater than that of the generator because of the reduction in field-winding turns, and therefore in field-winding inductance; (*b*) that, for the same output, the change in field-power input for the Amplidyne need only be 1 watt compared to 100 watts for the generator, and, therefore, that the Amplidyne operates from a low-power signal; and (*c*) that the amplification of the Amplidyne is 10,000 compared to 100 for the generator.

In the preceding discussion of the operation of an Amplidyne, only one field winding was considered. In many applications, the machine has two or more field windings, each of which may receive independent signals. In these applications, the operation of the Amplidyne is governed by the resultant field produced by the several windings.

19-10. THE ROTOTROL AND THE REGULEX

The standard **Rototrol** is similar in construction to a standard d-c generator except that it is provided with more field windings than the usual d-c generator. (Rototrols are also manufactured as multistage units to provide greater amplification, but only the standard Rototrol is considered in this book.) Basically, the Rototrol is a shunt or series generator with two additional field windings: namely, a **pattern-field winding** and a **pilot-field winding**.

A common connection of the Rototrol is as a series generator. Such

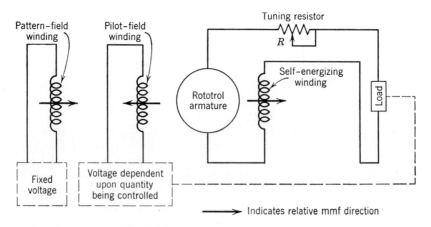

Fig. 19-21. Typical Rototrol connection.

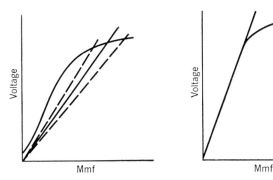

Fig. 19–22. Voltage versus mmf for d-c generators.

a connection in a control circuit is shown schematically in Fig. 19–21.

As indicated in Fig. 19–21, the pattern-field winding is connected to a fixed supply so that the mmf of the winding is constant and acts as a reference. The supply for the pilot-field winding has a magnitude dependent upon the quantity being controlled. (In control applications, the load connected to the Rototrol armature is a function of, or is related to, the controlled quantity.) The pattern field and pilot field are connected so that their mmf's are in opposition. The mmf of the series, or self-energizing field winding, is in the same direction as the pattern-field mmf.

The voltage output of a series generator, like that of a shunt generator, is given by the intersection of the field-circuit resistance characteristic and the saturation curve that is applicable to the generator. The usual d-c series or shunt generator has a field-circuit resistance characteristic and saturation curve as shown in Fig. 19–22a. In this example, the intersection of the two characteristics occurs at only one point, and thus the generator has a definite output voltage. A change in the field-circuit resistance causes the point of intersection, and therefore the output voltage, to change; however, a minor change in field-circuit resistance produces only a slight change in output voltage, as shown in Fig. 19–22a.

The Rototrol is so designed that the field-circuit resistance characteristic and the saturation curve coincide over a considerable portion of the curves, as shown in Fig. 19–22b. Therefore, the output voltage of the Rototrol is indeterminate and may be any value along the coincident curves. (The tuning resistor in the circuit of Fig. 19–21 is adjusted to make the field-circuit resistance characteristic coincident with the saturation curve.)

In actual operation, the value of the output voltage of the Rototrol is determined by the quantity being controlled and is so regulated that there is no difference between the mmf of the pattern field and the mmf of the pilot field. If the pilot-field mmf is smaller than the pattern-field mmf (indicating that the controlled quantity has deviated from the desired value), the net mmf of the two fields is in the same direction as the mmf of the self-energized field, and the output voltage of the Rototrol increases to restore the balance between the pattern and pilot fields. Similarly, if the pilot-field mmf is larger than the pattern-field mmf, the output voltage of the Rototrol decreases. Thus, although the output voltage of the Rototrol could be any value along the coincident curves, the actual output voltage remains constant at the proper value if the controlled quantity is constant at the desired value. Since only a slight difference in mmf's is necessary to cause the output voltage of the Rototrol to vary over a wide range, the Rototrol responds quickly to changes in the controlled quantity.

The preceding discussion of the operation of the Rototrol is based upon the series connection of the self-energizing field winding. The operation is fundamentally the same as that obtained when the self-energizing winding is shunt-connected. The shunt connection is the type commonly employed in the **Regulex,** an Allis-Chalmers product, which has the same basis of operation as the Rototrol but differs in design details. The discussion of the principles of operation of the Rototrol therefore also applies to the Regulex.

19-11. SOME ROTATING-AMPLIFIER APPLICATIONS

Each of the rotating amplifiers described in the preceding articles is widely applied in the control of voltage, current, speed, acceleration, position, power factor, tension, and power. No attempt is made here to list all the applications of such amplifiers, but the consideration of a few specific examples provides an insight as to their usefulness in control circuits. In the following examples, the Amplidyne is used as the amplifier, although, as previously indicated, any of the rotating amplifiers could perform the same functions.

Voltage Control. The connection of an Amplidyne to control the voltage output of a d-c generator is shown in Fig. 19–23.

In the connection shown in Fig. 19–23, the control-field winding of the Amplidyne is supplied by the voltage output of the generator that is being controlled, and the reference-field winding is connected to a separate d-c supply. The output voltage of the Amplidyne supplies the field winding of the d-c generator.

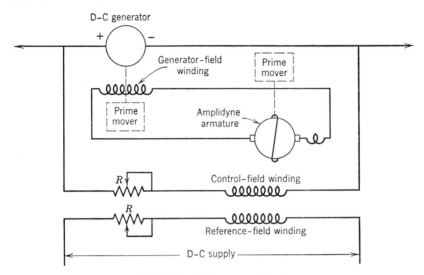

Fig. 19-23. Voltage control circuit.

When the d-c generator is producing the correct voltage, the Amplidyne field mmf's are such that the mmf of the reference field is larger than the mmf of the control field. In effect, therefore, the reference field governs the voltage output of the Amplidyne.

If, for some reason, the voltage of the d-c generator increases, the control-field-winding current and, therefore, the control-field mmf increase. The difference in mmf between the control field and the reference field therefore decreases, and the output voltage of the Amplidyne decreases. The decrease in Amplidyne voltage results in a decrease in generator field current and, therefore, generator voltage. Thus, for an increase in the voltage of the generator, the control circuit operates to lower the generator voltage.

An original decrease in generator voltage has the reverse effect; that is, the control circuit acts to raise the generator voltage.

The rheostats in the control-field and reference-field circuits are for the purpose of changing the value of the generator voltage to be maintained.

Current Control. A circuit employing an Amplidyne for the control of the armature current taken by a motor is shown as Fig. 19-24.

In Fig. 19-24, the Amplidyne is so connected in series with the motor field winding that the Amplidyne output voltage, which is in opposition to the supply voltage, governs the field current taken by the motor. Only one control field is used; it is so connected that the current through it depends upon the difference between a fixed reference voltage and a voltage produced by the motor armature current flowing through a fixed resistor.

When the armature current of the motor is at the value to be maintained, the reference voltage is larger than the voltage across the armature resistor, and the control-field current flows as indicated in Fig. 19–24, which also shows the polarity of the voltage generated in the Amplidyne.

If the load driven by the motor is increased, the current in the motor armature increases to produce an increased torque. The increase in armature current causes an increase in the voltage drop across the armature resistor and, therefore, a decrease in the control-field current. The decrease in control-field current causes a decrease in the Amplidyne voltage and, therefore, an increase in motor field current, which results in an increase in motor field flux. If the load torque remains constant after the initial increase in load, the increased field flux causes the armature current to decrease. Thus, the control circuit has operated to maintain the armature current at the correct value.

In maintaining the armature current constant by means of the control circuit shown in Fig. 19–24, the motor speed is varied; therefore, this

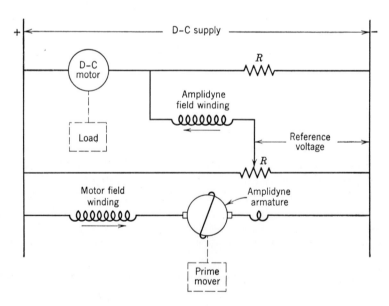

Fig. 19–24. Armature-current control circuit.

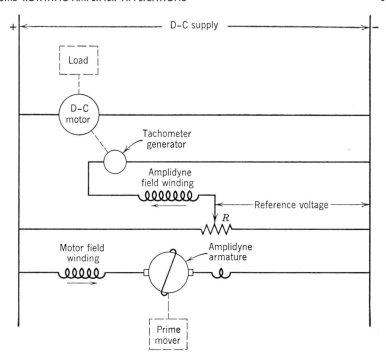

Fig. 19-25. Speed control circuit.

method of current control must not be applied where variations in motor speed are important to the motor application.

The value of the current to be maintained is changed by the adjustment of the rheostat in the reference-voltage circuit.

Speed Control. A circuit incorporating an Amplidyne to control the speed of a motor is shown as Fig. 19-25.

The Amplidyne connection in Fig. 19-25 is similar to that for current control. For speed control, the current through the single control-field winding depends upon the difference in voltage between a fixed-reference voltage and the output voltage of a tachometer generator connected to the motor shaft.

When the speed of the motor is at the value to be maintained, the reference voltage is larger than the output voltage of the tachometer generator, and the control-field current flows as shown in Fig. 19-25. If the speed of the motor increases, the control-field current decreases, which results in a decreased Amplidyne voltage. When the Amplidyne voltage decreases, the motor field current and, therefore, the motor field flux increase. The increase in motor field flux results in a decrease in motor speed; therefore, the control circuit has operated to maintain the speed at the desired value.

19–12. MAGNETIC AMPLIFIERS

The impedance of an iron-core coil connected in series with an a-c supply and a load affects the current, and therefore the power, taken by the load, and any change in this impedance causes the current to change. One method of changing the impedance of the coil is to change its inductance by changing the state of magnetization of its core material through the use of an auxiliary coil on the same magnetic circuit. Thus, by proper design of the magnetic circuit, a small change in the power input to the auxiliary coil results in a large change in the power to the load, and power amplification is accomplished. The main coil and auxiliary coil on a common magnetic circuit constitute a magnetic amplifier.

The basic circuit from which magnetic amplifiers are developed is shown in Fig. 19–26. The magnetic core with two coils, one in series with the a-c supply and the load (assumed to be resistance), and the other connected to a d-c supply, is known as a **saturable reactor**.

The voltage equation applying to the a-c circuit of Fig. 19–26 may be written as

$$e_{ac} = e_{ab} + e_{bc} \qquad (19\text{–}1)$$

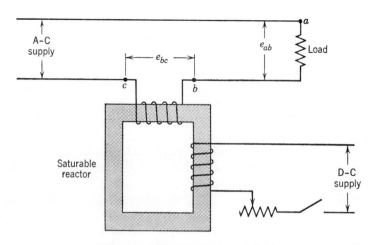

Fig. 19–26. Basic magnetic-amplifier circuit.

MAGNETIC AMPLIFIERS

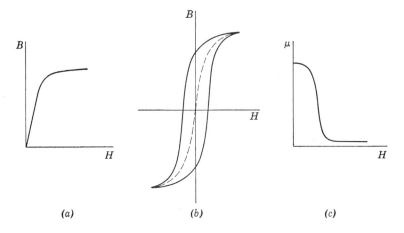

Fig. 19-27. Magnetization relationships.

Where e_{ac} is the instantaneous value of applied voltage, e_{ac} is the voltage across the load, and e_{bc} is the voltage across the main coil of the saturable reactor. If the applied voltage is sinusoidal and constant in frequency, and if the main coil is assumed to have negligible resistance, Eq. 19-1 may be written as

$$E_m \sin \omega t = \left(N \frac{d\phi}{dt} \times 10^{-8} \right) + iR \qquad (19\text{-}2)$$

where N is the number of turns of the main coil and R is the load resistance.

Because the normal magnetization curve for the core material of the saturable reactor has a nonlinear form similar to that shown in Fig. 19-27a, and because of the hysteresis effects shown in Fig. 19-27b, Eq. 19-2 cannot be solved by ordinary mathematical means, and graphical methods are usually employed. Modification of Eq. 19-2 so that it has the form

$$E_m \sin \omega t = L \frac{di}{dt} + iR \qquad (19\text{-}3)$$

does not eliminate the difficulties because the inductance L is not a constant, since it depends on the permeability μ of the magnetic circuit. If permeability is defined as the ratio of the change in flux density to the change in magnetizing force at a given magnetizing force (or the slope of the B–H curve at the given magnetizing force), the curve of μ vs. H for normal magnetization is as shown in Fig. 19-27c. Although the permeability curve shown does not take into account hysteresis effects resulting from a changing magnetomotive force, it is satisfactory for an initial analysis of the saturable reactor and, therefore, the magnetic amplifier.

If the current in the d-c winding of Fig. 19–26 is zero and the magnitude of the alternating voltage is such that the magnetic core is not saturated, the current in the load is small because of the large inductance of the main coil; that is, a large percentage of the applied voltage appears across the main coil, a small percentage across the load. The hysteresis loop applicable under these conditions is shown in Fig. 19–28, which also shows a line indicating the general slope of the hysteresis loop. This line may be considered to represent the average permeability for this condition of operation.

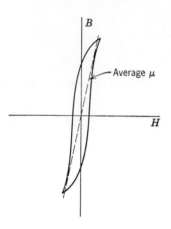

Fig. 19–28. Hysteresis loop for saturable reactor with no d-c magnetization.

If the current in the d-c winding is so adjusted that the magnetization with no a-c mmf applied to the core corresponds to point a in Fig. 19–20a, the permeability of the core, and therefore the inductance of the main coil, is at a very low value, and a larger alternating current flows to the load when the same alternating voltage is applied to the circuit. The hysteresis loop and the average permeability applicable under these conditions might be as shown in Fig. 19–29b.

The actual hysteresis loop obtained for the d-c magnetization conditions shown in Fig. 19–29b represents only one possibility. Thus, the

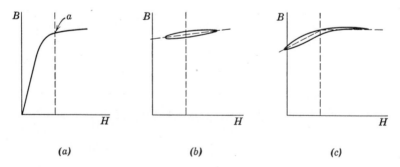

Fig. 19–29. Effects of d-c magnetization on saturable reactor.

MAGNETIC AMPLIFIERS

loop shown in Fig. 19–29c is equally possible and indicates one disadvantage of the type of saturable reactor shown in Fig. 19–26; that is, the loop is not so symmetrical about the original magnetization point as the loop shown in Fig. 19–29b. The result of the dissymmetry is that the current wave form for the positive portion of a cycle may be widely different from that of the negative portion because of the wide change in permeability. Another disadvantage of the saturable reactor shown is the fact that the d-c winding is linked by the alternating flux produced by the main coil, and, therefore, a large alternating voltage may be induced in the d-c winding. The two disadvantages mentioned can be eliminated by the construction shown in Fig. 19–30.

The saturable reactor shown in Fig. 19–30 consists of a three-legged magnetic circuit with a coil on each of the legs. The two identical coils on the outside legs are connected in series to form the main coil of the reactor, and the connection is such that the mmf's of the two coils are additive. Therefore, because of the symmetry of the magnetic circuit, when these coils are energized from an a-c supply, and the d-c winding is unenergized, no flux passes through the middle leg, and no alternating voltage is induced in the d-c winding.

Because of the connection of the magnetic circuit, the mmf of the d-c winding opposes that of one a-c coil while aiding the other during a given portion of the a-c cycle. The result is that one of the ouside legs of the magnetic circuit is saturated while the other may not be. The overall result of the energizing of the d-c winding, however, is a decrease in the combined inductance of the main coils and an increase in the effective value of current taken by the load. Because the main winding

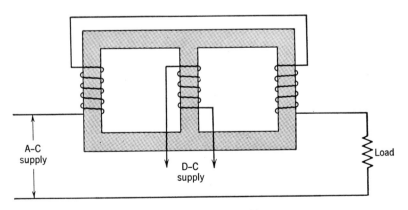

Fig. 19–30. Three-legged saturable reactor.

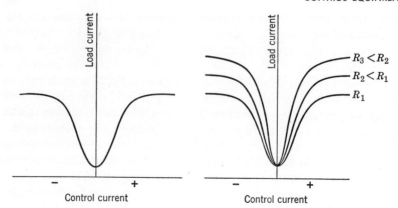

Fig. 19–31. Load current versus control current for saturable reactor.

consists of the two coils in series, the effect on one half of the cycle is the same as that on the next half, and a symmetrical current wave form results. (Similar results are achieved if the main coils are connected in parallel and the combination is connected in series with the load.)

A plot of the effective values of load current versus current in the d-c winding (called the control current) for a saturable reactor is shown as Fig. 19–31. The negative values of control current represent currents resulting from a change in the polarity of the voltage applied to the d-c winding. As indicated by the curves, such a reversal of polarity has no effect on the shape of the load-current characteristic.

For a practical reactor, the limiting value of load current shown in Fig. 19–31a is determined by the applied voltage, the main-coil resistance, and the load resistance. When the load current reaches the limiting value, the core is completely saturated, and the inductance of the main winding is at the minimum value. Under these conditions, the inductance may be considered to be negligible, and therefore the only impedance in the circuit consists of the main-coil resistance and the load resistance. A decrease in the load resistance for a given applied voltage results in a higher limiting value of load current, as indicated in Fig. 19–31b.

Although the saturable-reactor circuits shown in Figs. 19–26 and 19–30 have amplifying properties, the amplification is not large because of the large amount of power necessary in the control winding. Any decrease in the controlling power for a given controlled power results in an increased amplification, and such a decrease is made possible by the use of rectifiers in the load circuit to cause the reactor to become self-saturating.

19-13. SELF-SATURATING MAGNETIC AMPLIFIERS

The connection of a rectifier in the load circuit of the reactors previously considered results in a pulsating current that flows only during a portion of the applied voltage cycle; thus, the current is unidirectional and not alternating, as in the preceding discussion. Although reactors with rectifiers can be connected to produce an alternating current in the load, the first consideration here is to determine the effect of the rectifier upon the saturable-reactor circuit. The connection of a rectifier in the circuit of Fig. 19–26 is shown in Fig. 19–32.

For zero current in the control winding of Fig. 19–32 and a value of applied voltage that does not produce saturation, the hysteresis path during the first period of current flow is as shown in Fig. 19–33a. However, for subsequent periods the hysteresis path is as shown in Fig. 19–33b, because the end of the current flow during the first period left the core with residual magnetism, as indicated by Fig. 19–33a. Therefore, the average permeability during periods of current flow with a rectifier in the circuit is less than the average permeability when the rectifier is removed, which is indicated in Fig. 19–28. Thus, the effect of the insertion of the rectifier is to produce a higher average value of

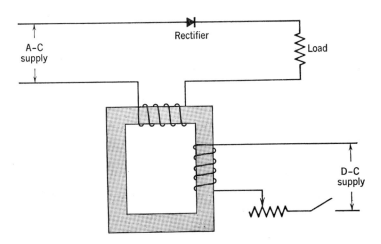

Fig. 19–32. Self-saturating saturable reactor.

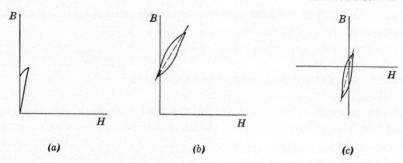

Fig. 19–33. Hysteresis effects in self-saturating saturable reactor.

current for the period when current flows, although the effective value may be reduced because current flow is blocked by the rectifier for a part of the voltage cycle. The increased average value for a portion of a cycle, however, is equivalent to the condition in the saturable reactor with no rectifier when a small value of current flows in the control winding. Thus, the tendency toward saturation is produced by the current in the load circuit rather than by current in the control winding, and the reactor is said to be self-saturating.

With the rectifier in the load circuit, the polarity of the control winding becomes important. With one polarity, the d-c mmf aids the saturating mmf produced by the main coil, and the core approaches saturation; with the reversed polarity, the d-c mmf opposes the saturating mmf produced by the main coil. Control current that produces an aiding

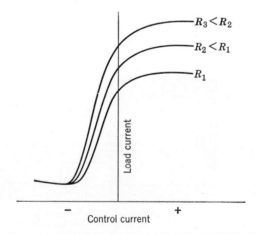

Fig. 19–34. Self-saturating saturable-reactor control characteristic.

mmf is designated as positive; current that produces an opposing mmf, negative. A negative control current may result in a hysteresis loop similar to that shown in Fig. 19–33c. The average permeability indicated for this loop is larger than that for the condition when no control-winding mmf is applied, and a lower average current results.

The complete control characteristic (plot of load current versus control current) for a saturable reactor with a rectifier is shown as Fig. 19–34. As for the saturable reactor with no rectifier, the limiting value of control current is determined principally by the load resistance, and the control characteristics for different values of load resistance are shown in Fig. 19–34.

19–14. MAGNETIC-AMPLIFIER CONNECTIONS

Practical magnetic amplifiers consist of combinations of saturable reactors and rectifiers, and the elements can be connected to supply either a-c or d-c loads. Although practical magnetic amplifiers usually have more than one control winding, the connections considered here show only one. For all the connections, control characteristics similar in form to those shown in Fig. 19–34 are applicable.

One connection for supplying an a-c load is shown in Fig. 19–35. As indicated, the main coils have separate cores and are connected in parallel. The rectifier connections are such that current flows through a given winding for only one half of a cycle, but the current through the load is alternating. The control windings are so connected in series that positive control current produces saturation for each of the main-winding coils.

Figure 19–36 shows two methods of controlling a d-c load by means of magnetic amplifiers. Figure 19–36a shows a bridge connection from

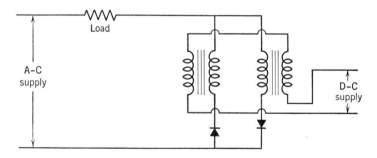

Fig. 19–35. Magnetic amplifier with a-c load.

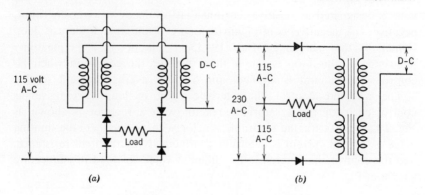

Fig. 19-36. Magnetic amplifiers with d-c loads.

a single-voltage supply; Fig. 19-36b is a bridge connection from a double-voltage supply.

The connections shown in Fig. 19-35 and 19-36 are representative of many possible connections; the particular connection chosen depends upon the application and the amplifier characteristics desired.

■ 20 High-Vacuum Tubes

20–1. INTRODUCTION

The tubes and semiconductor devices described in this and the following two chapters have several useful features that differentiate them from the circuit elements previously considered. Although those circuit elements were sometimes nonlinear, the nonlinearity was incidental to the functioning of the elements and was symmetrical; that is, it was independent of the direction of current through the element. The extreme nonlinearity of some electronic devices (as both tubes and semiconductors are called) is their most prominent and useful feature; that is, because they conduct current in one direction much more readily than in the other, they make efficient rectifiers for the conversion of a-c to d-c power. Also, the addition of a third element, or electrode, allows electric control of the current. This control is the basis for the many amplifier and other control circuits.

20–2. ELECTRON TUBES

An electron tube consists of a vacuum-tight glass or metal envelope in which there are two or more terminals, or **electrodes,** that carry or control the current within the tube. External terminals provide electric connections to the electrodes within the tube. The envelope of an electron tube may be highly evacuated, or it may contain an inert gas or metallic vapor at low pressure. Accordingly, electron tubes are classified as either **high-vacuum tubes** or **gas-filled tubes.** Tubes are also classified according to the number of electrodes. A two-electrode tube is known as a **diode;** a three-electrode tube, a **triode;** a four-electrode tube, a **tetrode;** and a five electrode tube, a **pentode.** A tube having more than five electrodes is generally referred to as a **multielectrode,** or **multielement, tube.**

The two principal electrodes in any tube are called the **anode** and the **cathode.** These electrodes carry the main current in the tube and are named with respect to the direction of that current: *conventional current flows from the anode to the cathode through the tube,* or the electron flow through the tube is from the cathode to the anode.

Since there are normally no free charges in either a vacuum or a gas, no current can flow through an electron tube unless the cathode of that tube emits electrons into the space between the electrodes. **Thermionic emission, photoelectric emission, secondary emission,** and **high-field emission** are processes by which electrons are liberated from the cathode. (These processes are treated separately in the following articles.) The operation of every electron tube depends upon one or more of these forms of electron emission.

A functional description of any electron tube can be made by classifying the tube in each of the three categories discussed in the preceding paragraphs: high-vacuum, or gas-filled; the number of electrodes; and the method for obtaining electron emission. Thus, one tube may be a high-vacuum thermionic triode; another, a mercury-vapor thermionic diode. However, it is common practice to apply special names to a few of the common tube types. For example, a gas-filled thermionic diode is called a **phanotron;** and any tube that employs photoelectric emission at the cathode is called a **phototube.** These names, and others that are defined in subsequent chapters, are used throughout this book.

20–3. THERMIONIC EMISSION

An electric conductor has many free electrons. However, these electrons are free only to the extent that they may transfer from one atom to another within the conductor; they cannot leave the conductor to provide electron emission. If an electron starts to leave a cathode, a positive charge is induced on the cathode. The force of attraction between this positive charge and the negative charge of the electron opposes the escape of the electron. This force is so great for small separations that, for practical purposes, no electrons can overcome it unless they receive additional energy from some external source. The manner in which additional energy is given to the electrons so that they may leave the surface of a cathode characterizes the type of emission.

Thermionic emission is the emission of electrons that results when the temperature of a cathode is raised. Much of the thermal energy of a heated cathode is, in reality, the kinetic energy of its free electrons and, if the cathode temperature is made high enough, an appreciable number of the free electrons gain the kinetic energy necessary to overcome the forces binding them to the cathode. The amount of thermionic emission increases rapidly as the cathode temperature is increased, as shown by Eq. 20–1, which relates the emission current density J to the absolute temperature T of the cathode:

$$J = aT^2 \epsilon^{-b/T} \tag{20-1}$$

The constants a and b depend upon the cathode material and upon the system of units. This expression was first derived on the basis of theoretical considerations, but it has since been verified experimentally.

Although all conductors are theoretically capable of thermionic emission, few materials emit in usable quantities at temperatures below the melting point. Because of its high melting point, tungsten is the only metal extensively used as a thermionic emitter; its emissivity is increased by the addition of a small amount of thorium.

For most tubes, pure-tungsten and thoriated-tungsten cathodes have been superseded by the **oxide-coated cathode.** This cathode consists of a thin layer of oxides (usually a combination of barium and strontium oxides) on a metallic structure which is necessary both for physical support and for making electric contact with the emitting oxides. The oxide-coated cathode not only is a more efficient emitter than a cathode of either pure or thoriated tungsten, but also operates at a lower temperature. The principal limitation of the oxide-coated cathode is that it cannot withstand high anode-to-cathode potentials. At less than a few thousand volts, the oxide-coated cathode is used exclusively.

20–4. CONSTRUCTION OF THERMIONIC TUBES

Thermionic cathodes are heated by either of two methods—by the passage of an electric current through the cathode, or indirectly from a current-carrying heater coil. A **directly heated cathode** is heated by the flow of electric current through the cathode itself and is therefore fabricated of a conducting wire or ribbon. Such a cathode may be made of tungsten wire, thoriated-tungsten wire, or an oxide-coated nickel wire or ribbon. Figure 20–1 illustrates schematically how connections to a cathode may be made.

A cathode heated from a separate heater is called an **indirectly heated cathode**. In this type of cathode, there is no heater current in the cathode to cause a voltage drop; therefore, the cathode is sometimes referred to as a **unipotential cathode**. The merit of the indirectly heated cathode is that an alternating heater current can be used without causing any noticeable alternation of the cathode temperature or emission. The emitting surface of an indirectly heated cathode is always oxide-coated.

Several practical thermionic cathode structures are shown in Fig. 20–2. The directly heated cathodes illustrated are (a) a wire and (b) a coil of edge-wound ribbon. The wire may be of tungsten (pure or thoriated), nickel, or a nickel alloy, but the ribbon must be nickel or a nickel alloy. If the directly heated cathode is oxide-coated—and only the tungsten wire operates without such a coating—the entire surface of the cathode structure is coated with the oxides.

The simplest indirectly heated cathode is a nickel sleeve with an oxide coating applied to its external surface. An indirectly heated cathode of the multicellular type, as shown in Fig. 20–2d, has a higher efficiency. The multicellular cathode consists of two concentric cylinders and several connecting fins between them. The emitting surfaces are the internal surfaces of the several sectors formed by the cylinders and fins. A central coil heats the entire structure.

The term **filament** denotes the element through which the cathode heating current flows; it refers to the heater of an indirectly heated cathode, or to the cathode itself in the case of the directly heated cathode.

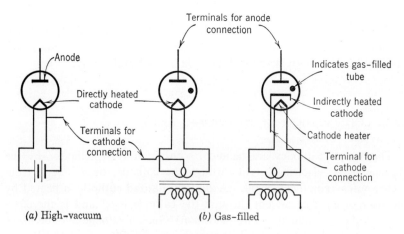

Fig. 20–1. Symbols for thermionic diodes.

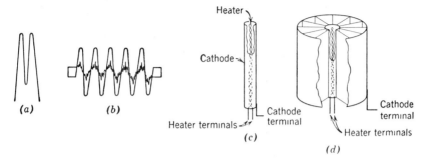

Fig. 20-2. Thermionic cathode structures.

To insure that the cathode of a thermionic tube operates at the temperature for which it was designed, the filament is always given a voltage or current rating. Operation at this rated voltage or current results in a cathode temperature sufficiently high for the necessary emission, but not so high as to damage the cathode by excessive evaporation of the emission material.

The function of an anode is to receive electrons that are emitted from the cathode; therefore, the construction of an anode depends somewhat upon that of the cathode with which it is used. The most common arrangement is a cylindrical piece of sheet metal surrounding the cathode. With the multicellular cathode and other heat-shielded cathodes discussed later, the anode may be simply a metal or carbon disk in front of the open end of the cathode.

20-5. CHARACTERISTICS OF HIGH-VACUUM THERMIONIC DIODES

The fact that there is electron emission at the cathode of a tube does not mean that the current through the tube is equal to the emission current, or that there is any current through the tube at all. The first requirement for current flow through a tube is that there be a complete external circuit connecting the anode and cathode. Thus, no current can flow through a diode that has no connection to its anode, even though its cathode is heated. Without a complete circuit, all emitted electrons eventually return to the cathode, and the net current is zero.

The current that flows through a high-vacuum diode depends upon the anode-to-cathode potential, since a positive anode attracts electrons and a negative anode repels them. A plot of anode current versus anode-to-cathode voltage for a typical diode is shown in Fig. 20-3. For all

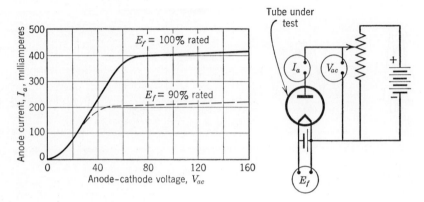

Fig. 20–3. Current–voltage characteristic for a high-vacuum diode having a pure tungsten cathode.

practical purposes, no current flows when the anode is at a negative potential with respect to the cathode. When the anode is made positive, the electrons are attracted to it, and a current flows through the tube and through the external circuit. As the anode is made more positive, the attractive force on the electrons is increased and the current through the circuit is increased. When all the electrons emitted at the cathode are attracted to the anode, a further increase in the anode potential produces no appreciable increase in the current.

If the filament voltage is below normal, the cathode temperature, and therefore the thermionic emission, is lower than normal. The effect of a reduction in the filament voltage is then a reduction in the maximum obtainable anode current, as indicated in Fig. 20–3. Tubes having oxide-coated cathodes exhibit no pronounced saturation such as that shown in Fig. 20–3. High currents apparently affect the oxide coating in such a manner as to increase the emission. Actually, the shape of the characteristic curve above the saturation point is of little importance, since tubes with oxide-coated cathodes are given a peak anode-current rating that is somewhat less than the normal emission current. The instantaneous value of anode current should never exceed this rating.

The most important feature of the current–voltage characteristic of a thermionic diode is that the current is zero for all negative values of anode voltage; the tube conducts only when the anode is at a positive potential with respect to the cathode. Thus, if it is desired that the current flow through a load in but one direction when only an a-c source is available, a thermionic diode may be placed in series with the load to allow current flow only in the desired direction.

20-6. THE HIGH-VACUUM DIODE AS A CIRCUIT ELEMENT

The simplest useful circuit involving a diode consists of a source of emf, an impedance, and the diode connected in series, as shown in the circuit of Fig. 20-4. In general, even this simple circuit requires the application of some indirect method for determining the current since the actual voltage across the diode is not known. Trial and error may be employed to find a current that produces the necessary voltage to satisfy the Kirchhoff voltage equation, or a graphical approach such as the load-line method described in Art. 3-5 may be used.

Figure 20-4 illustrates the application of the load-line construction to a diode-circuit problem in which the current–voltage characteristic for the diode is known, the values of both the supply voltage and the series resistance are known, and the current and the voltages of the circuit are to be determined. As indicated in Fig. 20-4, the load line is plotted to the same set of coordinate axes as for the tube characteristic. The load line intersects the voltage axis at a value equal to the voltage of the supply battery; it intersects the current axis at a value equal to the quotient of the battery voltage and the series resistance. The load line intersects the tube characteristic at a point corresponding to the current in the circuit and the voltage across the tube. The voltage across the series resistor is the difference between the battery voltage and the tube voltage.

In many applications, the voltage drop across a tube is so small in comparison with other voltages in the circuit that the tube drop can be neglected without appreciably affecting the accuracy of the circuit analysis. For the simple circuit of Fig. 20-4, neglecting the tube drop

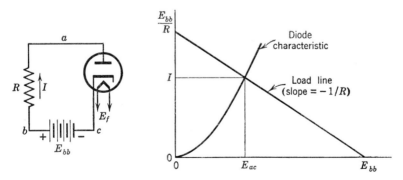

Fig. 20-4. Application of load-line construction to a diode circuit.

is equivalent to assuming that the voltage across the series resistor equals the battery voltage, or that the current in the circuit equals the battery voltage divided by the series resistance.

20-7. PHOTOELECTRIC EMISSION

Many materials emit electrons when exposed to light or other radiation. The emission resulting from such exposure is called **photoelectric emission.** The amount of photoelectric emission is directly proportional to the intensity of the radiation that is responsible for the emission; it is also a function of the frequency, or color, of this radiation. The manner in which photoelectric emission varies with frequency depends upon the emitting material. Some materials are most sensitive to ultraviolet radiation; others to blue, green, red, or even infrared radiation. The design data for any phototube generally include a plot of sensitivity versus the frequency of incident radiation.

Whereas thermionic cathodes may be designed to give an emission of several amperes, the emission obtainable from photoelectric cathodes is of the order of a few microamperes, or millionths of an ampere. To obtain as much current as possible, the cathode area is relatively large, the form of the cathode is such that most of its emitting surface can be illuminated, and the anode is so placed that it can receive electrons from the cathode without interfering with the illumination of the cathode. The construction illustrated in Fig. 20-5, which is typical of most phototubes, is a compromise among these various factors. The inner surface of the semicylindrical cathode is the light-sensitive emitter, and it is

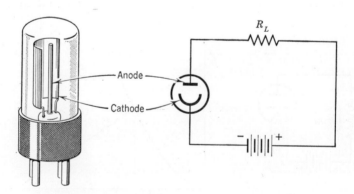

Fig. 20-5. Typical phototube construction and application.

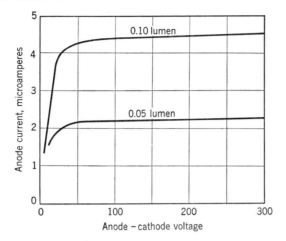

Fig. 20–6. Current–voltage characteristics of a high-vacuum phototube.

generally coated with some special material to give a desired frequency characteristic.

Although the characteristics of the high-vacuum phototube (Fig. 20–6) appear to be considerably different from those of the thermionic diode, the general features of the two are the same. Neither tube conducts when the anode is negative with respect to the cathode; and in both an increase in the anode potential results in an increase in the anode current until saturation is reached. The principal difference in the two characteristics is in the value of current at which saturation occurs. Photoelectric emission produces so few electrons that saturation is reached at low voltages and currents; in contrast to the thermionic tube, the phototube is generally operated beyond saturation. A change in the illumination of the phototube cathode changes the emission, and therefore the saturation current, in the same way that changes in the temperature of a thermionic cathode affect the saturation current in a thermionic tube.

Because current in a high-vacuum phototube is essentially independent of voltage in the normal operating range, approximate circuit calculations are easily made; the assumption is that the current depends only upon the intensity of the illumination. If greater accuracy is desired, a load line can be constructed on a plot of the tube characteristics. Actually, it is often impossible to make any calculations for phototube circuits because of the difficulty in determining the exact amount of light striking the cathode.

20-8. SECONDARY EMISSION

The free electrons of a metal may gain energy from external electrons or ions that strike the surface of the metal. These bombarding particles (if electrons, they are called **primary electrons**) give up some of their kinetic energy to the electrons within the metal. The emission resulting from electron or ion bombardment is called **secondary emission,** and the emitted electrons are called **secondary electrons.**

The factors influencing secondary emission are the emitting material, the energy of the bombarding particle, and the mass of that particle. In general, more secondary emission can be obtained from nonpure than from pure metallic surfaces. The emission is also greater for electron bombardment than for an equivalent ion bombardment, and, up to a certain point, the higher the energy of the bombarding particles, the greater the emission. Under ideal conditions, some specially prepared surfaces may emit ten times as many secondary electrons as there are primary electrons.

Because electrons strike the anode in any tube that is conducting current, it might then seem that secondary emission would always be present in any operating tube, and, in fact, this is the case. However, in most situations, the anode is the most positive electrode in the tube so that any secondary electrons immediately return to the anode. The net effect, as far as any external observations are concerned, is that there is no secondary emission.

Although secondary emission is often an undesirable source of emission from electrodes other than the cathode, the multiplier phototube takes advantage of the phenomenon. This tube utilizes secondary emission to multiply by several thousand times the minute current that can be obtained from photoelectric emission. It has a photosensitive cathode, an anode, and several electrodes called dynodes, whose surfaces are specially prepared to favor secondary emission. The electrons emitted from the cathode are directed to the first dynode, where perhaps five times as many secondary electrons are emitted. The electrons thus liberated are directed to a second dynode, and the process is repeated at each of the dynodes. The total emission from the last dynode goes to the anode and becomes the anode current. Thus, in a tube having eight dynodes and a secondary-to-primary ratio of 5 at each dynode, the current amplification, or the ratio of anode to cathode current, is 5^8, or 390,625.

In order for the multiplier phototube to function as outlined in the preceding paragraph, the cathode and dynodes are shaped and positioned to assist in directing the emitted electrons to the following electrode, and each dynode is maintained at a positive potential with respect to the preceding dynode. The total supply voltage required for this tube is about 1000 volts.

20-9. HIGH-FIELD EMISSION

The charge on an anode that is at a high positive potential with respect to the cathode may be capable of exerting a force of attraction great enough actually to pull some of the free electrons from the surface of the cathode. The same effect can be produced by a concentration of positive charges in the space adjacent to the cathode. The electron emission resulting from the attraction of a positive charge is called **high-field emission,** or simply **field emission.** Field emission may sometimes contribute to the emission from a thermionic or photoelectric cathode, but it is of practical significance only in the mercury-pool cathode, which is discussed in Chapter 24.

20-10. THE ELECTRIC FIELD

Just as the concept of the magnetic field is employed in the study of electric machinery, so the concept of the electric field is employed in the study of current flow in electron tubes. Thus, in predicting the current that will flow in a tube of given geometrical design for given applied voltages, a consideration of the electric field in that tube is essential. In addition to its value in quantitative analyses, the electric-field approach is helpful in gaining a qualitative understanding of many tubes.

There is a force of either attraction or repulsion on each of two separated charges; the magnitude of this force depends upon both the magnitude of the charges involved and the distance between them. A third charge in the vicinity of the first two experiences similar forces, or, specifically, it experiences a net force that is the resultant of the forces due to each of the two original charges. In general, the force on this third, or test, charge is dependent upon its position relative to the other charges. An electric field is said to exist wherever a stationary test charge experiences such a force.

The electric-field intensity at a point is defined as the force on a positive test charge of unit magnitude placed at the point. This quantity has both direction and magnitude, as has the force that defines it. Electric-field intensity is measured in volts per unit distance (volts per inch, volts per meter, etc.).

The source of both electric and magnetic fields is the electric charge. A magnetic field is produced as the result of the motion of electric charges, that is, as the result of an electric current, whereas an electric field is produced by the mere presence of electric charges.

The motion of electrons between the electrodes of a tube can be thought to be the result of forces due to the electric field in the space between the electrodes. The application of potential differences to the various electrodes causes an accumulation of electric charge on those electrodes, and the presence of the charges produces the electric field. An electron at any point in the interelectrode space is acted upon by an accelerating force proportional to the electric-field intensity at that point. Thus, if the field intensity is everywhere known, the path an electron will follow after being emitted from the cathode, and the time it will take to reach the anode, can be determined. A complication in most tubes is that the electron density in the interelectrode space may be great enough to make an appreciable contribution to the electric-field intensity.

The electric field in a region may be represented by a plot of field flux lines similar to those used to represent a magnetic field. Accordingly, the direction of the electric field at a point is indicated by the direction of the electric-field flux lines, and the magnitude of the electric-field intensity is indicated by the density of the flux lines. An electric field represents the force on a positive charge; therefore, electric flux lines originate on positive charges and terminate on negative charges. They do not form closed loops as magnetic flux lines do.

The simplest example of an electric field is that which exists between two parallel plates as the result of a difference in potential between the plates. A plot of the electric field for this example is given in Fig. 20–7a. If edge effects are negligible, charges are uniformly distributed over the inner surfaces of the plates—negative charges on one plate and positive charges on the other. This idealized condition produces a uniform electric field throughout the space between the plates. The direction of this field is normal to the surface of the plates, and its magnitude is equal to the difference in potential between the plates divided by the distance between them.

In most tubes, the electric-field pattern is more complex than that illustrated in Fig. 20–7a. However, three general statements can be made concerning the more complex fields: (a) If the two electrodes are concentric cylinders or spheres, the field is radial; its magnitude is inversely proportional to the distance from the axis of the cylinders, or inversely proportional to the square of the distance from the center, in

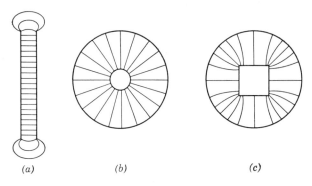

Fig. 20–7. Electric-field configurations.

the case of spheres. (*b*) If an electrode surface is irregular, the charge tends to concentrate on its outer, or projecting, areas. This effect, which is particularly noticeable with sharp projections, results in a greater value of electric-field strength in the vicinity of the projections. (*c*) The electric field at the surface of a conductor is always normal to that surface.

The design of cathodes for high-vacuum tubes is affected by the influence of surface irregularities upon the electric field, for it is this electric field that moves the electrons from the cathode to the anode. If the field is variable over the surface of the cathode, more electric current originates from the regions of higher electric-field strength; electron emission from hollows, crevasses, etc., is largely wasted. Therefore, such cathodes as the multicellular structure or the edge-wound ribbon (Fig. 20–2) are not efficient for high-vacuum tubes, which require cathodes that present a fairly regular emitting surface toward the anode.

20–11. POWER LOSS IN A DIODE

An electron in an electric field experiences an accelerating force; in traveling from a cathode to an anode, its velocity, and therefore its kinetic energy, increase continuously. The potential energy that the electron possessed because of its previous position with respect to the anode is converted into kinetic energy. This is analogous to a ball starting at the top of an incline and gaining kinetic energy as it rolls. If friction and windage are neglected, the analogy is even more complete. The kinetic energy that the ball gains is a function only of the weight of the ball and of the vertical displacement between the starting and end points; it is independent of the irregularities along its inclined path.

Similarly, the kinetic energy gained by an electron in traveling from cathode to anode depends only upon the electric charge of the electron and the potential difference between the anode and cathode; it is independent of the electric-field configuration along its path. The kinetic energy of an electron when it strikes the anode is proportional to the product of the anode-to-cathode voltage and the electron charge. Therefore,

$$\tfrac{1}{2}mv^2 = Ee \qquad (20\text{--}2)$$

where m is the mass of the electron in kilograms, v is the final velocity of the electron in meters per second, E is the potential difference in volts between the anode and cathode, and e is the charge of the electron in coulombs.

When electrons strike the anode, their kinetic energy is converted into thermal energy of the anode; that is, the current flowing through a diode heats the anode. (Any energy transferred to secondary electrons is recovered since all the secondary electrons return to the anode, the most positive electrode in the tube.) **Anode dissipation,** or the rate at which heat is dissipated at the anode, is equal to the product of the energy per electron and the number of electrons striking the anode per second. Thus,

$$P = Een \qquad (20\text{--}3)$$

where P is the anode dissipation in watts, and n is the number of electrons per second. The product en is the rate at which electric charge reaches the anode and is, therefore, the current through the tube in coulombs per second, or amperes. Anode dissipation is therefore given by the expression

$$P = EI \qquad (20\text{--}4)$$

Equation 20–4 indicates that the total power input to a diode (exclusive of cathode heating power) is spent in heating the anode.

20–12. TRIODES

As its name implies, the triode has three active electrodes: a plate (as its anode is more commonly called), a cathode, and a grid. For a tube having cylindrical symmetry, the grid is generally a spiral of fine wire mounted concentric with and between the plate and cathode; electrons must pass through the grid in traveling from the cathode to the plate. Symbolically the grid is shown as a broken line between the plate and cathode as, for example, in Fig. 20–8.

TRIODES

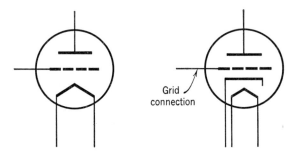

Fig. 20-8. High-vacuum triode symbols.

If the grid of a triode is at a negative potential with respect to the cathode, no electrons are attracted to it; however, some electrons are still capable of passing through the grid mesh to constitute a plate current. In any tube, the current that leaves the cathode is a function of the electric-field strength in the immediate vicinity of the cathode; for the triode, that field strength is a function of not only the plate potential but also the grid potential. In fact, the grid is in a much more favorable position to affect the electric field at the cathode. Thus, the plate current through the triode can be controlled by the control of the grid potential, and, if that grid potential is always negative with respect to the cathode, this control is achieved without the expenditure of any power. Therefore, in principle, the triode is capable of acting as an amplifier.

If the grid of a high-vacuum triode is held at cathode potential, the plate-current–plate-voltage characteristic is similar to that of a diode. This similarity is not surprising since, under these conditions, the presence of the grid cannot greatly alter the electric field at the cathode. However, if the grid is made negative with respect to the cathode, the passage of electrons through the grid is opposed, resulting in a reduction in the plate current for any fixed value of plate voltage; the more negative the grid, the smaller the plate current. These characteristics are best illustrated by curves of plate current versus plate voltage, one curve for each of several values of constant grid voltage. Such a family of curves is called the **plate characteristic.** The plate characteristic for a typical low-power triode is shown in Fig. 20-9.

For some special problems, it is convenient to use curves of plate current versus grid voltage for several constant values of plate voltage, or curves of plate voltage versus grid voltage for several values of plate current. Such curves contain no new information; they are simply a different presentation of information that can be obtained from the plate characteristic.

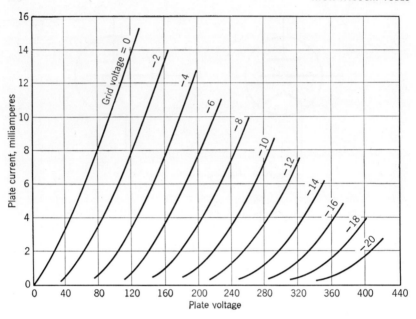

Fig. 20-9. Average plate characteristics for a type-6J5 triode. ($E_f = 6.3$ volts.)

20-13. TETRODES

A **tetrode** has two grids: a **control grid**, constructed in the same way as the grid in a triode of comparable ratings, and a **screen grid**, of approximately the same mesh size as the control grid and placed between the control grid and the plate. The presence of the additional grid gives the tetrode two important advantages over the triode: increased amplification, and reduced interelectrode capacitance. When the screen grid is operated at a fixed d-c potential (as it ordinarily is), it acts as an electrostatic shield between the plate and the grid. This reduces the effective grid-to-plate capacitance, which is important in high-frequency applications. The electrostatic shielding of the plate also accounts for the increased amplification possible with the tetrode. The plate is more effectively shielded from the cathode so that the plate potential has less effect upon the electric-field strength at the cathode, and hence less effect upon the plate current. The addition of the screen grid makes the plate potential less effective in controlling the plate current without affecting the effectiveness of the control grid, or, conversely, the effectiveness of the grid relative to the plate is increased.

Figure 20–10 illustrates the general shape of the plate characteristics of a high-vacuum tetrode. For plate potentials more positive than the screen-grid potential, the plate voltage has little effect on the plate current; for plate potentials less than the screen-grid potential, however, the current is definitely dependent upon the plate voltage. For some values of plate voltage, the plate current may even become negative, indicating electron emission at the plate. It is, in fact, secondary emission at the plate that accounts for most of the deviation of the plate current from a reasonably constant value at high plate potentials. Secondary emission occurs even at higher plate potentials, but, as long as the plate is more positive than any other electrode, all secondary electrons eventually return to the plate with no net effect on the external circuit. For plate potentials less than the screen-grid potential, the secondary electrons are attracted to the screen grid, thereby reducing the net plate current. If the number of secondary electrons going to the screen grid is greater than the number of primary electrons striking the plate, the plate current is negative.

For many purposes, a nonlinear plate characteristic is undesirable, and the operating range of the tetrode must be so limited that the plate potential is always greater than the screen-grid potential.

20–14. PENTODES

In addition to the control and screen grids, the pentode has a **suppressor grid,** which removes the undesirable effects of secondary emis-

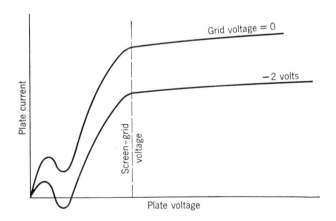

Fig. 20–10. Plate characteristics of a tetrode.

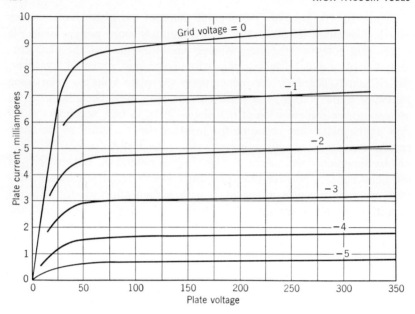

Fig. 20–11. Average plate characteristics for a type 6SJ7 pentode. ($E_f = 6.3$ volts, screen-grid voltage = 100 volts, and suppressor grid connected to cathode.)

sion from the plate. The suppressor grid is located between the screen grid and the plate and is constructed of an open mesh so as to interfere as little as possible with the electron flow to the plate. It is generally operated at cathode potential and is sometimes connected directly to the cathode within the tube.

If the suppressor grid is held at cathode potential, and if the plate is always positive with respect to the cathode, secondary electrons from the plate are repelled by the suppressor grid and return to the plate. Under these conditions, the plate characteristic of a pentode for all plate potentials should be similar to that of a tetrode with its plate more positive than its screen grid. However, at small plate voltages, the slope of the characteristic differs because more of the electrons that leave the cathode are attracted to the screen grid. The plate characteristic of a typical low-power pentode is shown in Fig. 20–11.

The pentode has all the merits of the tetrode, and, in addition, the range of useful operation is extended to lower plate potentials. As a result, the pentode has replaced the tetrode for most purposes.

20-15. BEAM-POWER TUBES

In tubes having a relatively high plate-current density, the function of the suppressor grid can be accomplished by deflecting plates mounted between the plate and the screen grid and electrically connected to the cathode. These plates concentrate the electrons into dense beams in the vicinity of the plate. The presence of this dense negative charge in the space adjacent to the plate is just as effective in repelling secondary electrons as a similar charge on the suppressor grid of a pentode. A tube that employs deflecting plates for the suppression of secondary emission is called a **beam-power tube.**

Figure 20–12, which shows the plate characteristic for a typical beam-power tube, indicates that the characteristic curves for a beam-power tube are similar to those for a pentode.

20-16. THE TRIODE AS A CIRCUIT ELEMENT

The important property of a triode or pentode is that its plate current can be controlled by varying its control-grid potential. Occasionally,

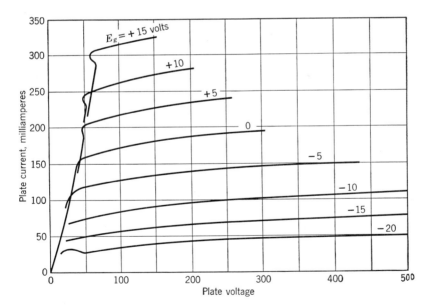

Fig. 20–12. Average plate characteristics for a type-6L6 beam-power tube. ($E_f = 6.3$ volts, screen-grid voltage $= 250$ volts.)

direct use is made of this feature; for example, a relay coil may be connected in the plate circuit of a tube so that the current through the coil, and hence the operation of the relay, can be controlled by variation of the voltage applied to the control grid.

Probably the most common application of vacuum tubes is in amplifier circuits. The amplifying property results from the fact that a change in the control-grid voltage changes the plate current, which in turn produces a change in the voltage across any load impedance through which it flows. If the circuit parameters are such that the change in voltage across the load impedance is greater than the corresponding change in the applied grid voltage, the circuit is a voltage amplifier. The relay application might also be termed an amplifier—a power amplifier—because a minute change in the grid power effects a much larger change in the power taken by the relay coil.

Regardless of the application, the basic tube circuit is that shown in Fig. 20–13. A triode is shown in the diagram, but the analysis of the

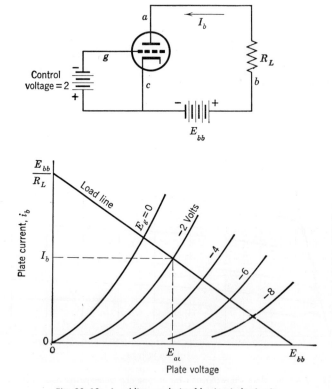

Fig. 20–13. Load-line analysis of basic triode circuit.

circuit is unchanged if the triode is replaced by a pentode with its additional grids maintained at constant potentials with respect to the cathode. Because the tube is a nonlinear device, graphical methods are employed to determine the voltages and currents that exist in the circuit.

One method of determining the plate current requires a plot of plate current versus grid voltage for the given circuit. The information for such a curve can be obtained from the plate characteristic of the given tube, but it is a tedious task. Once the curve is available, however, the plate current for any particular value of grid voltage may be taken directly from the curve.

A more versatile method of analyzing triode or pentode circuits is the load-line method, illustrated in Fig. 20–13. The interpretation of this load line is the same as for a diode, except that the appropriate current–voltage characteristic depends upon the value of the grid voltage. Thus, for a grid voltage of -2 volts in the circuit of Fig. 20–13, the intersection of the load line with the corresponding plate characteristic gives the plate voltage E_{ac} and the plate current I_b directly. The voltage across the load resistance is

$$E_{ba} = E_{bb} - E_{ac} = I_b R_L \qquad (20\text{–}5)$$

The load-line method is equally applicable to the determination of instantaneous values of voltage and current for a continuously varying control voltage. For this application, it is generally necessary that the effects of inductance and capacitance be negligible.

PROBLEMS

20–1. A diode is connected in series with a 150-volt battery and an 800-ohm resistor. The electrical characteristic of the diode is given in Fig. 20–3, and rated voltage is applied to its filament.
  (a) What current flows in this circuit? *Ans.* 150 ma.
  (b) What error would result from neglecting the tube drop in this calculation? *Ans.* 25%.
  (c) What is the anode dissipation? *Ans.* 4.8 watts.

20–2. To what value could the source voltage of Prob. 20–1 be raised without exceeding an anode dissipation of 15 watts? *Ans.* 290 volts.

20–3. The tube of Prob. 20–1 is operated at 90% of rated filament voltage and is connected in series with a 300-ohm resistor and an a-c source of 110 volts (rms). Plot one complete cycle of current versus time for this circuit.
  (a) What is the average value of this current? *Ans.* 80 ma.
  (b) What is the average value of voltage across the series resistor?
  Ans. 24 volts.

(c) Explain why the average power dissipated in the resistor is not equal to the product of the average values of voltage and current.

(d) How is the current wave shape altered by raising the filament voltage to its rated value?

20–4. A diode having an electrical characteristic as shown in Fig. 20–14 is connected in series with a variable resistance and a 250-volt battery. Approximately what value of series resistance will result in an anode dissipation of 6.0 watts? *Ans.* 1100 ohms.

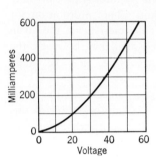

Fig. 20–14.

20–5. The tube of Prob. 20–4 is connected in series with an a-c source and a 1000-ohm resistor. What is the maximum rms value of source voltage that can be applied without exceeding an instantaneous current of 400 ma? *Ans.* 305 volts.

20–6. A phototube having the characteristic of Fig. 20–6 is connected in series with a 250-volt battery and a 10-megohm resistor.

(a) With an illumination of 0.10 lumen, what error would be introduced in the determination of the phototube current by the assumption that the full battery voltage is applied across the phototube terminals? *Ans.* 1%.

(b) What is the change in the voltage across the 10-megohm resistor when the phototube illumination is reduced to 0.05 lumen? *Ans.* 22 volts.

(c) Approximately what magnitude of series resistance will result in the maximum change in voltage across that resistor for the change in illumination specified in (b)? *Ans.* 60 megohms.

20–7. A relay that has a coil resistance of 20,000 ohms and requires a coil current of 5.0 ma for operation, is to be the plate load for a type-6J5 tube. What is the minimum value of plate-supply voltage that can be used in this circuit to assure that the relay will operate when the grid voltage is −2.0 volts? *Ans.* 195 volts.

20–8. A type-6L6 beam-power tube is to be used as a variable-resistance d-c load for a 250-volt circuit. Its screen grid is to be connected to the plate. How must the control-grid voltage of this tube be varied to change the load resistance from 5000 ohms to 1000 ohms? *Ans.* −18 to +6 volts.

20–9. In machines that take information from punched cards, it is necessary to operate a relay when electric contact is made through a hole in the card. If the contacts carrying the entire relay current are opened by the card, the resulting sparking might burn the cards. This difficulty can be overcome by employing the circuit of Fig. 20–15. The contacts C_t act in a low-current circuit to affect the grid voltage of the tube, whose plate current flows through the relay coil.

(a) Explain the reaction of this circuit to the opening and closing of the contact C_t.

(b) Specify a value of E_{cc} that will allow the circuit to operate satisfactorily, and a value of R that will limit the current in the contacts to 0.1 ma. *Ans.* $E_{cc} =$ at least 4.0 volts.

Data on relay, *CR*:

Coil resistance = 10,000 ohms
Contacts close at coil current = 5.0 ma
Contacts open at coil current = 2.0 ma

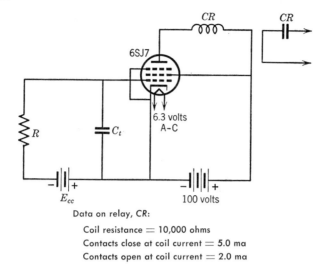

Fig. 20–15.

20–10. A response to the angular position of a shaft may be obtained by mechanically coupling the shaft to a rheostat that is connected in the grid circuit of a tube, as shown in Fig. 25–16. The rheostat has a total resistance of 1.0 megohm and can turn through 300°.

(*a*) Specify the supply voltage E_{bb} required to cause the relay contacts to close at $\theta = 200°$. *Ans.* 245 volts.

(*b*) At what value of θ will the relay contacts open? *Ans.* 100°.

Data on relay, *CR*:

Coil resistance = 10,000 ohms
Contacts close at coil current = 8.0 ma
Contacts open at coil current = 4.0 ma

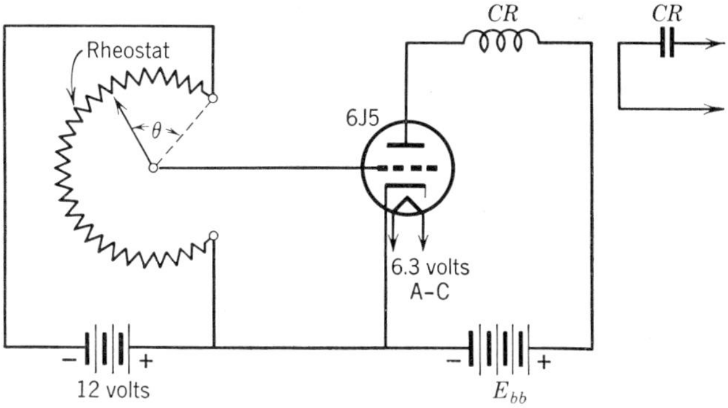

Fig. 20–16.

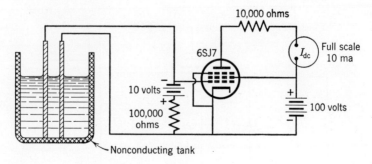

Fig. 20–17.

20–11. The circuit of Fig. 20–17 is employed to obtain a meter indication of the water level in a tank. For a water depth of 10 in., the resistance between the two probes is 10,000 ohms. What is the depth of the water when the meter indicates 2 ma? 6 ma? *Ans.* 1.7 in.; 8.1 in.

20–12. The current in a high-vacuum phototube is much too small to operate any ordinary magnetic relay but, by using a triode as shown in Fig. 20–18a, the phototube can indirectly operate the relay. Assuming that the relay is identical with the one in Fig. 20–15, that the phototube characteristics are given by Fig. 20–6, and that the initial illumination on the phototube is 0.10 lumen, determine the reaction of the circuit to a complete removal of illumination. How small a plate-supply voltage for the 6J5 could be employed and still achieve the same reaction? *Ans.* 160 volts.

20–13. Another possible circuit for the application of the phototube to relay operation is given in Fig. 20–18b. In this circuit, an increase in illumination is required to energize the relay coil. What value of E_{cc} will cause the relay contacts to close at an illumination of 0.05 lumen? At approximately what value of illumination will the contacts reopen?
Ans. 22 volts; 0.035 lumen.

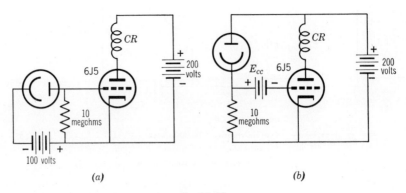

Fig. 20–18.

■ 21 Gas-Discharge Tubes

21-1. INTRODUCTION

Gas-filled tubes are made for a wide variety of uses and in a wide range of sizes. The range extends from small phototubes that control a few millionths of an ampere to large mercury-arc rectifiers that deliver several thousand amperes. In all these tubes, the gas atoms in the interelectrode space contribute to the flow of current between the electrodes. (A high-vacuum tube contains some gas, since a perfect vacuum is an impossibility, but the gas atoms in such a tube are so far apart that they have no noticeable effect on the current through the tube.) When the current between two electrodes depends upon the presence of the gas in the interelectrode space, the phenomenon is called a **gas discharge**. The appearance and electrical characteristics of gas discharges are much different from those of high-vacuum tubes and take many different forms, depending upon the type and pressure of the gas, the magnitude of the current, and the physical arrangements of the electrodes.

21-2. STARTING A GAS DISCHARGE

A gas is generally considered to be an excellent electrical insulator, since there are few free electrons in a gas. It is possible, however, to produce an ample supply of free electrons in a gas by the process of ionization and thus cause the gas to become a conductor. **Ionization** of a gas atom occurs when one or more of the outer electrons of the atom

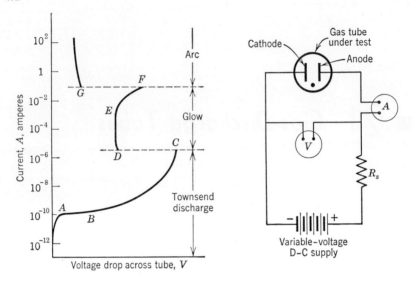

Fig. 21-1. Current–voltage characteristic for a gas discharge.

are freed from the binding force of the nucleus, and may be accomplished in any of several ways. An electron in motion may collide with a gas atom and knock one of the outer electrons from that atom, or one of the outer electrons may absorb sufficient radiant energy (ultraviolet, visible light, X rays, cosmic rays, etc.) to allow it to leave the atom. Regardless of the source of the ionizing energy, the result is an additional free electron and a positively charged ion for each atom that is ionized. Since both these particles carry electric charges, the motion of either may contribute to the current through the gas in a tube. Many atoms must be ionized if the gas is to become a good conductor.

A circuit that can be employed to determine the electrical characteristic of a gas discharge is shown in Fig. 21–1. The characteristic is obtained by measuring corresponding values of the current through, and the voltage across, a cold-cathode gas-filled diode as the d-c supply voltage is gradually increased from zero. Included in Fig. 21–1 is a plot of the characteristic that might be obtained from such a test on a typical tube. This plot is made to a semilogarithmic scale in order to present on a single set of coordinate axes the detail of the characteristic at both low and high currents.

If a sufficiently sensitive ammeter is used in the circuit of Fig. 21–1, a minute current is observable for small applied potentials. There is always a slight amount of ionization because of light or cosmic radiation, and the resulting free electrons and positive ions are attracted to the positive and negative electrodes, respectively; the movement of these charges constitutes the observed current through the tube. This source of free charges is so meager that saturation occurs for a small applied voltage, as indicated by point A on the curve of Fig. 21–1.

As the voltage applied to the circuit of Fig. 21–1 is increased, the energy gained by each electron in traversing the interelectrode space is increased, and, at a voltage corresponding to point B on the curve, some of the free electrons gain sufficient energy to ionize the gas atoms with which they may collide; the resulting free electrons and positive ions add to the current through the tube. The voltage at which such ionization first takes place is called the **ionization potential** of the gas in the tube. For the common gases the value of the ionization potential is between 10 and 25 volts.

As the applied voltage is increased above the ionization potential, more free charges are produced, and the current through the tube increases, as indicated between points B and C on the curve of Fig. 21–1.

The gas discharge represented by the curve to point C is called a Townsend discharge. In this discharge, all current ceases if the radiation responsible for the original ionization is removed; for this reason, the Townsend discharge is frequently referred to as a **non-self-maintaining discharge.**

At the higher voltages in a Townsend discharge, an electron may have more than one ionizing collision, and the electrons thus liberated may also cause ionization. In addition, positive ions bombarding the cathode liberate secondary electrons. Together, these processes cause a cascading ionization which, at a voltage corresponding to point C on Fig. 21–1, becomes uncontrolled and fills the interelectrode space with ions and free electrons. The resulting increase in current would proceed without limit if the series resistance were not in the circuit. Furthermore, the appearance and characteristics of the discharge are changed markedly. The discharge has become visible and is now a **self-maintaining discharge** because it is no longer dependent upon the external ionizing agent that was so essential to its formation. (Both the glow and arc discharges indicated in Fig. 21–1 are self-maintaining discharges. They are treated separately in the following articles.)

The voltage required to produce a self-maintaining discharge is called the **breakdown potential,** or the **sparking potential.** It is a function of the cathode material, the shape and spacing of the electrodes, the gas in the tube, and the gas pressure. Although all these factors are important in the design of electron tubes, gas pressure, because of its dependence upon temperature, is the only one that is subject to variation in the application of a given gas tube. The effect of gas pressure on the breakdown potential is shown in Fig. 21–2. The general shape of this curve is characteristic of all such curves.

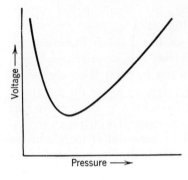

Fig. 21–2. Variation of break-down voltage with gas pressure.

The formation of a self-maintaining discharge fills the space between the electrodes with many free electrons, but an appreciable voltage is necessary to maintain the ionized condition; a continuous supply of electrons must be liberated from the cathode, and the electrons must be given sufficient energy to produce ionization. The means by which the electrons are liberated from the cathode characterize the two types of self-maintaining discharges: the **glow discharge** depends upon secondary emission; the **arc discharge,** upon thermionic emission.

21–3. GLOW DISCHARGE

The cathode of a tube designed to operate with a glow discharge is not heated. In such a tube, the discharge depends upon positive-ion bombardment of the cathode to produce the secondary emission of electrons. The voltage drop between the electrodes of a glow discharge is that required to give the ions sufficient energy to produce the secondary emission and is, therefore, a function of the gas filling the tube and of the secondary emitting properties of the cathode. In glow tubes filled with neon or argon, the voltage drop may vary between 50 and 200 volts and, for a given tube, is reasonably constant for a wide range of currents, as indicated in Fig. 21–1.

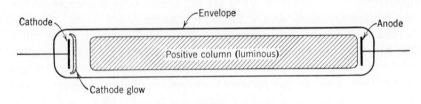

Fig. 21–3. Glow discharge in a long tube.

A glow discharge between electrodes at opposite ends of a tubular envelope has an appearance similar to that shown in Fig. 21–3. All or part of the cathode is covered with a velvety glow, called the **cathode glow**. (In the operating range for which the voltage drop of the glow discharge is constant, the area of the cathode glow is directly proportional to the magnitude of the discharge current.) Adjacent to the cathode glow is a relatively dark region, and the remainder of the interior of the tube is luminous. This luminous region of the discharge is known as the **positive column**; it may be absent in a glow discharge between closely spaced electrodes.

21–4. ARC DISCHARGE

If, by increasing the applied voltage to the circuit of Fig. 21–1, the current through the tube is made greater than that required to cause the cathode glow completely to cover the cathode surface, the voltage drop across the tube is no longer constant. As indicated between points E and F on the curve of Fig. 21–1, there is a definite increase in the voltage drop across the tube as the tube current is increased. Since the energy of the bombarding ions is proportional to the voltage drop, and since some of this energy is transferred to the cathode in the form of heat, the increase in the voltage drop across the tube causes an increase in the temperature of the cathode. If the current and voltage become high enough, the cathode provides thermionic emission.

At a voltage drop corresponding to point F on the characteristic curve, the thermionic emission is adequate to supply the needs of the discharge, and the discharge becomes an arc. The arc concentrates on a small spot on the cathode and keeps that spot sufficiently hot to maintain the necessary thermionic emission. Under these conditions, little energy is required to liberate electrons from the cathode; the voltage between the electrodes needs to be only slightly greater than that necessary to give the electrons ionizing energy. In other words, the arc voltage is approximately equal to the ionization potential of the gas filling the tube and is not appreciably affected by the magnitude of the discharge current.

Except in the region surrounding the cathode, the appearance of an arc is similar to that of a glow discharge. There is nothing recognizable as a cathode glow in an arc discharge, but the hot spot on the cathode is visible because of its high temperature.

The cathodes of many gas-filled tubes are similar to vacuum-tube cathodes in that they are heated by external means—either directly or indirectly. The operating characteristics of a gas-filled tube with an externally heated cathode are essentially the same as those for a tube in which the cathode is heated by the discharge itself. (The voltage drop may be somewhat less in a tube with an externally heated cathode since no energy is taken from the discharge for cathode heating.) The only important difference is that a high breakdown potential is not necessary to initiate the arc; thermionic emission is already available because the cathode is externally heated. An applied voltage equal to, or slightly greater than, the ionization potential of the gas is sufficient to start the arc.

21–5. GAS-FILLED PHOTOTUBES

Although the Townsend discharge is essential to the formation of any glow discharge, the only important applications of the Townsend discharge, as such, are the gas-filled phototube and certain radiation detectors. The Townsend discharge in a gas-filled phototube differs from that described in Art. 21–2 in that the source of the original free electrons is photoelectric emission at the cathode, instead of ionization due to stray radiation. The electrical characteristics are no different, however; at higher voltages the ionization of the gas provides the same increase in free electrons and, therefore, the same increase in current.

In Fig. 21–4, the current–voltage characteristics of a gas-filled photo-

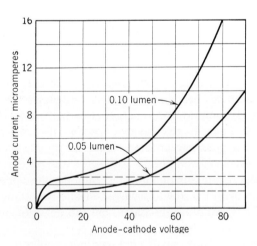

Fig. 21–4. Current–voltage characteristic for a gas-filled phototube. (Approximate characteristics for the same tube with no gas are shown with broken lines.)

tube are superimposed on the characteristics that the tube would have without the gas. At voltages less than the ionization potential of the gas filling the tube, the characteristics are identical; at higher voltages, ionization provides a greater current if gas is present. (The higher current for a given amount of light is the only advantage of the gas-filled phototube.) Structurally, the two types of phototubes are identical, except for the presence of some gas in the gas-filled tube.

The formation of a self-maintaining discharge in a gas-filled phototube would make the tube insensitive to changes in illumination and might permanently damage the photosensitive cathode surface. Therefore, the anode supply voltage for these tubes is generally limited to 90 volts. If the voltage rating is not exceeded, there is no possibility of forming a glow discharge. Care should be exercised in interchanging phototubes since high-vacuum tubes having the same dimensions as gas tubes are commonly operated at anode voltages of several times 90 volts.

21-6. GLOW TUBES

A glow discharge has several important properties. For example, both neon signs and glow lamps depend upon glow discharge to produce light. In a glow lamp, the electrodes are so arranged that the positive column is eliminated and the cathode glow is easily visible. When an alternating voltage is applied to this lamp, the electrodes act alternately as cathodes and anodes, and at a frequency of 60 cycles or higher both appear to glow continuously. As in all self-maintaining discharges, a series impedance is necessary to limit the current in a glow lamp and a resistor is frequently built into the base of the lamp for this purpose.

A neon sign utilizes a glow discharge between cold electrodes at opposite ends of a long tube. The positive column fills most of the tube and is the principal source of light. The neon sign is supplied from a high-voltage high-reactance transformer. The high voltage, of the order of several thousand volts, is necessary to start the discharge; the high reactance, to provide the current-limiting series impedance needed after the discharge has been started.

Another useful characteristic of the glow discharge is the relatively constant voltage drop that it exhibits over a relatively wide range of currents. The operation of a voltage-regulator tube depends upon this characteristic; for this reason, voltage-regulator tubes have large cathode surfaces, as shown in Fig. 21-5. Voltage-regulator tubes are available for operation at constant voltages between 75 and 150 volts, and for currents up to 50 milliamperes.

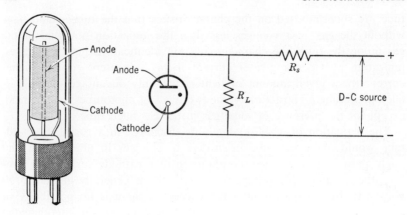

Fig. 21–5. Construction and application of a voltage-regulator tube.

Investigation of the circuit of Fig. 21–5 indicates the possibility of a constant voltage across the load resistance R_L for a considerable variation of either the source voltage or the resistance of the load. As long as the tube current is between certain limits, the load voltage (which is identical with the tube voltage) remains constant. The rating of a voltage-regulator tube includes current limits between which the variation in tube voltage is no greater than a specified amount—usually 1 or 2 volts.

21–7. ARC-DISCHARGE LAMPS

The most common application of the arc discharge is to lighting. Ordinary fluorescent lamps, sodium-vapor lamps, high-pressure mercury-vapor lamps, and some ultraviolet light sources are all simple arcs between filamentary electrodes, either of which may act as anode or cathode. Except for the fluorescent lamp, the source of light is the positive column, and the color of the light depends only upon the type and pressure of the gas in the lamp. In the fluorescent lamp, a mercury-vapor arc acts as a source of ultraviolet light which excites the coating on the bulb to luminescence; therefore, the color of the light depends upon the material used for the fluorescent coating.

After the discharge has been established, the cathodes of most arc-discharge lamps are heated by the discharge itself. It is common, however, to heat the cathodes externally during the starting period to reduce the breakdown potential and to prevent damage of the cathode due to

ARC-DISCHARGE LAMPS

high-voltage ion bombardment. Figure 21-6 shows a simple fluorescent lamp circuit that includes a means of preheating the electrodes.

When voltage is first applied to the circuit of Fig. 21-6, current flows through the series inductor, the two lamp electrodes, the normally closed contacts C_H, and the small heater H. The inductor limits the current to a reasonable value for heating the electrodes to emitting temperature. Together, H and C_H form a small thermal switch that acts to open the circuit after current flows through the heater for a short time—2 or 3 seconds. When the contacts do open, a high-voltage surge is developed between the electrodes of the lamp owing to the breaking of a current-carrying inductive circuit; this high voltage initiates an arc discharge between the lamp electrodes. (If the switch contacts open at a time when the current in the circuit is zero or nearly zero, the induced voltage may be insufficient to start the discharge. If the lamp fails to start, the relay heater cools, the contacts reclose, and the starting process is repeated.) The arc discharge in the lamp completes the circuit otherwise broken by the opening of the switch contacts, and current continues to flow through H to hold the switch contacts open.

If the circuit of Fig. 21-6 is supplied from a d-c source, the inductor is of little value for limiting the current, either during or after starting; therefore, an additional series resistor must be inserted. The inductor is still required, however, to provide the high-voltage surge for starting.

The starting method outlined in the preceding paragraph is but one of many possible schemes. The common glow-switch starter for fluorescent lamps achieves the same result and requires no power to hold the contacts open after starting. Another method is the direct application of a voltage high enough to start a self-maintaining discharge even though the electrodes are initially cold. When a lamp is started with cold electrodes, the formation of the discharge proceeds as outlined in Art. 21-2: that is, a Townsend discharge is formed first; the ionization builds up to allow the formation of a glow; and the glow heats the

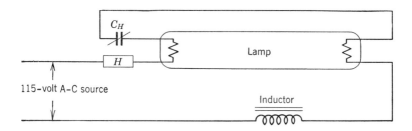

Fig. 21-6. Fluorescent lamp circuit.

electrodes to establish the arc. In a lamp designed for this type of starting, there is but a fraction of a second from the application of the voltage to the establishment of the arc.

21–8. PHANOTRONS

A **phanotron** is a two-electrode gas-filled tube having an externally heated cathode. This tube is capable of the same functions as a high-vacuum diode, except that it cannot be constructed to withstand the high voltages for which some vacuum tubes are designed. At the more common voltages, however, the phanotron has the distinct advantage that its voltage drop is small, even for large currents. The voltage drop is important since, for a given tube current, anode dissipation is proportional to the anode-to-cathode voltage drop. Phanotrons are always preferred to vacuum tubes except in (*a*) low-power applications, where efficiency is less important than the first cost, and (*b*) extremely high-voltage applications, where the phanotron might be caused to conduct equally well in either direction.

The power loss in a phanotron is low, not only because of the lower anode dissipation, but also because the nature of a gas discharge allows the use of a more efficient cathode. The energy required to hold a cathode at a desired operating temperature is approximately equal to the thermal energy radiated from that cathode. For a given emission surface, cathodes in the form of the edge-wound ribbon or the multi-cellular structure (Fig. 20–2) present little area for heat radiation and are therefore efficient. The efficiency is increased by surrounding the cathode with a reflecting heat shield to reduce the amount of radiation. These heat-conserving schemes are not applicable to vacuum tubes because emission from the concave surfaces would not be available.

Most of the space between the electrodes of a phanotron is filled with electrons and positive ions in approximately equal numbers. Except for a thin layer adjacent to the cathode, the space between the electrodes exhibits no net charge and can be thought of as a metallic conductor. In effect, the anode is but a fraction of a millimeter from the cathode and is shaped to follow all the cathode contours. Thus, regardless of the shape of the cathode, there is a uniform electric field over the entire emitting surface, and the emission from all parts of the cathode is available.

The operation of a phanotron at lower-than-normal cathode temperature results in a higher voltage drop across the tube. Under these conditions the increased energy of the bombarding ions may damage the cathode and shorten the life of the tube. For this reason, voltage should not be applied to the anode of a phanotron until the filament power has heated the cathode to the proper temperature; the rating of a phanotron includes the heating time required. The heat-conserving cathodes of most phanotrons are slow to reach operating temperature; a 5-minute heating time is not uncommon.

21-9. THE PHANOTRON AS A CIRCUIT ELEMENT

For practical purposes, there is no current in a phanotron for negative anode-to-cathode voltages, or for positive voltages less than the ionizing potential of the gas. When the anode voltage becomes equal to, or slightly greater than, the ionizing potential, an arc is formed, and, regardless of the magnitude of the current through the tube, the voltage drop across the tube can be assumed to be constant. The constant tube drop is approximately equal to the ionizing potential of the gas filling the tube. The magnitude of this tube drop is all that need be known about a phanotron in order to predict the behavior of any circuit of which it is a part.

Two examples of circuits that include a phanotron are shown in Fig. 21-7. If it is desired to find the current that flows in the circuit of Fig. 21-7a, the Kirchhoff voltage equation gives the answer directly. Thus,

$$E_{bb} = E_{TD} + IR \qquad (21\text{-}1)$$

or

$$I = \frac{E_{bb} - E_{TD}}{R} \qquad (21\text{-}2)$$

where E_{TD} is the anode-to-cathode tube drop. In any circuit in which

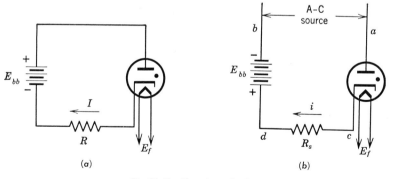

Fig. 21-7. Phanotron circuits.

the phanotron conducts, the phanotron may be considered to be replaced by a battery having the appropriate polarity and an emf equal to the constant tube drop.

Inspection of Eq. 21–2 reveals that the tube alone is incapable of limiting current; that is, with no series resistance ($R = 0$), the current in the circuit becomes infinite. Therefore, there must always be some impedance in series with any arc or glow-discharge tube. (Note the resistance in series with the tube in Fig. 21–1.)

The addition of an a-c source to the circuit of Fig. 21–7a converts it to a practical circuit that can be used to charge a battery as illustrated in Fig. 21–7b. The phanotron passes current in the direction for charging the battery whenever the supply voltage e_{ab} exceeds the sum of the battery voltage and the tube drop; no current flows at other times. If inductance and capacitance effects are negligible, the method used for the analysis of the d-c circuit of Fig. 21–7a can be applied for the instantaneous values of voltage and current in the a-c circuit, and a plot of a full cycle of either quantity may be obtained by a point-by-point calculation.

An investigation of the behavior of this circuit requires a determination of the time in the cycle at which conduction begins. Conduction begins when the actual tube voltage e_{ac} becomes equal to the normal tube drop E_{TD} and may be determined from the relation

$$e_{ac} = e_{ab} + e_{bd} + e_{dc} \tag{21-3}$$

The drop across the series resistance e_{dc} equals zero while no current flows in the circuit. Therefore, for zero current, and for a sinusoidal a-c supply voltage, the tube voltage becomes equal to the normal tube drop when

$$E_{TD} = e_{ab} + e_{bd} = E_m \sin \omega t - E_{bb} \tag{21-4}$$

Solving for ωt,

$$\omega t = \sin^{-1}\left(\frac{E_{bb} + E_{TD}}{E_m}\right) \tag{21-5}$$

If E_m is great enough to give a solution to Eq. 21–5, that is, great enough so that the tube is certain to conduct, there are two values of ωt that satisfy the equation. One value corresponds to the time at which conduction begins; the other, to the time at which conduction ceases. During the interval between the two, the tube voltage equals E_{TD}, and the magnitude of the current can be found by dividing the voltage across the resistor by the value of the resistance. Thus, since

$$i = \frac{e_{cd}}{R_s} = \frac{e_{ab} + e_{ca} + e_{bd}}{R_s}$$

THYRATRONS

during conduction, it becomes

$$= \frac{E_m \sin \omega t - E_{\text{TD}} - E_{bb}}{R_s} \qquad (21\text{-}6)$$

Equation 21–6 illustrates that, even though both the battery and the tube are acting as loads, neither serves to limit the current. With no series resistance, the current would be unlimited. Every circuit must contain at least one element capable of limiting current: that is, at least one element whose voltage drop increases with an increase in the current through it.

21–10. THYRATRONS

A **thyratron** is a gas-filled thermionic triode or tetrode. In several respects concerning both construction and operating characteristics, the thyratron is similar to the phanotron; in other respects the two differ radically. The greatest difference between the operating characteristics of these two types of tubes has to do with the requirements for the establishment of an arc. A phanotron starts to conduct when its anode-to-cathode voltage becomes equal to, or slightly greater than, the normal tube drop; but the magnitude of the voltage necessary to start conduction in a thyratron can be controlled. This control is made possible by the presence of a third electrode, the **control grid**, the potential of which determines the anode-to-cathode voltage required for breakdown.

When the anode-to-cathode voltage of a phanotron is made equal to, or greater than, the ionizing potential of the gas filling the tube, electrons emitted from the cathode gain sufficient energy before reaching the anode to ionize some of the gas atoms with which they collide. These ionizing collisions are essential to the formation of an arc. If a control grid is placed between the anode and cathode, and if this grid is maintained at a sufficiently negative potential with respect to the cathode, electrons from the cathode are repelled by the grid and do not gain the energy necessary to produce ionization. For a more positive anode potential, the control grid must be more negative to prevent electrons from getting through the control grid and ionizing the gas. This is the principle upon which thyratron operation depends. The thyratron is essentially a phanotron to which a third electrode has been added. Control of the potential of this third electrode offers a means of control over the anode potential at which the tube starts to conduct.

Cross sections of the electrodes of several typical thyratrons are shown in Fig. 21–8. The anodes and cathodes for each of the structures

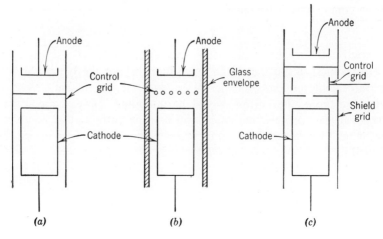

Fig. 21-8. Thyratron electrode arrangements.

shown are similar to those of a phanotron of comparable current-carrying capacity; only the grids of the three structures differ.

The grid of Fig. 21-8a is arranged to shield the anode from the cathode except for a small opening along the axis; therefore the potential of this grid controls the anode potential required to establish an arc through the opening. The shape and large area of this grid are necessary if the grid is to be capable of preventing the formation of an arc along any external path, as well as controlling the formation of an arc along the axis. There are disadvantages, however, to the large grid area. Ions and electrons are always present in the space between the cathode and grid—even a few strays before any arc has been formed—and the presence of these charged particles allows grid current to flow whenever a difference of potential exists between the grid and cathode; the greater the grid area, the greater the grid current. A large grid area also results in a large grid-to-cathode capacitance and a relatively large alternating current when there is an alternating voltage in the grid circuit.

Before the establishment of an arc, the grid current in any thyratron is only a fraction of a microampere, but even this current may have undesirable effects. If the grid current flows through a high impedance or, in other words, if the grid circuit is supplied from an extremely low-power source, the impedance drop may cause the actual grid-to-cathode voltage to differ considerably from the expected value. Once the cathode-to-anode arc is established, grid currents may be as high as 100 milliamperes.

The structures of Fig. 21-8b and c are schemes for reducing the control-grid area and the grid-to-cathode capacitance, thereby reducing the grid current, without interfering with the ability of the grid to exert control over the formation of an arc. In effect, the greater portion of the control grid of Fig. 21–8a has been replaced in Fig. 21–8b by the glass envelope, and in Fig. 21–8c by a fourth electrode, the **shield grid**. In both these arrangements, the only function of the control grid is the control of breakdown; shielding to prevent arcs along external paths is provided by the glass wall in one and by the shield grid in the other. The latter tube is called a **shield-grid**, or **double-grid, thyratron**.

21–11. THYRATRON CONTROL CHARACTERISTICS

A plot of the anode potential required to initiate conduction versus grid-to-cathode potential is known as the **control**, or **breakdown, characteristic** of a thyratron. An example of a control characteristic is given in Fig. 21–9. Together with the normal anode-to-cathode arc drop, the control characteristic of a thyratron provides all the information needed to predict the behavior of the thyratron in an electric circuit. If the point representing a certain combination of grid and anode voltages

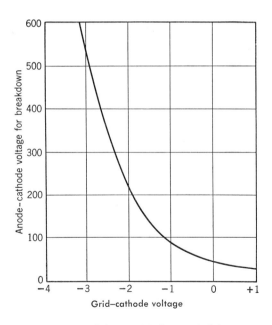

Fig. 21–9. Control characteristic for a typical thyratron.

lies above and to the right of the control characteristic, an arc is formed from the anode to the cathode; if below and to the left, no arc is formed.

In a mercury-vapor thyratron, an increase in the temperature of the condensed mercury causes a rise in the vapor pressure in the tube and results in a lower breakdown potential. Because of this dependency upon temperature, the published control characteristics for many mercury-vapor thyratrons include several curves, one for each of several operating temperatures. Some characteristics are plotted as a broad band to include variations of individual tubes, as well as variations of temperature within the rated range.

For some applications, the variation of the control characteristic with a change of ambient temperature is undesirable. Such applications require thyratrons filled with an inert gas rather than with mercury vapor. There is no appreciable change in the control characteristic of an inert-gas-filled tube as the temperature is changed. In the normal range of temperatures, the fact that gas pressure is directly proportional to absolute temperature does not produce an appreciable change in the breakdown potential of the tube.

It is possible to change the control characteristic of a shield-grid thyratron by altering the potential of the shield grid. When no use is to be made of this feature, the shield grid of a double-grid thyratron is connected to the cathode.

21–12. THYRATRON AS A CIRCUIT ELEMENT

The operation of a thyratron in an electric circuit can be explained with reference to the circuit of Fig. 21–10. As indicated in Art. 21–2, there must always be a current-limiting impedance in series with a self-maintaining discharge. In this respect, the thyratron is no different from a phanotron or any other gas tube. It is also common practice to include a resistor (R_g of Fig. 21–10) in series with the grid circuit to prevent excessive grid current. The grid resistor is not absolutely necessary unless there is a possibility that the grid-to-cathode voltage may become positive. If the grid does become positive, an arc may form between the cathode and grid, in which case the grid-current-limiting resistor is as essential as the resistor in the anode circuit.

If the grid potential of the tube of Fig. 21–10 is made so negative that the tube does not conduct when anode voltage is first applied, then no current can flow through the load resistor R_L. With no voltage drop across the load, the anode-to-cathode voltage equals E_{bb}, the terminal voltage of the anode supply battery. If the grid voltage is now made less

THYRATRON AS A CIRCUIT ELEMENT

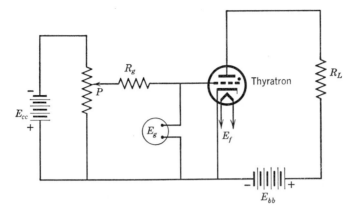

Fig. 21-10. Basic thyratron circuit.

negative by adjusting the rheostat P, a voltage is reached at which the tube starts to conduct. The exact value of the grid voltage at which conduction begins can be predicted from the control characteristic for the thyratron under test. Thus, for an anode supply voltage of 200 volts applied to a thyratron having a control characteristic as shown in Fig. 21-9, conduction starts at a grid voltage of approximately -2.0 volts. Actually, this is the method for experimentally determining the control characteristic of a thyratron. A variable anode supply voltage is required, however, in order to obtain more than one point on the characteristic.

Once the thyratron starts to conduct, the grid loses control of the anode current. Its electrical characteristics are then identical with those of a phanotron; that is, the anode-to-cathode tube drop is small and, for practical purposes, independent of the current through the tube. To stop the anode current, the anode circuit must be opened, or the anode supply voltage must be reduced to a value less than the normal tube drop. If an alternating voltage serves as the anode supply, the anode current is automatically stopped at the end of each positive half-cycle, and the grid regains control over the firing time for the following cycle. The use of an a-c anode supply is discussed in detail in Arts. 23-10, 11, and 12.

The thyratron in the circuit of Fig. 21-10 acts as a one-way relay, or switch. A small amount of power supplied to the grid can cause the anode circuit to conduct and thus close a relatively high-power circuit. Many thyratron applications depend upon this relay action. Small signals involving almost no power can be employed to "trigger" a thyratron that has a d-c motor, an electric light, or some other electric device as its load. The load is often a relay coil that, when energized, closes contacts in a circuit of still greater current and power. A few simple applications of thyratrons as relays are considered in Chapter 28.

21-13. COLD-CATHODE TRIGGER TUBE

Although the cold-cathode trigger tube is not a thyratron and bears little physical resemblance to a thyratron, its general functions are similar to those of a thyratron. The tube is a gas-filled cold-cathode triode in which the initiation of a glow discharge between the cathode and anode is controlled by means of the voltage applied to a third electrode, called the trigger.

Because the trigger tube operates with a glow discharge instead of an arc, its operating and control characteristics differ from those of a thyratron. In general, the glow discharge of the trigger tube requires a higher breakdown voltage, the normal operating voltage for the glow discharge is considerably higher than that of an arc, and the trigger tube is limited to relatively small currents—a few milliamperes. The advantage of the trigger tube is that no cathode heating power is needed. This feature makes the tube applicable to portable battery-operated devices that require the occasional relaying, or triggering, action of a thyratron, but which cannot afford continuous power for cathode heating.

Figure 21-11 shows a possible arrangement of the electrodes for a trigger tube. The anode and cathode are far enough apart so that, without the trigger, several hundred volts would be required to start

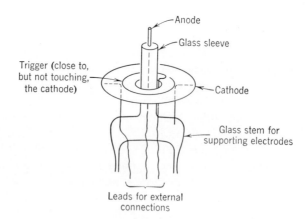

Fig. 21-11. Electrode structure of a typical cold-cathode trigger tube.

COLD-CATHODE TRIGGER TUBE

a discharge. However, a voltage of 80 to 100 volts between the trigger and the cathode produces a glow discharge between the trigger and cathode, and, if the anode-to-cathode voltage is then 80 volts or more, a discharge is established between the anode and the cathode. The initiation of a glow discharge between the cathode and anode of a trigger tube normally requires the presence of a glow discharge between the trigger and cathode; the trigger-to-cathode discharge may be produced by either a positive or a negative trigger potential.

PROBLEMS

21-1. A 10-megohm voltmeter is connected in series with a phototube and a 90-volt battery. The electrical characteristics of the phototube are given in Fig. 21-4. What is the approximate illumination on the phototube when the meter reads 65 volts? *Ans.* 0.22 lumen.

21-2. The voltage-regulator tube in the circuit of Fig. 21-5 has a constant voltage drop of 75 volts for currents between 5 and 50 ma. The series resistance is 1000 ohms.

(a) For a source voltage of 125 volts, determine the limits between which the load resistance may be varied and still maintain the constant load voltage of 75 volts. *Ans.* 1670 ohms to ∞.

(b) The load resistance is next held constant at 3000 ohms. Between what limits may the source voltage be varied and still maintain the constant load voltage? *Ans.* 105 and 150 volts.

(c) If the voltage-regulator tube requires a breakdown voltage of 95 volts to establish the glow discharge, are all the above answers necessarily valid? Explain.

21-3. The supply voltage in the circuit of Prob. 21-2 is adjusted to 150 volts and the load resistance to 3000 ohms. How would the load voltage be affected by a failure of the voltage-regulator tube, that is, by an absence of conduction in the tube?

21-4. A 100-volt (rms) a-c source, a phanotron having a normal tube drop of 10 volts, and a 100-volt battery are connected in series with sufficient resistance to limit the peak charging current to 15 amperes. What value of peak current would you expect if the connections to the phanotron were reversed? Repeat for a reversal of the battery connections.
Ans. 110.2 amperes.

21-5. In the circuit of Fig. 21-7b, the rms value of the a-c source voltage is 30 volts, the battery potential is 24 volts, and the phanotron tube drop is 10 volts. Neglect the internal resistance of the battery.

(a) What value of series resistance must be used to limit the peak value of the tube current to 20 amperes? *Ans.* 0.425 ohm.

(b) Plot one complete cycle of the tube current.

(c) What is the average value of the tube current if the series resistance is 0.425 ohm? *Ans.* 2.7 amperes.

(d) What is the maximum negative value of the anode-to-cathode voltage?

Ans. 66.5 volts.

(e) What is the average anode dissipation in the phanotron?

Ans. 27 watts.

(f) How would you determine the necessary power rating for the series resistor?

21-6. A certain relay operates when the direct current through its 10,000-ohm coil equals 10 ma. This relay coil is to be connected in the anode circuit of a thyratron so that grid control of this thyratron will control the operation of the relay. The control characteristic for the thyratron is given in Fig. 21-9.

(a) What d-c anode supply voltage should be used? (Assume normal arc drop is 15 volts.)

(b) What is the minimum negative grid-to-cathode potential that will prevent the thyratron from firing?

(c) What is the reaction of the circuit to an increase (less negative) in the grid potential, and to a subsequent decrease?

21-7. Figure 21-12 is typical of many of the thyratron control characteristics that are published, in that it is plotted as a band to include variations among individual tubes, changes with the age of the tube, and differences

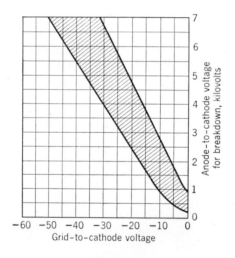

Fig. 21-12.

due to differences in temperature. A thyratron of the type to which this characteristic applies is employed in a circuit with an a-c anode supply of 4000 volts (rms).

(a) What is the smallest (negative) grid-to-cathode voltage that should be used if no conduction is desired?

(b) What is the smallest change in grid voltage that should be used to insure conduction?

PROBLEMS

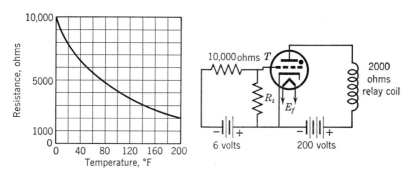

Fig. 21-13.

21-8. The circuit shown in Fig. 21-13 is so arranged that the relay coil is energized when the ambient temperature rises above a certain value. The control characteristic for the thyratron T_1 is given in Fig. 21-9, and the relationship between the resistance of the temperature-sensitive resistor R_t and the ambient temperature is given by the curve of Fig. 21-13. At what temperature will the relay operate? *Ans.* Approximately 90° F.

21-9. Draw a circuit diagram in which a phototube and thyratron are employed to energize a relay coil when illumination on the phototube is interrupted. Using the same thyratron, relay, and anode supply as in Prob. 21-8 and a phototube whose characteristics are given by Fig. 21-4, specify all other components so that circuit will function properly if the initial illumination is 0.05 lumen.

■ 22 *Semiconductor Devices*

22–1. INTRODUCTION

Most of the characteristics of the various electron tubes, and therefore most of the functions performed by electron tubes, can be duplicated by one or another of the semiconductor devices. There are 2- and 3-element semiconductor devices that are quite analogous to high-vacuum tubes; analogies for some of the gas tubes are found; and semiconductor diodes that are sensitive to light are easily made. The purpose of this chapter is to provide a physical basis for understanding the operation of the more common semiconductor devices and to describe their electrical characteristics so that their advantages and disadvantages with respect to electron tubes are evident.

The outstanding feature of all semiconductor devices is that no cathode heating power is required. The fact that there is no cathode heating power to dissipate means that equipment using semiconductors instead of tubes can be made smaller as well as lighter, and accounts in part for the semiconductor device itself being much smaller than an equivalent tube.

Semiconductors also suffer several shortcomings. Their low operating temperature is desirable but it also represents a limitation. For germanium, the most commonly used semiconductor, temperatures above 100° C result in radical changes in the operating characteristics, and even at that temperature the power-handling capabilities are greatly reduced. Silicon is somewhat better than germanium in this respect, but both exhibit temperature dependence that could be of importance even at room temperature. Similarly, the sensitivity to light can be either desirable or undesirable. A device that is not completely enclosed in metal should be shielded from any strong light if its behavior is to be independent of that light.

At present no very high-power semiconductor devices are constructed. However, diodes that can pass several amperes and withstand back voltages of several hundred volts are available, and devices with greater capacities will undoubtedly be developed. The power limitation is partially alleviated by the relative simplicity with which semiconductors can be connected in series or parallel to meet greater voltage or current requirements.

22–2. CONDUCTION IN PURE GERMANIUM AND SILICON

The common metallic conductors consist of an aggregate of atoms that, except for the valence electrons, are fixed in position. The valence electrons—one, two, or three per atom—are essential to the electrical neutrality of an atom or to a region of the metal, but are not so tightly bound to particular atoms as are the valance electrons involved in the formation of chemical bonds. At any temperature above absolute zero, these electrons have some random motion, but this motion is not completely unrestricted, for an external force would be required to cause an accumulation of charge in any particular region. If an external electric field is applied, however, and if electrons can be supplied at one end and removed at the other end of the metal, then there can be a net drift of the electrons without upsetting the neutrality of the metal. This drift constitutes an electric current.

The chemical elements having a valence of 4 form crystals in which the situation is quite different. The four-valence electrons of each atom are shared, one each with its four nearest neighbors, to form covalent bonds similar to those involved in chemical combinations. The result is a chemically and physically stable solid (diamond, the crystalline form of carbon, is an example) and one that is a poor electric conductor, since even the valence electrons are tightly bound to particular atoms or to particular pairs of atoms.

Even the purest crystals of germanium or silicon, the important valence-4 elements for semiconductors, have some conductivity. Just as thermal energy causes a random motion of the valence electrons in ordinary metals, it can disrupt an occasional covalent bond in a semiconductor. A given bond does not stay broken for long, but there is a continuous breaking and re-forming of bonds. An electron that is broken free of its bond is as free as a valence electron in an ordinary metal and, like it, can contribute to an electric current if an electric field is applied. In addition, for each freed electron a positive **hole** is left behind; an atom is left with a net positive charge and a further need for an electron to complete one of its covalent bonds. The situation is somewhat like that in a gas discharge when an atom is ionized producing a free electron and a positive ion, both of which can take part in an electric current. In the semiconductor, not only is the freed electron capable of motion, but also the positive hole is. The motion of a hole may seem to be an indirect description of what actually happens: a neighboring electron, under the influence of thermal agitation or an applied electric field, moves to fill in the hole and leaves behind it a new hole. Actually, only an electron has moved, but effectively it is a positive hole that has moved—in a direction opposite to the motion of the electron. Viewed externally, it is a region of positive charge that has moved, and any test that may be devised will so indicate. Furthermore, the energies involved in moving an electron from a covalent bond to an unfilled bond are different than those required for moving an excess electron about in a crystal lattice. Thus, for purposes of this discussion, the hole is a reality, and the use of this concept is almost essential to an understanding of the operation of the various semiconductor devices.

An idea of the density of the thermally broken bonds can be obtained from a comparison of the resistance of a semiconductor with that of some more common materials. Pure germanium at room temperature has a resistivity of approximately 0.6 ohm-meter, whereas that of copper is 1.724×10^{-8} ohm-meter, and that for most ordinary insulating materials between 10^8 and 10^{14} ohm-meters. The resistivity of germanium lies midway between that for a good conductor and a good insulator—hence the name, semiconductor.

22–3. EFFECT OF AN IMPURITY IN SEMICONDUCTORS

The presence of minute quantities of impurities can greatly affect the resistivity of a semiconductor, and it is the addition of carefully controlled amounts of impurities that produces the useful semiconductor devices. The impurities used have a valence of either 3 or 5; that is, they have either one too many or one too few valence electrons to ideally replace semiconductor atoms in the crystal lattice. When atoms of arsenic or antimony, which have a valence of 5, take their place in the lattice structure of a semiconductor crystal, each provides a free electron over and above what is required to complete the covalent bonds. Impurities having a valence of 5 are therefore called **donors.** An atom of gallium or indium, which are commonly used impurities of valence 3, lacks an electron to complete the four covalent bonds with its nearest neighbors and must therefore accept an electron from another atom. These impurities are called **acceptors.** In the process of completing its covalent bonds, an acceptor atom creates a hole that is free to move and therefore to contribute to an electric current.

Most of the conduction in semiconductors devices results from the free electrons and holes produced by the added impurities. A limitation of semiconductors is that, at high temperatures, the number of electrons and holes produced thermally overshadows the number produced by the impurities, and the semiconductor becomes little different from an ordinary metal.

If both donor and acceptor impurities are present in the crystal, they tend to neutralize each other. The neutralization would be complete only if the same number of atoms of each were present; otherwise, the character of the material is determined by the more abundant impurity. A semiconductor that has only donor impurities or that has a surplus of donors, is called **N-type;** the current carriers are negatively charged. Similarily, those materials with a surplus of acceptor atoms are known as **P-type** materials because the current carriers—holes in this case—are positive.

22–4. P–N JUNCTIONS

There is nothing outstanding about the characteristics of a piece of semiconductor. Its resistance is considerably less than that of the pure germanium or silicon by an amount dependent upon the amount and type of impurity, and this change in resistance is perhaps more dramatic than the changes that occur in the formation of most of the common alloys, but, otherwise, a piece of semiconductor considered alone has a definite resistance that is not out of the ordinary. The interesting and useful properties of semiconductors are those exhibited by the junction between some N-type and some P-type material.

The P–N junctions under consideration are not formed by simply placing a piece of one type material in contact with a piece of the opposite type. In order to achieve a junction of useful size with the intimate contact required, the junction is usually built into a single crystal of the germanium or silicon. That is, the amount of impurity is made to vary so that one part of the crystal has a surplus of donor atoms, and another part a surplus of acceptor atoms. Ideally, the transition would be a single atomic layer; actually, it is several but it is still effective.

Many methods of producing P–N junctions have been developed, two of which are (1) altering the amount or type of impurity in molten germanium as a crystal is being formed and withdrawn from the bath, and (2) diffusion. To form a diffusion junction, it is only necessary to deposit some acceptor material on an N-type crystal and heat the combination until enough of the acceptor atoms have diffused into the crystal to form a region of P-type material. The desired junction then exists between this formed region of P-type and the main body of the crystal, which is N-type.

22–5. ELECTRICAL CHARACTERISTICS OF A P–N JUNCTION

If the applied voltage across a piece of ordinary metal is varied, and the temperature is held constant, the resulting current remains proportional to the voltage, regardless of its direction or magnitude. Figure 22–1, which shows the current–voltage characteristic for a typical P–N junction illustrates that no such proportionality exists for the P–N junction. The principal feature of this characteristic is that, for moderate

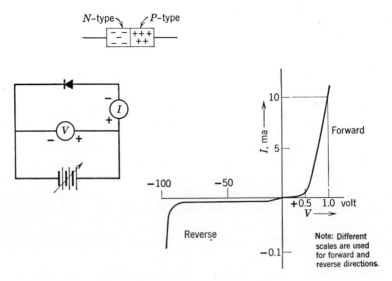

Fig. 22–1. Current–voltage characteristic for a typical P–N junction.

voltages, the junction conducts readily in one direction and poorly in the other. When the applied voltage is of the polarity shown in the circuit diagram of Fig. 22–1, both the electrons in the type-N material and the holes in the type-P material are forced toward the junction where they neutralize each other. At the same time, the external circuit supplies more electrons to the N-type and takes electrons from the P-type to form more holes in the P section. Thus a continuous flow of electric current is possible in the circuit. This direction of current flow, that for low resistance, is spoken of as the **forward** direction.

If the applied voltage across the junction is reversed, all available current carriers are drawn away from the junction, and so current can flow only as a result of the formation of electron-hole pairs at the junction by thermal agitation, photoexcitation, or some other external means. Unless some effort is made to enhance this reverse current, it is smaller than the forward current by several orders of magnitude.

If the reverse voltage across a P–N junction is increased sufficiently, the high-resistance character of the junction breaks down (at about 100 volts for the example shown in Fig. 22–1.) In the reverse direction, practically all the voltage appears across the thin junction region to produce relatively high electric fields. At the breakdown voltage, the field strength is great enough to create electron-hole pairs by ripping valence electrons loose or by so accelerating the few free electrons present that upon collision they do the same thing. The situation is very much the same as in the breakdown of a gas discharge, both as to the electrical characteristics and the consequences. In many applications, the breakdown may be a limitation; in others, the external circuit is arranged to limit the current, and use is made of the constant-voltage characteristic. Semiconductor diodes can be used in the same manner for voltage regulation as glow tubes are. (See Art. 21–6).

22–6. OTHER SOLID-STATE JUNCTIONS

In the original crystal diode, the P–N junction was not used, at least not wittingly; instead, the junction between a pointed wire and a crystal was used. The point-contact diode, as this arrangement is called, is still very much in use, and, as a detector of extremely high radio-frequency signals, it has almost no competition. The operation of the point-contact diode is much less clearly understood than that of the P–N junction, but its electrical characteristics are essentially the same. The differences are simply those arising from the difference in junction area. The point contact necessarily has a small area and therefore a small effective capacitance across the junction; this is a desirable feature for high-frequency applications. On the other hand, only the *P–N* junction can be constructed with a large enough area to accommodate currents greater than a few milliamperes.

Several other solid-state junctions that exhibit essentially the same characteristics as the *P–N* junction are known variously as contact rectifiers, blocking-layer rectifiers, or metallic rectifiers and include, as specific examples, the copper-oxide and selenium rectifiers. Again, the theory of operation is not as well understood as it is for the *P–N* junction. The junctions of interest are the boundary between copper and a layer of copper oxide in the one case, and between some pure selenium and an impure layer that is formed on its surface in the other. Both are produced in disk shape—one quarter of an inch to several inches in diameter, depending on the current to be passed—and as many disks as are necessary to withstand the applied voltage are then assembled in series by simply stacking them with suitable contacting material between and bolted together. This stacking is almost always necessary because the individual junctions can withstand only 6 to 8 volts for the copper-oxide type and 25 volts for the selenium.

The symbol used for the *P–N* junction in Fig. 22–1 is also used to represent any of the solid-state diodes.

22–7. APPLICATION OF SEMICONDUCTOR DIODES

Semiconductor diodes have been constructed to serve all the functions provided by high-vacuum or gas-filled diodes. Their most important characteristic is rectifying ability; that is, they conduct more readily in one direction than in the other. The way in which this is utilized to convert a-c to d-c power is discussed in some detail in Chapter 23. The possibility of using a semiconductor diode as a voltage regulator has been mentioned in Art. 22–5.

A *P–N* junction that is arranged so that it can be illuminated is called a phototransistor, and is capable of reacting to light just like a phototube. A reverse voltage is normally applied to a phototransistor so that only a minute current flows in the absence of some external excitation. Light of the proper frequency can supply this excitation; that is, photons can break covalent bonds to produce electrons and holes in or near the junction that then act as current carriers, and current flows through the junction. The proper connection for a phototransistor is shown in Fig. 22–2. For a given frequency of illumination, the rate at which electron-hole pairs are produced, and therefore the current, is propor-

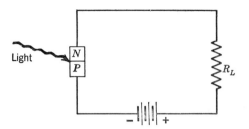

Fig. 22-2. Connections for a phototransistor.

tional to the intensity of the light striking the junction. The frequency for maximum response for a germanium phototransistor is in the visible range.

22-8. TRANSISTORS

A **transistor** in its simplest form is a three-terminal solid-state or semiconductor device in which the current through one P–N junction is made to control the current through a second junction. Any of the methods for forming P–N junctions are applicable to the construction of junction transistors, and some transistors are made using two closely spaced point contacts on a single crystal. The latter, or **point-contact transistor** as it is known, is the form in which the transistor was first

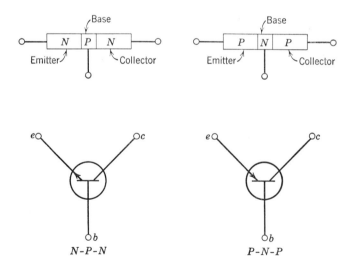

Fig. 22-3. Diagrams of transistors and the corresponding circuit symbols.

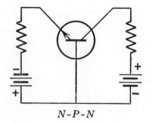 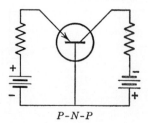

Fig. 22–4. Proper polarity for transistor supplies.

constructed. At present its use is limited to some special applications; the junction transistor has replaced it for most amplifier applications. Only the junction transistor will be discussed here.

No matter how the junctions are produced, a junction transistor consists of a thin layer of one type of semiconductor sandwiched between two layers of the opposite type. This arrangement is shown diagrammatically in Fig. 22–3 for the two possible configurations along with the corresponding electrical symbols for the two devices. The two configurations are named for the order in which the semiconductor types appear—*NPN, PNP*. The thin center section is known as the base, and the other two as the emitter and collector. Theoretically, the transistor is symmetrical, and so there is no difference between the emitter and collector; actually, the difference is a matter of usage in the circuit. The direct voltage, or bias, applied to the collector is such as to cause a reverse current through the collector-base junction, while the emitter-base bias is in the forward direction. Either a preference is built into the transistor, or whichever junction in a particular transistor gives the better reverse characteristics is labeled the collector-base junction.

The proper polarity of the bias voltages for both types of transistors is shown in Fig. 22–4. The arrow on the emitter of the transistor symbol indicates the direction of conventional current in the emitter when it is properly biased, and as such distinguishes between an *NPN* and a *PNP* transistor. For example, on the symbol for the *NPN* transistor, the arrow is directed outward, since a forward current through the emitter-base junction would flow from the base to the emitter and to the external circuit.

22–9. TRANSISTOR CHARACTERISTICS

The useful characteristics of the transistor derive from the fact that current through the emitter-base junction affects the current through the collector-base junction. When the emitter current is zero, that is, when the emitter circuit is opened, the collector-base circuit is an uncomplicated *P–N* junction and exhibits all the characteristics of one. In normal use, this junction is biased in the reverse direction so that only a small current is present. For the *NPN* transistor, this means that the electric field is in the direction to carry electrons from the base to the collector, or holes in the opposite direction. The current is very small because these carriers do not exist where they are needed except for the few produced by thermal excitation.

If current is now made to flow through the emitter-base junction, carriers become available in the base region that can contribute to current through the collector-base junction. Specifically, in the *NPN* transistor, with conduction in the forward direction, electrons pass from the emitter into the base region. Because the base is a relatively good conductor, the field strength is small, and many of these electrons diffuse towards the collector, where they are immediately swept to the collector

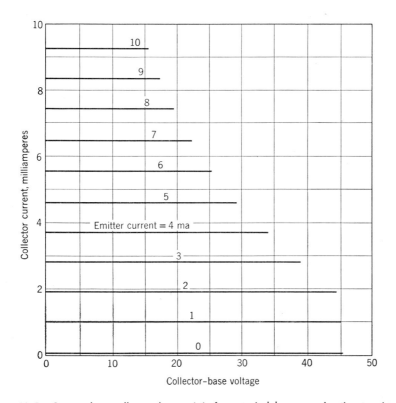

Fig. 22–5. Common-base collector characteristic for a typical low-power junction transistor.

by the field that exists across the collector-base junction. The collector characteristics for a typical *NPN* transistor are shown in Fig. 22–5 and illustrate the effect of emitter current just described. A study of these curves indicates not only that some of the electrons from the emitter diffuse over to the collector but also that most of them do. Increasing the emitter current by one milliampere increases the collector current by almost as much. The ratio of change in collector current to change in emitter current for a constant collector voltage is known as the **current-amplification factor**; typical values of the current-amplification factor range between 0.9 and 1.00.

The operation of the *PNP* transistor is the same as that of the *NPN* except that the carriers from the emitter that contribute to the current through the base-collector junction are positive holes instead of electrons. Furthermore, in order that these currents be reversed from what they are in the *NPN* transistor, all applied voltages are reversed as indicated in Fig. 22–4.

As described, it appears that the emitter of a transistor is the control electrode and therefore corresponds to the grid of a high-vacuum triode. Actually, there is no restriction as to which is considered the control electrode since all three carry current: in fact, the analogy between the transistor and the high-vacuum triode is very close if the emitter is made to correspond to the cathode, the base to the grid, and the collector to the plate. In the high-vacuum triode, electrons emitted from the cathode pass through the grid to the plate, and the number of electrons getting through, or the magnitude of the plate current, depends upon the potential of the grid. In the case of the transistor, electrons or holes are emitted from the emitter (injected into the base region is the way in which it is usually expressed), and most of them pass through the base region to the collector. Since the fraction of the emitter current that gets through to the collector—the current-amplification factor—is relatively constant, changes in the base current produce corresponding changes in the collector current; therefore, the magnitude of the collector current depends upon the current through the base terminal. The analogy is complete except for the fact that in one case the control is by voltage and in the other by current. In both, however, the control electrode is the low-current electrode; grid current in the tube is usually negligible, and base current in the transistor is much smaller than the other currents.

22-10. TRANSISTOR AS A CIRCUIT ELEMENT

The analogy of the transistor to the high-vacuum triode carries over to applications, for the most common transistor circuits employ the base as the control electrode, just as the most common tube circuits employ the grid as the control electrode. For the analysis of such a circuit one uses the common-emitter collecter characteristics as illustrated in Fig. 22-6. The difference between this characteristic and the common-base characteristic of Fig. 22-5 is simply in the reference electrode, or grounded electrode, from which the collector voltage is measured, and the variable parameter for which the family of curves is drawn. The voltages and currents that exist in a given circuit can be determined by application of the load-line method to the appropriate set of curves. Thus, for the circuit of Fig. 22-7, the load line, drawn on the characteristics of Fig. 22-6, intersects the $I_b = 0.20$ milliampere curve at a

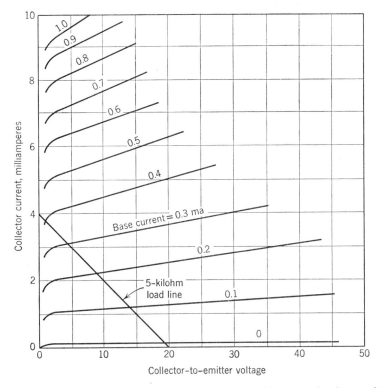

Fig. 22-6. Common-emitter collector characteristics for a typical low-power junction transistor.

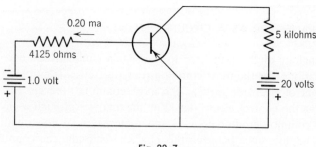

Fig. 22-7.

collector current of 2.3 milliamperes and a collector-to-emitter voltage of 9 volts. The effect of a change of input current can be determined by taking a second point on the load line corresponding to a second base current and noting the change in collector voltage or current.

The analysis of the preceding paragraph did not determine the base voltage. The corresponding quantity in the vacuum tube, grid current, is considered to be negligible. The base-to-emitter voltage is small since that junction has a forward voltage impressed across it, but it is not always negligible. To determine the base voltage, or to determine the base current when the base voltage is known, requires additional information given by the base characteristic. An example of the base characteristic for a typical transistor is shown in Fig. 22-8. Except that the ordinate of this curve is only a fraction of the current through the junction, it is the forward characteristic of the emitter-base junction and is not greatly affected by the collector voltage or current. In practical

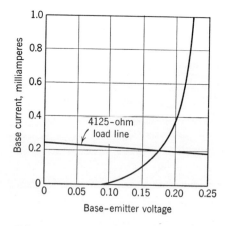

Fig. 22-8. Common-emitter base characteristic for a typical low-power junction transistor.

situations, there is often a large enough external resistance in series with the base so that the base current is obtainable from the applied voltage and that resistance without recourse to the base characteristic. For a more exact solution, or in situations where there is not a large external resistance, the base current is determined by the construction of a load line on the base characteristic corresponding to the applied voltage and series resistance; then that value of base current is used to evaluate the collector voltage and current. As an example, if it is desired to determine the current through the 5000-ohm resistor in the collector circuit of Fig. 22–7, and if the base characteristic is assumed to be independent of the collector current, a load line is constructed on the curve of Fig. 22–8 intersecting the voltage axis at 1.0 volt and having a slope equal to minus the reciprocal of 4125 ohms. The load line intersects the characteristic at 0.20 milliampere, the actual base current. The proper collector characteristic of the family shown in Fig. 22–6 is now known, from which the collector voltage and current—9 volts and 2.3 milliamperes, respectively—are determined as previously described.

PROBLEMS

22–1. The current–voltage characteristic of a diode similar to that of Fig. 22–1 is to be measured, using a d-c source that is continuously variable from 0 to 250 volts. What series resistance should be used if the power dissipated in the diode must never exceed 50 mw? *Ans.* 300 kilohms.

22–2. A diode, whose current–voltage characteristic is given by Fig. 22–1, is connected in series with a 20-kilohm resistor and a 50-volt sinusoidal source.
(*a*) What are the peak values of the forward and reverse currents?
Ans. 3.5 ma; 0.01 ma.
(*b*) Approximately what average power is dissipated in the resistance?
Ans. 60 mw.

22–3. The diode of Fig. 22–1 is to be used with a series resistance to provide a regulated supply of 100 volts to a 1-megohm resistance load. If the source voltage is 150 volts, what value of series resistance should be used to limit the diode dissipation to 5.0 mw? *Ans.* 333 kilohms.

22–4. Indicate the current directions in each of the three branches of both circuits shown in Fig. 22–4. For each circuit relate the magnitudes of the branch currents and the current-amplification factor.

22–5. If all of the batteries in the circuit of Fig. 22–7 are replaced with 9-volt batteries, what values of resistances should be used to cause the collector current and collector-emitter voltage to be 3.2 ma and 5.0 volts, respectively? (Assume that the characteristics of Fig. 22–6 apply.)
Ans. 30 kilohms; 1350 ohms.

22–6. Draw a circuit diagram showing a d-c source and a phototransistor connected in series between the base and emitter terminals of a transistor. Indicate all necessary information concerning polarities to make certain that light on the phototransistor will cause current through the base-emitter junction in the proper direction.

22–7. The collector circuit of the transistor in Prob. 22–6 is completed by a 2000-ohm relay coil and a 20-volt battery connected in series between the collector and the emitter. Assuming that the transistor characteristics are given by Fig. 22–6, how is the relay coil current affected by an increase in the phototransistor current from 0 to 0.5 ma?

Ans. Increase from 0.2 to 5.5 ma.

22–8. Figure 22–9 illustrates a possible connection whereby a decrease in illumination of the phototransistor is required to increase the relay coil current. Select a value for R such that, with the other components given, the relay current is essentially zero when the phototransistor current is 0.8 ma. To what value must the phototransistor current drop to obtain a relay current of 6 ma?   *Ans.* 11 kilohms, 0.23 ma.

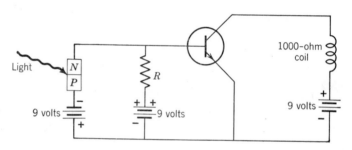

Fig. 22–9.

■ 23 Single-Phase Rectifiers

23-1. INTRODUCTION

A rectifier is a device that permits the use of an a-c source to supply d-c power to a load. Since most of the devices considered in the preceding three chapters conduct current readily in but one direction, any of them can act as a rectifier; that is, one or more of these devices can be connected in a circuit with a load and an a-c source so that current flows through the load in only one direction. Actually, a rectifier is not required to prevent current flow in the reverse direction, but it must cause an average current in the desired direction through the load.

The simplest possible rectifier circuit is one in which a single rectifying element is placed in series with the load and the a-c source, as shown in Fig. 23-1. Since current can flow through the tube only in the

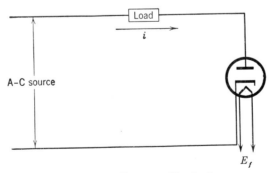

Fig. 23-1. Half-wave rectifier circuit.

direction indicated, current flows through the load in that direction only. This rectifier is known as a half-wave rectifier because current flows in the load for only one half of each cycle of the supply voltage.

Other circuits require more than one rectifying element, some require special transformers, and a few require capacitors. In order to compare the various single-phase rectifier circuits, it is necessary to understand their operation and to determine the following characteristics of each:

(*a*) Average load current or voltage obtainable from a given supply voltage, or the supply voltage needed to produce a desired load voltage.

(*b*) Load-voltage wave form. The function of a rectifier is to provide d-c power. Any departure from pure direct current then represents an a-c, or ripple, component that is unwanted. The smaller the ripple, the more efficient the rectifier.

(*c*) Minimum ratings for all the circuit elements.

23-2. RATINGS FOR RECTIFIER ELEMENTS

All rectifier elements are given the following ratings:

(A) PEAK CURRENT. For a thermionic tube, this rating is equal to, or somewhat less than, the total emission from the cathode when rated filament voltage is applied. If the instantaneous anode current exceeds the rated value, the cathode may be damaged.

(B) AVERAGE CURRENT. Current through any diode heats it. Under the conditions most commonly encountered, the equilibrium temperature is a function of the average current through the device. Exceeding this rating of a rectifier tube causes the anode to operate at a higher temperature than that for which it was designed. Such overheating of the anode may liberate gases from the anode structure, thereby altering the characteristics of the tube, or it may crack the glass envelope of the tube.

Exceeding the average-current rating for a semiconductor diode causes it to overheat, which if carried far enough causes the diode to lose its rectifying properties.

(C) PEAK INVERSE VOLTAGE. An inverse voltage on a rectifying element is one that tends to make current flow in the reverse, or unwanted, direction. The peak inverse voltage rating is the maximum instantaneous value of inverse voltage that can be applied to the rectifying element without danger of breakdown between the electrodes with consequent current flow in the reverse direction. Such a breakdown may damage the diode in question, other diodes in the circuit, or the transformer supplying the rectifier.

23-3. HALF-WAVE RECTIFIER

The simple circuit of Fig. 23-1 is seldom used without refinements, but an analysis of that circuit provides a convenient introduction to the more important circuits. In the discussion of this rectifier, and most of the other circuits covered in this chapter, tubes are shown as the rectifying elements. In principle, there is no reason why the tubes cannot be replaced with semiconductor diodes; the same simplifying assumptions are applicable, and the characteristics of a rectifier are independent of the rectifying element used.

The analysis of this and other rectifier circuits is greatly simplified by assuming that the rectifying elements are perfect: that is, that they conduct currents in the forward direction with zero voltage drop, and that they pass no current in the reverse direction. The assumption of negligible tube drop causes no appreciable error in the calculations for an efficient rectifier. Since the same current flows through the tube and load of the half-wave rectifier, the ratio of power loss in the tube to power output equals the ratio of tube drop to load voltage whenever there is current. High efficiency requires that the losses be small in comparison with the output, or that the tube drop be small in comparison with the output voltage.

In the analysis of a circuit involving an appreciable tube drop, neglecting that tube drop gives a value of load voltage somewhat greater than the actual value. Such an analysis also results in conservative values for the tube ratings.

In the circuit of Fig. 23-2, the tube conducts when the source voltage e_{ab} is positive, and, if the tube drop is negligible, the load voltage is equal to the source voltage during this positive half-cycle. During the negative half-cycle, no current flows; therefore, the load voltage is zero, and the

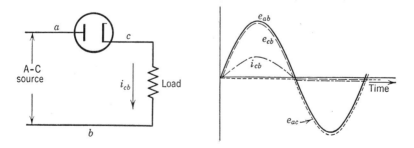

Fig. 23-2. Voltage and current wave forms for a half-wave rectifier.

tube voltage must equal the source voltage. The resulting wave forms are plotted in Fig. 23–2. Inspection of these curves indicates that the Kirchhoff voltage equation, $e_{ab} = e_{ac} + e_{cb}$, is satisfied at all times during the cycle.

The plot of load current in Fig. 23–2 is obtained by the application of Ohm's law; that is,

$$i_{cb} = \frac{e_{cb}}{R} \tag{23-1}$$

The necessary tube ratings for this rectifier can be found by a consideration of the voltage wave forms for the circuit and by the application of Eq. 23–1. The peak inverse voltage (PIV) is equal to the peak of the source voltage; therefore, if E is taken as the rms value of the sinusoidal source voltage,

$$\text{PIV} = \sqrt{2}\,E \tag{23-2}$$

The peak anode current equals the peak value of the load voltage divided by the resistance of the load. Since the peak values of the source and load voltages are equal (neglecting tube drop), the peak anode current is

$$I_{pk} = \frac{\sqrt{2}\,E}{R} \tag{23-3}$$

The average anode current is found by integrating the instantaneous current over one complete cycle and then dividing the result by the period of the alternations. The anode current in the half-wave rectifier is zero for one half of the cycle and is equal to the quotient of the source voltage and the load resistance for the other half-cycle. Thus,

$$I_{avg} = \frac{1}{2\pi} \int_0^\pi \frac{\sqrt{2}\,E \sin \omega t \, d(\omega t)}{R} = \frac{\sqrt{2}\,E}{\pi R} \tag{23-4}$$

(ωt is used as the variable of integration instead of t, to simplify the calculations.) The peak value of the anode current under discussion is π times the average value of that current. The same relationship exists between the peak and average values of any voltage or current having the same wave shape.

23–4. HALF-WAVE RECTIFIER WITH CAPACITOR FILTER

The simple half-wave rectifier previously considered is unsatisfactory in several respects. For a given supply voltage, the average load voltage of the half-wave rectifier is small, and the load current and load voltage are both zero for half the time. In general, a less fluctuating load voltage is desirable.

The rectifier circuits discussed in the following articles all produce a more satisfactory load-voltage wave form than that of the simple half-wave rectifier. It is possible to effect considerable improvement and still retain the simple one-tube circuit, however, by the addition of a filter to the circuit. Any combination of capacitance and inductance that is used to "smooth out" the wave form of the output voltage of a rectifier is called a **filter**.

A single capacitor connected in parallel with the load of a single-phase half-wave rectifier, as shown in Fig. 23–3, acts as a filter. While the source voltage is positive and increasing, the tube conducts, and, if the tube drop is negligible, the capacitor charges to maintain its voltage equal to the supply voltage. However, when the source voltage starts to decrease, the capacitor can discharge only through the load; the tube does not conduct in the direction necessary to allow the capacitor to discharge through it. The rate at which the capacitor discharges, or the rate at which the load voltage decreases, depends only upon the load resistance and the value of the filter capacitance. Specifically, the rate is inversely proportional to the product of the load resistance and the filter capacitance. Thus, the greater the load resistance, or the greater the filter capacitance, the less rapid the decrease in load voltage after the tube ceases to conduct. This discharge of the capacitor continues until the source voltage becomes equal to, or slightly greater than, the load voltage (time t_1 on Fig. 23–3), at which time the tube can again conduct to recharge the capacitor.

If, for some particular values of load resistance, filter capacitance, and source voltage, the curves corresponding to those plotted in Fig. 23–3 are available, the necessary tube ratings and the average load voltage

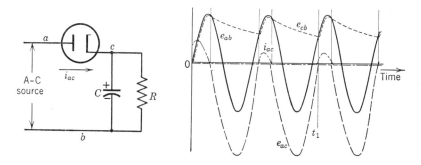

Fig. 23–3. Half-wave rectifier with capacitor filter.

can be readily determined. The peak inverse voltage is equal to the negative peak value of the anode-to-cathode voltage; the peak anode current is the maximum value indicated by the anode-current curve; and the average current can be determined by a graphical averaging process. The average load voltage is the product of the average load current and the load resistance. Without the current and voltage wave forms, however, it is a tedious analytical process to determine the four quantities just discussed.

When the product of the load resistance and the filter capacitance (ohms × farads = seconds) is large in comparison with the period of the voltage variations, the capacitor does not discharge appreciably between successive positive peaks. If there is negligible discharge, the load voltage is constant and equal to the peak value of the source voltage; the peak inverse voltage equals twice the peak of the source voltage; and the average anode current equals the load voltage divided by the value of the load resistance (average anode current equals average load current since the capacitor conducts no average current). The peak anode current is not readily determined, but it must be several times the average value since current flows through the tube only during short intervals, when the source voltage is most positive.

Although the capacitor filter offers a simple means of obtaining a steady d-c load voltage, it is practical only for low-current loads. For high-current low-resistance loads, the magnitude of the capacitance required for adequate filtering is generally too great.

For a constant load resistance, increasing the filter capacitance increases the average value of the load voltage and hence increases the average load current and average tube current. At the same time, the period during which the tube conducts is decreased. To supply a higher average current in a shorter time requires a much higher instantaneous anode current. These peak currents may become excessive as the load-voltage wave form approaches a straight line. A high-vacuum tube is capable of limiting the value of peak current so that its cathode is not damaged, but a phanotron possesses no such current-limiting ability. Therefore, phanotrons are never employed in rectifiers with capacitor filters.

23–5. VOLTAGE DOUBLER

The voltage-doubler circuit, as shown in Fig. 22–4, consists of two half-wave rectifiers supplied from a common source. Each rectifier has a capacitance load and therefore acts like a half-wave rectifier that has

FULL-WAVE RECTIFIER

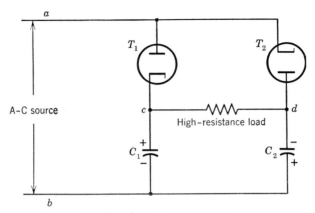

Fig. 23-4. Voltage-doubler circuit.

a large capacitance filter. The source voltage e_{ab}, the tube T_1, and the capacitor C_1 constitute one of the rectifiers. Current flows through T_1 in the direction to charge C_1 with the indicated polarity, and, if the load current is negligible, the capacitor charges to a voltage equal to the peak value of the source voltage. The source, the tube T_2, and the capacitor C_2 constitute the other rectifier. On alternate half-cycles, T_2 conducts to charge C_2 to the peak value of the source voltage with the polarity indicated. The output of the voltage doubler is taken across the two capacitors in series, and, since the relative polarities are additive, the output voltage is double that produced by each of the half-wave rectifiers. For small load currents, the average load voltage is twice the peak of the source voltage.

With no transformer, or with a given transformer, a higher output voltage can be obtained with a voltage-doubler circuit than from any of the other rectifiers.

23-6. FULL-WAVE RECTIFIER

The full-wave rectifier circuit is shown in Fig. 23-5. Although this circuit requires two diodes, it is a more popular circuit than the half-wave rectifier because current flows in the load throughout the cycle, resulting in a more satisfactory output-voltage wave form.

Like the voltage doubler, the full-wave rectifier is made up of two half-wave rectifiers: one consists of tube A, the load, and one half of the transformer secondary; the other, of tube B, the same load, and the other half of the transformer secondary.

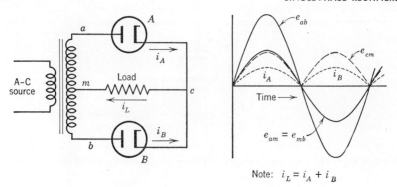

Fig. 23–5. Single-phase full-wave rectifier.

Point m on the transformer secondary is the center tap on a continuous winding between points a and b; therefore, the voltage e_{am} is 180 degrees out of phase with the voltage e_{bm}, and the two tubes conduct on alternate half-cycles. The direction of the current through the load is the same for both halves of a cycle. The load is supplied by the two half-wave rectifiers, which produce the load-voltage wave form indicated in Fig. 23–5.

The peak inverse voltage on either tube of the full-wave rectifier can be determined from a plot of the tube voltage (for tube A, $e_{ac} = e_{am} - e_{cm}$). A shorter method incorporates the assumption that the voltage across the conducting tube is zero. Accordingly, the voltage across tube A must equal the total transformer secondary voltage while tube B conducts. The peak inverse voltage is then the peak value of the voltage e_{ab}.

One half of the average load current is supplied by each of the two tubes of a full-wave rectifier. The average tube current is, therefore, one half of the average load current. The peak current is an instantaneous value, however, and, since only one tube conducts at a time, the peak tube current must equal the peak load current, or $1/R$ times the peak value of e_{am}.

Because of the higher average load voltage for a given peak load voltage, the full-wave rectifier is more satisfactory than the half-wave rectifier. It is also preferred to the half-wave rectifier with a capacitor filter for any but very light loads. There are at least two reasons for this preference: the load-voltage wave form of a half-wave rectifier with a capacitor filter is affected by changes in the load resistance; the full-wave rectifier conducts on both halves of every cycle and therefore takes a balanced (zero average value) current from the a-c source.

BRIDGE RECTIFIER

Several tubes have been designed specifically for low-power full-wave rectifiers. For this purpose, two diodes with a common cathode are contained in a single envelope. Inspection of the circuit diagram of the full-wave rectifier indicates that a common cathode is consistent with the requirements of this circuit, since the cathodes of the two diodes are normally connected together. This tube, called a double diode, is widely used in the rectifiers that supply d-c power to radio receivers.

23-7. BRIDGE RECTIFIER

The bridge rectifier produces the same output-voltage wave form as the full-wave rectifier, and for that reason it is sometimes referred to as a full-wave rectifier. As indicated by the circuit diagram of Fig. 23-6, the bridge requires twice as many tubes as the full-wave rectifier of Fig. 23-5. The bridge circuit has some advantages, however, the most obvious being that a special transformer is not required; in fact, as far as the operation of the circuit is concerned, no transformer is needed. A transformer is shown in Fig. 23-6, however, since one is generally required to provide the proper load voltage.

The operation of the bridge rectifier can be understood by a consideration of the circuit of Fig. 23-6. When the source voltage e_{ab} is positive, current flows through tube A, the load, and tube C; the direction of this current is as indicated. When the polarity of the source voltage reverses, current flows through tube B, the load, and tube D. Regardless of the path taken by the current, its direction through the load is the same. For every half-cycle, the load voltage e_{cd} is the positive half of a sine wave. The resulting load-voltage wave form is identical with that of the full-wave rectifier of Fig. 23-5.

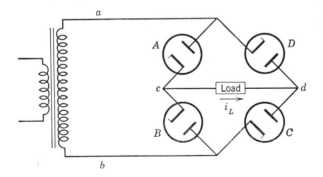

Fig. 23-6. Bridge-rectifier circuit.

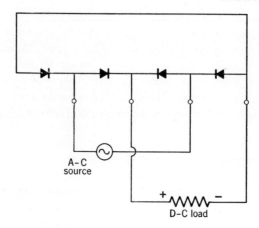

Fig. 23–7. Arrangement of selenium or copper-oxide rectifier stack to form a bridge.

The bridge rectifier is important as a source of high-voltage d-c power. With a fixed transformer voltage, the bridge produces twice the average load voltage produced by the full-wave rectifier of Fig. 23–5. There are several disadvantages to the bridge rectifier, however: (*a*) It requires four tubes instead of two. (*b*) There are always two tubes conducting, and this increases the total tube drop and power loss. (*c*) The cathodes of the rectifier tubes are at different potentials; this eliminates the possibility of using double diodes and necessitates a separate heater supply for each of the four tubes.

It may appear that the disadvantages of the bridge circuit outweigh its advantages. However, for large output voltages, the tube drop is negligible in comparison with the output voltage, and the savings resulting from the lower-voltage supply transformer may make the bridge circuit more economical than the two-tube full-wave rectifier. When the tubes are replaced by semiconductor rectifying elements, which require no filament supply, the bridge circuit is also practical at low voltages.

Since copper-oxide and selenium rectifiers are frequently stacked to achieve the desired voltage rating (see Art. 22–6), they are available with the four arms of a bridge rectifier all mounted in a single stack and bolted together. Figure 23–7 illustrates this arrangement and shows how an a-c source and a load are connected to the stack. For a bridge that is to be operated directly from a 110-volt a-c line, each arm is subjected to a peak inverse voltage of approximately 150 volts. Thus, each arm would require at least 6 elements of selenium rectifiers or a total of 24 elements.

23-8. RECTIFIER FILTERS

It is a characteristic of a capacitor that time is required to change the electric charge on its plates (or to change the voltage between its terminals, since this voltage is proportional to the charge on the capacitor). It was shown in Art. 23–4 that, because of this characteristic, a capacitor can act as a simple filter to reduce load-voltage fluctuations. Although it was shown to act as a filter for a half-wave rectifier, the capacitor is equally effective as a filter for any rectifier circuit.

Inductors are also applicable to filters, since time is required to change the current through an inductor, or, as sometimes stated, an inductor tends to oppose any change in the current through it. An inductor connected in series with a rectifier load tends to oppose changes in the load current; for pure resistance loads, this tendency is equivalent to an opposition to changes in the load voltage.

Instead of consisting of a single inductor or a single capacitor, most filters are made up of combinations of series inductors and shunt capacitors. Several practical arrangements are shown in Fig. 23–8.

The function of any filter is to reduce the magnitude of the a-c component of the rectifier output without seriously interfering with the d-c component. The degree to which any of the circuits satisfies this objective can be determined analytically by determining the a-c and d-c components of the load voltage that result from a given input voltage to the filter. The input voltage (output of the rectifier) consists of a d-c component, an a-c component of the fundamental frequency,* and

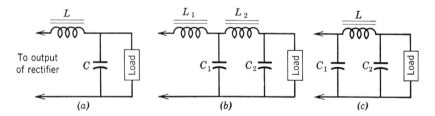

Fig. 23–8. Rectifier filters. (a) Single-section L, or choke input. (b) Two-section choke input. (c) Single-section π, or capacitor input.

---

* The frequency of the fundamental a-c component is equal to the frequency of the supply for a half-wave rectifier and is equal to twice the supply frequency for the full-wave rectifiers.

higher harmonics, all of which can be found by a Fourier analysis of the input-voltage wave form. For each of these components of the input voltage, there is a corresponding component in the load voltage. These components are found by assuming that each acts independently of the others. This may seem like a tedious task, but it can be greatly simplified.

An inspection of the circuits of Fig. 23–8 indicates that only the resistance of the series inductors and the resistance of the load are involved in the determination of the d-c component of the load voltage. Furthermore, only the fundamental a-c component need be considered, since the higher-frequency components are filtered out even more completely than the fundamental. The validity of these simplifying assumptions is demonstrated for the filter of Fig. 23–8a in the following paragraphs.

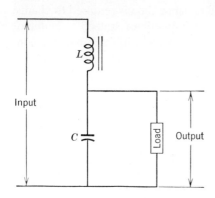

Fig. 23–9. Choke-input filter circuit.

The circuit of Fig. 23–8a is redrawn as Fig. 23–9 to show its similarity to the voltage-divider circuit.

The d-c component of the output voltage in the circuit of Fig. 23–9 differs from the corresponding component of the input voltage only by an IR drop due to the resistance of the series inductor. When considering the a-c components, however, it is convenient to think of the input voltage as being impressed across the series combination of the inductor and capacitor, and the output voltage as being that part of the input that appears across the capacitor. In a practical filter, the capacitive reactance is much less than the resistance of the load; therefore, the effect of the load resistance can be neglected in determining the voltage division between the inductor and the capacitor. The ratio of output voltage to input voltage then equals the ratio of the impedances across which they appear. Thus,

$$\frac{E_{\text{out}}}{E_{\text{in}}} = \frac{\frac{1}{2\pi f C}}{2\pi f L - \frac{1}{2\pi f C}} \qquad (23\text{–}5)$$

For satisfactory filtering, most of the a-c component of voltage should appear across the series inductor; that is, a practical filter requires that the reactance of the series inductor be much larger than that of the shunt capacitor. Thus, Eq. 23–5 may be simplified and rearranged as follows:

$$E_{\text{out}} = E_{\text{in}} \frac{1}{(2\pi f)^2 LC} \tag{23-6}$$

Equation 23–6 indicates that, if a filter reduces the fundamental component of a voltage to one tenth of its original value, the second-harmonic component is reduced to one fortieth of its original value; therefore a consideration of only the fundamental a-c component is justified. Equation 23–6 also indicates that filtering is improved by the use of larger shunt capacitors or larger series inductors.

Analysis of the other filter circuits indicate that (*a*) a given amount of inductance and capacitance is more effective if split up into two sections (Fig. 23–8*b*) than if applied in a single section (Fig. 23–8*a*); (*b*) higher output voltage is obtained with capacitor-input filters than with choke-input filters, but the voltage regulation is poorer. (The same restriction on the use of phanotrons with capacitor filters applies to all capacitor-input filters, such as the one illustrated in Fig. 23–8*c*.)

Filters can be designed to provide any desired degree of filtering. The filters in the d-c power supplies of most radio receivers are capable of reducing the rms value of the a-c component from 70 per cent of the d-c component at the filter input, to a fraction of 1 per cent of the average voltage across the load.

23–9. VIBRATOR RECTIFIERS

Occasionally, a direct voltage of several hundred volts is required when the only available source is a d-c battery of low voltage. An example is an automobile radio; the radio requires a d-c source of about 200 volts, and the only available supply is a 6-volt battery. There are at least two practical solutions to this problem. One solution is the use of a small dynamotor—in effect, a low-voltage d-c motor driving a high-voltage d-c generator—but this is too expensive for most applications. A more economical solution is to convert the direct voltage to an alternating voltage which can then be transformed and rectified to provide the required direct voltage. A vibrator is capable of both conversion and rectification.

In the circuit of Fig. 23–10*a*, the vibrator provides only an alternating voltage. With the armature in its normal position, as shown, current flows from the battery through one half of the transformer primary; a smaller current flows through the other half because of the connection

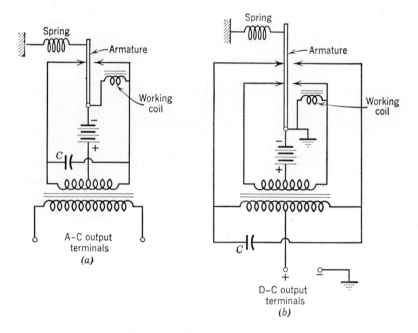

Fig. 23–10. Vibrator power supplies.

of the working coil. The current in the working coil provides enough magnetic force to pull the armature from its normal position and connect it to the other stationary contact. Current ceases to flow in the first half of the transformer primary, but a large current now flows through the second half, linking the transformer core in the opposite direction. At the same time the working coil is de-energized, and the spring returns the armature to the start position. This cycle is repeated continuously, applying an alternating mmf to the transformer core. The vibrator frequency depends upon the mechanical and electrical characteristics of the entire apparatus and is generally between 60 and 180 cycles per second.

The circuit shown in Fig. 22–10b is similar to that of Fig. 22–10a but has an additional pair of contacts. These contacts make and break in synchronism with the first pair and act to switch the secondary in the same manner as the two diodes of a two-tube full-wave rectifier. This vibrator is therefore capable of stepping up a direct voltage. The capacitors shown in both diagrams serve to reduce sparking at the contacts, and in conjunction with the inductance of the transformer they form a resonant circuit to improve the output wave form.

23-10. GRID-CONTROLLED RECTIFIERS

A thyratron meets all the requirements of a rectifier tube and, in addition, can be controlled. In diode rectifiers, the average load voltage can be altered only by a change in the magnitude of the a-c supply voltage. Variation of the control-grid voltage of a thyratron rectifier offers a more flexible means of control and requires the expenditure of negligible power.

Figure 23-11 shows a half-wave thyratron rectifier with a variable d-c grid voltage as the control. (Other forms of grid control are discussed in following articles.) For any given value of grid voltage, the anode voltage necessary to start conduction can be read directly from the thyratron control characteristic. Then, knowing the manner in which the anode supply voltage varies with time, one can determine the time in the cycle at which conduction starts. While the tube conducts, the anode-to-cathode voltage equals the constant tube drop, and the load voltage equals the source voltage minus the tube drop. Since the grid loses control as soon as conduction begins, the tube continues to conduct until the supply voltage becomes less than the tube drop. This operation is repeated every cycle. The resulting wave forms are plotted in Fig. 23-12.

If the grid-to-cathode voltage in the circuit of Fig. 23-11 is made more negative, conduction starts later in the cycle; therefore, the period of conduction is decreased. Since the load voltage is independent of the grid voltage during conduction, the shorter period of conduction results in a smaller average load voltage. Grid-voltage variation therefore offers a means of controlling the average value of the load voltage.

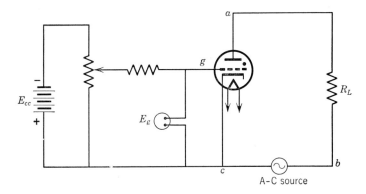

Fig. 23-11. Half-wave grid-controlled rectifier.

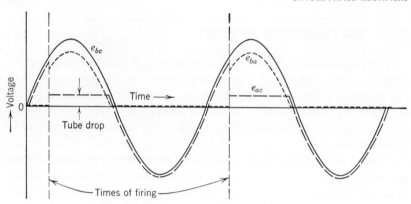

Fig. 23–12. Voltage wave forms for the rectifier of Fig. 23–11.

For any particular set of constants in the circuit of Fig. 23–11, a plot of critical grid voltage versus time is an aid to an understanding of the effect of grid-voltage variations. The thyratron control characteristic gives the critical grid voltage as a function of the anode voltage, and, if the anode voltage is known as a function of time, the two relationships can be combined to obtain the desired curve. This is done graphically in Fig. 23–13. For each of several values of time (t_1 through t_5), the anode supply voltage is found directly from the supply-voltage curve, and corresponding values of the critical grid voltage are found from the control characteristic. These values of critical grid voltage are then plotted against the same time scale as the plot of anode supply voltage.

The significance of the curve of critical grid voltage versus time is that the thyratron starts to conduct whenever the actual grid voltage becomes less negative than the critical value. If the actual grid voltage is plotted to the same scale as the critical-grid-voltage curve, their intersection indicates the time at which conduction starts. However, the intersection must be such that the grid voltage is becoming more positive than, or less negative than, the critical grid voltage. An intersection in the other direction would be meaningless, for the tube would already be conducting.

The actual grid voltage for the circuit of Fig. 23–11 would appear as a horizontal line on the critical grid-voltage–time plot, and, as the magnitude of this d-c grid voltage is varied, the horizontal line representing it would move up and down accordingly. For zero grid voltage, the intersection of the two curves and the beginning of conduction follow soon after the start of the cycle. As the grid voltage is made more negative, conduction starts later in the cycle. However, if the grid voltage is made any more negative than that required to delay the start of conduction until the anode supply voltage reaches its positive peak value, the curves do not intersect, and there is no conduction.

There are two important disadvantages to this method of control, both of which are evident from a consideration of the construction of Fig. 23–13. First, as indicated in the preceding paragraph, the time of firing can be controlled only during the first quadrant of each cycle. The effect of this limitation is that the average load voltage may be varied only between its maximum value and one-half that value. Second, the average load voltage is sensitive to changes in the thyratron characteristic. A change in the control characteristic is equivalent to a vertical displacement of the critical-grid-voltage curve; therefore, a slight change in temperature may change the characteristic sufficiently to stop conduction entirely when the tube has been firing at, or just before, the time of maximum anode supply voltage.

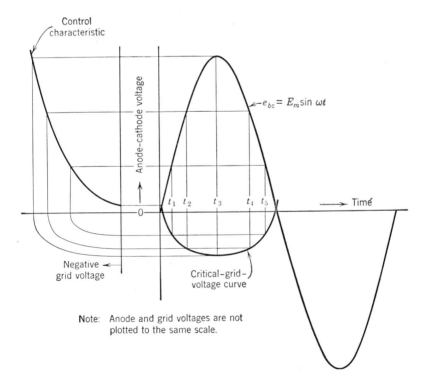

Fig. 23–13. Critical grid voltage versus time for a sinusoidal supply voltage.

23-11. PHASE-SHIFT CONTROL

The disadvantages of the variable d-c grid-voltage control are overcome by employing an a-c grid voltage of fixed magnitude and controllable phase shift. This type of control is illustrated in Fig. 23–14, which shows a plot of an a-c grid voltage for each of two possible phase positions relative to the critical-grid-voltage curve. The times at which conduction begins, labeled t_1 and t_2, are determined from the intersection of the corresponding grid-voltage curves with the critical-grid-voltage curve.

By properly shifting the phase of the grid-voltage curve relative to the critical-grid-voltage curve, the start of conduction can be varied throughout a full half-cycle. To accomplish this, the grid voltage must lag the anode supply voltage by an amount that is controllable between 0 and 180 degrees. Such grid-voltage control would allow the average load voltage of a rectifier to be varied smoothly from zero to the maximum possible value. In addition, if the peak value of the grid voltage is large in comparison with the maximum (negative) value on the critical-grid-voltage curve, the slope of the grid-voltage curve is steep at its intersection with the critical-grid-voltage curve. Therefore, variations in the thyratron characteristic (which have the effect of moving the critical-grid-voltage curve up or down) have little effect on the time at which the thyratron fires.

There are several means of obtaining an alternating voltage of constant

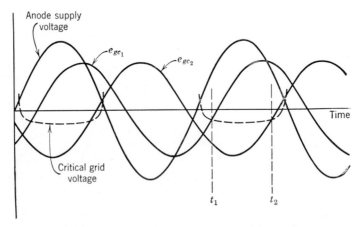

Fig. 23–14. Voltage wave forms for phase-shift control.

BIAS-PHASE CONTROL

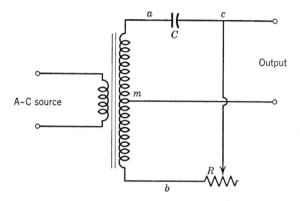

Fig. 23-15. Phase-shift circuit.

magnitude and variable phase. One circuit for this purpose is shown in Fig. 23-15. Since the grid voltage must bear a definite phase relationship to the anode supply voltage, the source for this phase-shift circuit must be the same as that for the anode supply. The voltage e_{cm} is equal in magnitude to that of the voltage across each half of the transformer secondary. Changing the resistance of the rheostat R shifts the phase of e_{cm} without affecting its magnitude. To obtain a phase shift of 0 to 180 degrees, R must be continuously variable from 0 to ∞. This is obviously impossible, but an adequate phase shift (168 degrees) can be obtained with a rheostat whose resistance can be varied from zero to ten times the reactance of the capacitor C.

23-12. BIAS-PHASE CONTROL *

Another method of grid control is illustrated in Fig. 23-16. The grid-to-cathode voltage for bias-phase control is made up of two components: an alternating voltage of fixed magnitude and lagging the anode supply voltage by a fixed displacement, about 90 degrees; and a d-c component of variable magnitude and polarity. Curves are plotted in Fig. 23-16 for three values of the d-c component: one positive, one negative, and one zero.

A study of the curves of Fig. 23-16 reveals that the bias-phase method of control has all the merits of the variable phase-shift control. It requires a slightly more complicated circuit, however; therefore, for

---

* The term "grid bias," or simply bias, denotes a d-c component of the grid-to-cathode voltage.

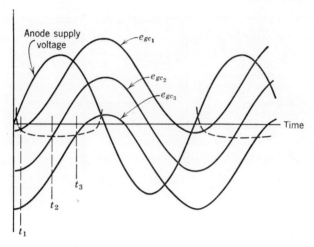

Fig. 23-16. Voltage wave forms for bias-phase control.

an application involving only manual control, the phase-shift method is simpler. Bias-phase control readily lends itself to automatic-control circuits in which the quantity exercising the control is a varying direct voltage. The choice between these two control systems must be made on the basis of the most desirable method of activating the control: that is, whether by varying a resistance or by varying a direct voltage.

PROBLEMS

23-1. A 100-volt battery is charged from a 120-volt 60-cycle line through a phanotron and a 50-ohm limiting resistor. Assuming a 10-volt drop across the phanotron when conducting, calculate the peak current and peak inverse voltage to which the phanotron is subjected.

Ans. 1.2 amperes, 270 volts.

23-2. A half-wave rectifier supplied from a 120-volt a-c source uses a type-35Z5 diode having the following ratings: peak anode current, 600 ma; average anode current, 100 ma; peak inverse voltage, 700 volts.

(a) What is the minimum value of load resistance without exceeding either of the current ratings? *Ans.* 540 ohms.

(b) How high a source voltage could be applied to this rectifier without exceeding the PIV rating of the tube? *Ans.* 495 volts.

23-3. Two identical diodes are available, and they may be used in a voltage-doubler circuit or connected in series for a simple half-wave rectifier. In either case, a capacitor filter will be used and the load current will be negligible.

PROBLEMS

(a) What is the maximum load voltage that can be developed in each case without exceeding the 2500-volt PIV rating of each tube?

Ans. 2500 volts.

(b) Why is one circuit to be preferred over the other?

23-4. It is necessary to purchase new tubes for a certain full-wave rectifier. This rectifier is supplied from a center-tapped transformer secondary, having a total voltage of 2000 volts. The load is a pure resistance of 1000 ohms.

(a) Specify minimum ratings of I_{pk}, I_{avg}, and PIV for each of the two required tubes. *Ans.* 1.41 amperes; 0.45 ampere; 2830 volts.

(b) Qualitatively, how would each of these required ratings differ if a capacitor filter were connected in parallel with the load? Explain.

23-5. Selenium rectifier cells are available that can withstand a PIV of 25 volts each and an average current of 100 ma in the forward direction. It is desired to use some of these cells for a bridge rectifier that can be used with a 115-volt a-c source, and that will be capable of delivering an average current of 200 ma to a pure resistance load. Cells may be connected in series and/or parallel to obtain higher voltage and/or current ratings for the various arms of the bridge. What is the minimum number of cells that will be needed if none of the ratings are to be exceeded, and how should those cells be arranged? *Ans.* 28 cells.

23-6. A certain average voltage across a fixed resistance is to be obtained through the use of the selenium cells described in Prob. 23-5. These cells may be connnected as a half-wave, a full-wave, or a bridge rectifier. How does the number of cells required for the different circuits compare? Assume that transformers of any desired ratings are available.

23-7. A plot of load voltage versus time for a single-phase half-wave rectifier with a capacitor filter is shown in Fig. 23-17. To what peak inverse voltage is the rectifier tube subjected?

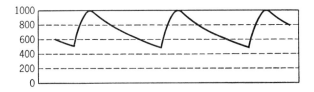

Fig. 23-17.

23-8. A plot of load voltage versus time for a two-tube full-wave rectifier with capacitor filter is shown in Fig. 23-18. To what peak inverse voltage is each of the tubes subjected? Assuming that the same load-voltage curve applies to a bridge rectifier, what is the peak inverse voltage to each of the four tubes?

23-9. A simple half-wave rectifier consists of an a-c source, a phanotron, and a pure resistance load. How would its efficiency be affected by each of the following changes:

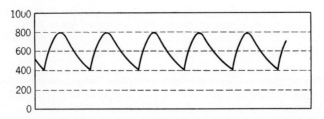

Fig. 23-18.

(a) An increase in the load resistance?
(b) An increase in the supply voltage?

23-10. A resistance-loaded bridge rectifier employs four tubes each having the following ratings:

> Peak anode current: 1.00 ampere
> Average anode current: 0.40 ampere
> Peak inverse voltage: 5000 volts

(a) Neglecting tube drop, what is the maximum rms value of voltage that can be safely applied to this rectifier? *Ans.* 3535 volts.

(b) With a source voltage of 3535 volts, what is the maximum d-c power ($E_{avg}I_{avg}$) that can be delivered without exceeding either of the current ratings of the tubes? *Ans.* 2030 watts.

23-11. A single-phase full-wave rectifier uses two diodes, each of which is rated as follows:

> Peak inverse voltage: 5000 volts
> Peak anode current: 1.0 ampere
> Average anode current: 0.25 ampere

The rectifier load is a pure resistance of 2000 ohms, and the transformer supplying the rectifier is rated at 115/4000 volts (2000 volts each side of a center tap). What is the greatest voltage (rms) that can be applied to the transformer primary without exceeding any of the rectifier-tube ratings?
Ans. 64 volts.

23-12. A full-wave rectifier with a single-section choke-input filter is supplied from a 60-cycle source. The filter consists of a 10-μf capacitor and a 30-henry choke having a d-c resistance of 300 ohms. The load is a pure resistance of 3000 ohms. The average load voltage is 300 volts, and the voltage wave form at the input to the filter can be expressed by the following Fourier series:

$$e = K(1 - \tfrac{2}{3}\cos\theta - \tfrac{2}{15}\cos 2\theta - \tfrac{2}{35}\cos 3\theta - \cdots)$$

(a) What is the average value (K) of the voltage at the input to the filter?
Ans. 330 volts.

(b) What is the rms value of the load voltage ripple? (Ripple and a-c component are synonymous terms.) State any assumptions you make to simplify this calculation. *Ans.* 0.9 volt.

23–13. Repeat Prob. 23–12 for a 2-section filter (Fig. 23–8b) having the same total capacitance, resistance, and inductance. For this calculation, assume that the presence of the second section does not affect the a-c operation of the first section. *Ans.* (a) 330 volts. (b) Approx. 0.4 volts.

23–14. The thyratron in a half-wave grid-controlled rectifier has the characteristics given by Fig. 21–9. The rectifier is supplied from a 120-volt a-c source. The grid voltage is adjusted so that the thyratron fires when the anode supply voltage is at its maximum positive value. Neglecting tube drop, determine the minimum value of load resistance that can be used without exceeding either of the following tube ratings: $I_{avg} = 100$ ma and $I_{pk} = 500$ ma. *Ans.* 340 ohms.

23–15. Draw a vector diagram for the circuit of Fig. 23–15 with $R = X_c$. Assume a transformer polarity, and determine the phase of the output voltage relative to the source voltage. Prove that this circuit would be just as effective if the capacitor were replaced by an inductance. (Actually, a combination of fixed resistance and variable inductance is sometimes used in control circuits. The variable inductance is provided by a two-winding reactor, called a saturable-core reactor, in which the direct current in one of the windings controls the effective inductance of the other.)

23–16. Draw a circuit diagram for a single-phase full-wave rectifier that is controlled by means of a variable-phase-shift alternating voltage on the grids of each of the two thyratrons. The tubes must be controlled simultaneously, and, since their anode voltages are 180 degrees out of phase, their control grid voltages must also be displaced by 180 degrees.

23–17. Repeat Prob. 23–16, using bias-phase control.

23–18. The control characteristic for the thyratron in a single-phase half-wave rectifier is given by Fig. 21–9. The rectifier is supplied from a 230-volt a-c source. The grid-to-cathode voltage is sinusoidal, has an rms value of 4.0 volts, and lags the anode supply by 150 degrees.

(a) Plot one cycle of the load voltage. Neglect tube drop.

(b) What are the peak and average values of anode current if the load resistance is 2000 ohms? *Ans.* 130 ma; 10 ma.

24 Power Rectifiers

24-1. INTRODUCTION

Electronic rectifiers that supply more than about 1 kilowatt have several features that distinguish them from the low-power rectifiers of Chapter 23. First, since efficiency is important, gas-filled tubes are used unless the voltage is too high. The low-voltage drop of the gas tube means a low-power loss in the tube and, consequently, a relatively high efficiency. Second, it is impractical to construct thermionic cathodes having a peak anode-current rating of more than about 50 amperes. This limitation requires the introduction of a different cathode, the mercury-pool cathode, whenever higher currents are encountered. Third, three-phase circuits, or circuits derived from three-phase, are used to improve the efficiency.

24-2. MERCURY-POOL CATHODES

If an arc is established between a pool of mercury and a positive electrode in a tube, the mercury acts as an excellent cathode. There is practically no limit to the emission current that can be drawn from this cathode, and condensed mercury returns to the pool at the bottom of the tube so that the emission material of the cathode is never exhausted. In addition, the presence of the mercury assures a supply of gas (mercury vapor) in the tube. When an arc is established, a bright spot, called the **cathode spot,** is formed on the surface of the mercury pool and is in continuous motion, appearing to dance about on the mercury. This spot is the actual cathode, the source of the electron emission that sustains the discharge.

The exact process by which the cathode spot supplies electron emission is not too clearly understood, but several facts concerning it are known. Although the spot appears hot, its temperature is definitely not high enough to provide adequate thermionic emission to account for the extremely high currents that have been observed. It is generally agreed that the bulk of the emission must be caused by high-field emission. This seems reasonable, for, as in other arcs, most of the 10- to 15-volt anode-to-cathode drop occurs within a short distance of the cathode, making the voltage gradient (the electric field) high at the surface of the cathode.

The application to the electrodes of a mercury-cathode tube of a voltage greater than the normal arc drop is not sufficient to initiate an arc; a cathode spot must be formed. In general, the formation of a cathode spot requires a spark on the surface of the mercury. The earliest mercury-pool tubes were so mounted that the tube could be tilted to allow the mercury to make contact with an auxiliary energized electrode, thus permitting current to pass through the mercury. When the tube was restored to its original position, the mercury contact was broken, causing a small spark from which the desired cathode spot would develop if the anode were at a sufficiently positive potential.

Other, more practical methods of starting an arc in a mercury tube have since been developed. One method commonly used in large tank rectifiers requires a movable carbon rod which is normally in contact with the mercury but which can be raised by means of a solenoid to form a small arc between the rod and the mercury. This low-current d-c arc is often operated continuously to insure the presence of a cathode spot. A spark can also be formed by a high-voltage spark coil. This method, somewhat similar to the ignition system of an internal-combustion engine, allows precise control over the time at which the arc is established.

24-3. IGNITRONS

An **ignitron** is a mercury-cathode tube in which an arc is started by still another method, one that allows the same precise control that is possible with a spark coil. The ignitron has an auxiliary electrode (called the **ignitor**) of high-resistance material which is partially immersed in the mercury. When a sufficiently high current is passed

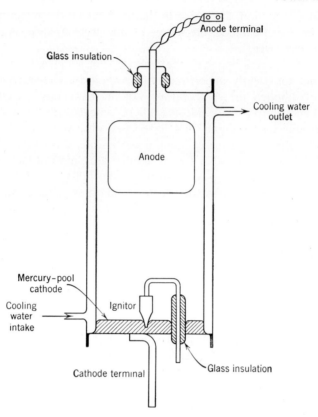

Fig. 24–1. Cross section of a low-voltage sealed-off ignitron.

through the ignitor into the mercury, a spark is formed at the contact, and a cathode spot can develop from this spark.

Figure 24–1 illustrates the construction of a typical ignitron. The envelope is usually made of stainless steel (not subject to attack by mercury) and is surrounded by a water jacket for cooling. Ignitrons designed to carry more than a few hundred amperes generally require a continuously operated pump to maintain the necessary degree of vacuum.

The action of the ignitor in initiating an arc depends upon the high resistance of the ignitor and upon its not being wetted by the mercury. When current flows through the ignitor, there is a voltage drop along the length of the ignitor, but the mercury remains essentially at a uniform potential throughout because of its high electric conductivity. As a result, there is a potential difference across the gap produced by the mercury meniscus. If the ignitor current is great enough (of the order of 10 to 20 amperes), this voltage causes a spark to the mercury, and therefore a cathode spot may form. Thus, control of the firing of an ignitron is accomplished by controlling the ignitor current.

If an ignitron is used as a half-wave rectifier, or as one element of a more complex rectifier, the arc is extinguished at the end of each positive half-cycle, and the cathode spot must be re-formed in every cycle.

If no control is desired, starting can be accomplished by employing a single phanotron (or contact rectifier) for each ignitron. A circuit that illustrates this method of starting applied to a half-wave rectifier is shown in Fig. 24–2. When the supply voltage becomes equal to the breakdown potential of the phanotron, the phanotron conducts and current flows through the ignitor and the mercury pool. The ignitor current increases as the supply voltage increases until a spark forms between the mercury and the ignitor. A cathode spot is immediately formed, and current flows from the anode to the cathode. (When conditions are such that the phanotron can conduct, the polarity of the voltage across the ignitron is also proper for conduction.) While the ignitron conducts, the sum of the voltages across the resistor R_s, the phanotron, and the ignitor equals the arc drop of the ignitron (normally 12 to 15 volts). The small current that flows through the ignitor circuit as a result of this voltage is negligible. For practical purposes, the ignitor current is limited to a short pulse at the beginning of each conduction period. This limitation is desirable, for it prevents overheating and consequent damage to the ignitor.

To control the time at which an ignitron fires, it is necessary only to replace the phanotron in the circuit of Fig. 24–2 with a thyratron, and to provide a means of controlling conduction in the thyratron. The ignitron starts to conduct only after the thyratron fires, so that any method of grid control applied to the thyratron indirectly controls the firing of the ignitron.

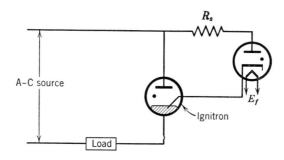

Fig. 24–2. Ignitron circuit.

24–4. MULTIANODE TANK RECTIFIERS

Most rectifier circuits require more than one rectifying element, and in many of these circuits the cathodes of the various tubes are electrically connected. For rectifier circuits with a common cathode connection, it is possible to combine all the rectifying tubes into a single unit having but one cathode and as many anodes as the circuit requires. Such a rectifier, shown schematically in Fig. 24–3, is contained in a steel tank (1 to 10 feet in diameter). A mercury pool at the bottom of the tank serves as the common cathode for all the anodes (usually 6 or 12) which are supported from the top of the tank and insulated from it.

The anodes of a tank rectifier are often enclosed by control grids similar to those of thyratrons. With these grids, the whole structure becomes functionally equivalent to several thyratrons with their cathodes connected. Any of the methods for grid control of thyratrons is applicable to the control of a tank rectifier equipped with control grids.

As discussed in Art. 24–2, an arc to a movable carbon rod may provide the cathode spot in a large tank rectifier. When the tank rectifier has no grid control, there is always current flowing from one of the anodes. This current maintains the cathode spot so that the starting arc is needed only when the rectifier is first started. However, with grid

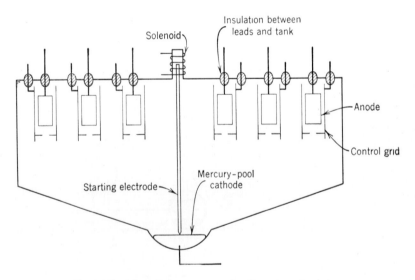

Fig. 24–3. Schematic arrangement of multianode tank rectifier.

control, there may be a delay after the current from one anode ceases before current starts flowing from the next, and the auxiliary arc (sometimes called the "keep-alive arc") must be kept in continuous operation.

24–5. THREE-PHASE HALF-WAVE RECTIFIER

In the analysis of the polyphase rectifier circuits discussed in this chapter, several simplifying assumptions are made. First, the voltage drop across the rectifying element while it conducts is neglected. Second, only a pure resistance load is considered. Third, the transformers are treated as ideal transformers having neither leakage reactance nor coil resistance and requiring no exciting current. The significance of this last assumption is that the transformer primary current can be considered to have the same wave shape as the a-c component of the secondary current, and the ratio of the primary current to the secondary current (a-c components) is inversely proportional to the turns ratio of the transformer.

The simplest of the polyphase rectifiers is the three-phase half-wave rectifier shown in Fig. 24–4.* This rectifier is made up of three single-phase half-wave rectifiers that send current in the same direction through a common load. The source voltages for the individual tubes are balanced three-phase voltages. Each tube conducts for one third of a cycle, and at any given time only that tube having the most positive supply voltage can conduct. The validity of the last statement can be demonstrated by a consideration of the curves of Fig. 24–4. At time t_1, both e_{an} and e_{bn} are positive, but e_{an} is the larger of the two. If tube A is conducting and its tube drop is negligible, the anode-to-cathode voltage of tube B is

$$e_{bo} = e_{bn} - e_{on} = e_{bn} - e_{an} \qquad (24-1)$$

At the time under consideration, e_{an} is greater than e_{bn}. Equation 24–1 therefore indicates that the anode of tube B is negative with respect to the cathode, and tube B cannot conduct. Further investigation applying the same reasoning would show that only one tube can conduct at any time.

Since tube drop is neglected, the load voltage e_{on} is at all times equal to the supply voltage of the conducting phase, as indicated on the curves

---

* For convenience only, phanotrons are indicated as the rectifying elements. Tank rectifiers or ignitrons are more likely to be used in any but relatively low-power installations.

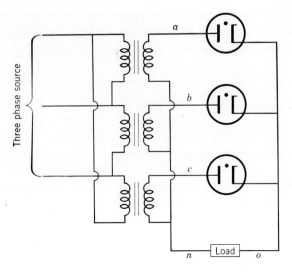

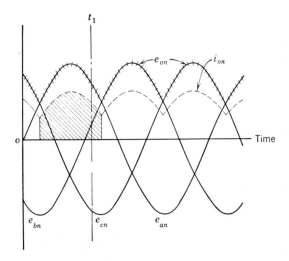

Fig. 24–4. Three-phase half-wave rectifier circuit.

of Fig. 24–4. If the supply voltage for phase A is taken as

$$e_{an} = E_m \sin \omega t$$

the load voltage, for values of ωt between $\pi/6$ and $5\pi/6$, is

$$e_{on} = E_m \sin \omega t \tag{24-2}$$

ZIGZAG CONNECTION

Since the wave form of the load voltage repeats itself every third of a cycle, the average value of the load voltage may be found by a consideration of only that third for which Eq. 24–2 applies. Thus,

$$\text{Average load voltage} = \frac{3}{2\pi}\int_{\pi/6}^{5\pi/6} E_m \sin \omega t \, d(\omega t)$$

$$= \frac{3E_m}{2\pi}\left[-\cos \omega t\right]_{\pi/6}^{5\pi/6} = \frac{3E_m \cos \pi/6}{\pi} = 0.827 E_m \quad (24\text{–}3)$$

For a resistance load, the average load current equals the average load voltage divided by the resistance of the load. Thus, since each tube carries one third of the average load current,

$$\text{Average tube current} = \frac{1}{3} \times \frac{0.827 E_m}{R_L} = 0.276 \frac{E_m}{R_L} \quad (24\text{–}4)$$

The current carried by tube A is indicated by the shaded portion under the load-current curve of Fig. 24–4.

24–6. ZIGZAG CONNECTION

One disadvantage of the three-phase half-wave rectifier of Fig. 24–4 is that the currents in each of the transformer secondary windings have a large average value (or d-c component). No proportional primary current flows to cancel the magnetizing effect of these direct currents, with the result that the cores of the transformers are likely to become saturated. The prevention of excessive primary currents in the transformers of this rectifier requires either an oversized transformer, or a special arrangement of the transformer connections to eliminate the magnetizing effect of the direct currents.

The zigzag connection of Fig. 24–5 is one possible arrangement for eliminating d-c saturation of the transformer cores. Each of the three transformers required for this circuit has one primary and two secondary windings. The three windings of each transformer are separated in Fig. 24–5 to simplify the connections, but they are shown parallel to each other for identification. Each set of three windings is shown 120 degrees from each of the other two sets. This arrangement of the windings on the diagram is identical with the vector diagram for the circuit, if each winding is replaced by the corresponding voltage vector.

The supply voltage for each tube of the zigzag rectifier is the sum of

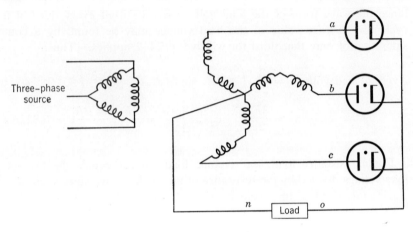

Fig. 24–5. Zigzag transformer connection for a three-phase half-wave rectifier.

the voltages of two windings, one from each of two different transformers. The connections are such that the two secondary windings of each transformer conduct current in opposite directions with respect to the core, resulting in no net d-c component of magnetizing force, and no d-c saturation of the core.

As far as the tubes and load are concerned, the zigzag circuit is the same as that of Fig. 24–4. The three tubes are supplied by three voltages equal in magnitude and 120 degrees apart in phase. Therefore, the wave forms of Fig. 24–4 are equally applicable to this rectifier.

To obtain the same output voltage, the total number of secondary turns required is greater for the zigzag connection than for the three-phase half-wave rectifier of Fig. 24–4. Although this might seem to be an added expense, it is negligible in comparison with the expense of transformers large enough to exhibit no saturation with the d-c magnetizing force present in the three-phase half-wave rectifier of Fig. 24–4.

24–7. SIX-PHASE HALF-WAVE RECTIFIER

Except for the number of phases, the operation of the six-phase half-wave rectifier (circuit diagram shown in Fig. 24–6) differs little from that of the three-phase half-wave rectifier. There is the same symmetrical arrangement of the phase voltages, each supplying one tube. There are six phases, however, so that each tube conducts for only one sixth of each cycle, and, as indicated in Fig. 24–6, the resulting load-voltage wave form is much smoother. The average value of this load

ZIGZAG CONNECTION

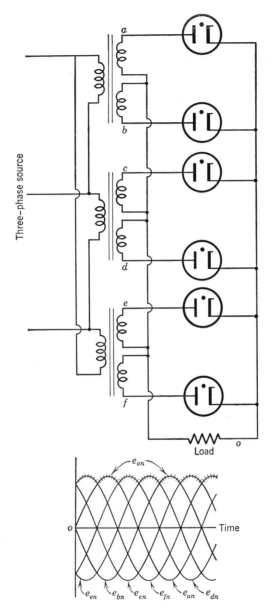

Fig. 24-6. Six-phase half-wave rectifier circuit.

voltage is 0.955 times the peak value of the supply voltage for one phase.

There is no d-c saturation of the transformer cores in the six-phase rectifier since each transformer supplies two phases, and the magnetizing effects of the d-c components in these two phases cancel.

24-8. THREE-PHASE FULL-WAVE RECTIFIER

The three-phase full-wave rectifier circuit, shown in Fig. 24–7, is the three-phase counterpart of the single-phase bridge circuit; in fact, it is sometimes called a three-phase bridge rectifier. As in the single-phase bridge, there must always be two tubes conducting to complete the circuit, and current flows through each of the voltage sources in both directions. Different pairs of tubes conduct to allow this reversal of

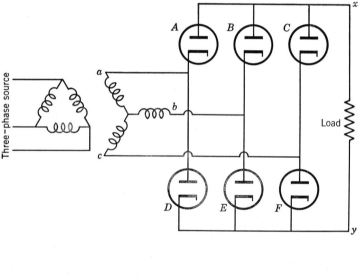

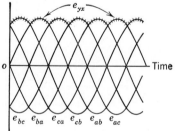

Fig. 24–7. Three-phase full-wave rectifier circuit.

current in the source without reversing the direction of the current in the load. For instance, when the voltage e_{ab} of Fig. 24–7 is at its maximum positive value, current flows through tube D, the load, and tube B; one half-cycle later, e_{ab} is at its maximum negative value, and the current flows through tube E, the load, and tube A. The direction of the current through the load is always from y to x.

The load of Fig. 24–7 has six supply voltages, the phases of which differ by 60 degrees, as indicated by the curves. The path taken by the current at any time is determined by the most positive of these six voltages. For example, when the voltage e_{cb} is the most positive, current must flow through tube F, the load, and tube B. If the tube drop is negligible, the load voltage is always equal to the most positive of the six supply voltages, resulting in a load-voltage wave form identical with that of the six-phase half-wave rectifier, as shown in Fig. 24–7.

No use is made of the neutral point of the wye-connected transformer secondary windings in the circuit of Fig. 24–7. The transformers supplying a three-phase full-wave rectifier may be connected either in wye or in delta. In fact, the circuit would function properly with no transformers.

The three-phase full-wave rectifier has all the advantages of the six-phase circuit—no d-c saturation, and a smoother output-voltage wave form—and, in addition, makes more effective use of the transformer windings. Each winding conducts for two thirds of the time instead of for one sixth of the time.

The only disadvantage of the three-phase full-wave rectifier is that the cathodes of the six tubes are not electrically connected. Lack of a common cathode connection precludes the possibility of using this circuit with a tank rectifier. Even with phanotrons as the rectifying elements, the problem of insulation between the various cathode heater circuits is the same as for the single-phase bridge rectifier.

At modest power levels, this circuit is very attractive for use with any of the semiconductor rectifying elements; no transformer is required, the output-voltage wave form is as smooth as that of any circuit considered (without a filter), no direct current is drawn from the source, and the lack of a common cathode connection is of no consequence with the semiconductor diodes.

24–9. DOUBLE-WYE CONNECTION

The double-wye connection is a popular rectifier circuit combining most of the advantages of the other rectifiers discussed in this chapter.

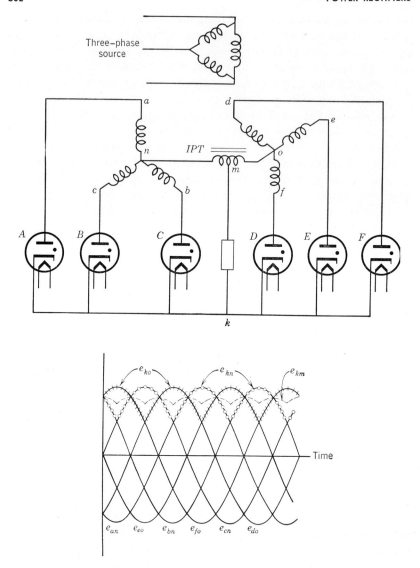

Fig. 24–8. Double-wye-connected rectifier circuit.

The load-voltage wave form is the same as for all other six-tube circuits; the circuit has the advantage of the common cathode connection; there is no d-c saturation of its transformer cores; and better use is made of the transformers than in any of the other circuits with the exception of the three-phase full-wave rectifier.

The circuit diagram of Fig. 24–8 indicates that the double-wye-connected rectifier consists of two three-phase half-wave rectifiers operating in parallel and that it differs from the six-phase rectifier circuit only by the addition of an interphase transformer (IPT on the circuit diagram of Fig. 24–8). The presence of the interphase transformer is necessary to allow conduction in two tubes, one in each set of three, at any given time. Each tube conducts for one third of the time, which results in more effective use of the transformer windings than in the six-phase half-wave rectifier.

An analysis of the double-wye circuit may be made by considering it to be two three-phase half-wave rectifiers. The load voltage for the half-wave rectifier that includes tubes A, B, and C of Fig. 24–8 is e_{kn}, and for the rectifier that includes tubes D, E, and F the load voltage is e_{ko}. In each case, the load voltage is found as for any other three-phase half-wave rectifier.

The interphase transformer, which serves to isolate the two half-wave rectifiers, is simply an inductor with a connection made to the midpoint m of its winding. Since the two halves of the winding are linked by the same core (and flux), the voltages across the two halves must always be equal. Therefore, the potential of the point m is always midway between the potentials of the two end terminals, n and o, of the inductor. The load voltage of the entire double-wye rectifier e_{km} must then be midway between the two voltages e_{kn} and e_{ko}. The resulting curve of voltage e_{km} plotted in Fig. 24–8 indicates a wave form identical with that of the other six-tube circuits.

24–10. RECTIFICATION EFFICIENCY

The load voltage or current of any rectifier includes, in addition to the desired d-c component, a ripple or a-c component. In any load requiring a rectifier, this a-c component is of no value; it constitutes a loss. For instance, if a rectifier is used to charge a storage battery, only the d-c component of the current contributes to the charging; the ripple component serves merely to heat the battery. This loss depends upon the magnitude of the ripple and therefore upon the load-current wave form.

For power rectifiers, the loss due to the presence of a ripple in the load voltage or current is of considerable concern and is the basis for two different figures of merit for comparing rectifier circuits. One figure of merit, the **ripple factor,** is the ratio of the rms value of the a-c component to the average value of the load voltage or current. Values of ripple factor for some common rectifier circuits with pure resistance loads are listed in Table 24–1.

TABLE 24–1

COMPARISON OF THE MERITS OF RESISTANCE-LOADED RECTIFIER CIRCUITS

Circuit	Ripple Factor	Rectification Efficiency	Ratio of Total Transformer Volt-Amperes to D-C Power	
			Secondary	Primary
One-phase full-wave	0.48	0.810	1.75	1.23
One-phase bridge	0.48	0.810	1.23	1.23
Three-phase half-wave (zigzag connection)	0.18	0.965	1.73	1.22
Three-phase full-wave	0.04	0.998	1.05	1.05
Six-phase half-wave	0.04	0.998	1.81	1.28
Double-wye	0.04	0.998	1.50	1.06

The same information on loss due to ripple can also be expressed as an efficiency. **Rectification efficiency** is defined as the ratio of the d-c power output ($E_{dc}I_{dc}$ or $I_{dc}^2 R_L$) to the total power transferred to the load. Rectification efficiency is the efficiency obtained when only those losses that are inherent in the particular rectifier circuit are considered; that is, losses in the tubes and transformers are not included.

From mathematical expressions for the definitions of ripple factor and rectification efficiency, it is possible to develop the relationship between these two quantites. Thus,

$$\text{Ripple factor} = f = \frac{I_{ac}}{I_{dc}}$$

or

$$I_{ac} = f I_{dc} \qquad (24\text{–}5)$$

The d-c power in the load is

$$P_{dc} = I_{dc}^2 R_L \qquad (24\text{–}6)$$

and the loss due to the ripple

$$P_{ac} = I_{ac}^2 R_L = f^2 I_{dc}^2 R_L \qquad (24\text{–}7)$$

Since rectification efficiency is the ratio of d-c power to the sum of the d-c power and losses, Eqs. 24–6 and 24–7 can be combined to give

$$\text{Rectification efficiency} = \frac{I_{dc}^2 R_L}{I_{dc}^2 R_L + f^2 I_{dc}^2 R_L} = \frac{1}{1 + f^2} \qquad (24\text{–}8)$$

TRANSFORMER UTILIZATION

Equation 24–8 indicates that ripple factor and rectification efficiency are merely different ways of expressing the same information, and that either one may be determined if the other is known.

24-11. TRANSFORMER UTILIZATION

The cost of the transformers for a large electronic rectifier is a large percentage of the total first cost of the rectifier. Furthermore, the cost of a transformer is largely determined by its volt-ampere rating. Because of different current wave forms, the rms transformer currents (which are the basis of its volt-ampere rating) needed to produce a certain d-c load current may differ for the different rectifier circuits. As a result, the total transformer volt-amperes required to produce a given d-c power may differ considerably for the different circuits. An indication of the relative cost of the transformers required for a certain rectifier application may therefore be obtained from a comparison of the ratio of the total transformer volt-amperes required to the d-c power delivered. This information is included in Table 24–1.

In several of the rectifier circuits, the primary-current wave form differs from that of the secondary. For such circuits, the volt-ampere rating of the primary differs from that of the secondary.

The reciprocal of the ratio of transformer volt-amperes to d-c power is called the **transformer utilization factor**. The utilization factor (always less than unity) for any circuit gives the maximum d-c power that can be delivered by the rectifier in terms of the total volt-ampere rating of the transformers supplying that rectifier.

PROBLEMS

24-1. Prove that the average load voltage for a six-phase half-wave rectifier equals 0.955 times the peak value of the supply voltage for one phase.

24-2. The peak current in one tube of a three-phase half-wave rectifier is 100 amperes. What are the average and rms values of this current?
Ans. 27.6 amperes; 48.5 amperes.

24-3. A three-phase half-wave rectifier uses three phanotrons, and its average load current is 15.0 amperes. What are the peak and average values of the anode current in one of the phanotrons.
Ans. 18.1 amperes; 5.0 amperes.

24-4. The average load voltage for the rectifier of Prob. 24–3 is 600 volts. Plot one cycle of the anode-to-cathode voltage for one of the phanotrons, and determine its peak inverse value. Neglect tube drop. *Ans.* 1260 volts.

24–5. Repeat Prob. 24–3 for a three-phase full-wave rectifier using six phanotrons. *Ans.* 15.7 amperes; 5.0 amperes.

24–6. A three-phase full-wave rectifier delivers 1000 watts (d-c power) to a 100-ohm resistance load. Plot one cycle of the voltage across one of the tubes, and from that plot determine the peak inverse voltage. *Ans.* 330 volts.

24–7. Two different phanotron types have been suggested for replacement in a polyphase rectifier. Their costs are the same and their anode current ratings are as follows:

$$\text{Type } A: I_{pk}, 10.0 \text{ amperes}; I_{avg}, 3.0 \text{ amperes}$$
$$\text{Type } B: I_{pk}, 15.0 \text{ amperes}; I_{avg}, 2.5 \text{ amperes}$$

Assuming that you wish to obtain as large a current output as possible, which type would you choose for replacement in (*a*) a three-phase half-wave rectifier, and (*b*) a six-phase half-wave rectifier? Explain. *Ans.* (*a*) *A*; (*b*) *B*.

24–8. The average load current in a three-phase double-wye-connected rectifier is 100 amperes. What is the average tube current? Explain why the peak tube current in this rectifier is not equal to the peak load current. *Ans.* 16.7 amperes.

24–9. Six selenium-rectifier units are to be connected to form a three-phase bridge rectifier.

(*a*) What is the maximum rms value of line voltage that can be applied to this rectifier without exceeding a peak inverse voltage of 150 volts? *Ans.* 106 volts.

(*b*) With an applied voltage of 106 volts and a load resistance of 1000 ohms, what average value of current would you expect in each of the rectifier units? *Ans.* 47.7 ma.

24–10. Sketch the load-voltage wave form that would be obtained from the three-phase full-wave rectifier of Fig. 24–7 if one of the transformer secondary windings became open-circuited.

24–11. A polyphase rectifier is to be designed that is capable of delivering 20 amperes at 1250 volts to a pure resistance load. For each of the circuits studied, which of the following tubes would you use, and what are the voltage and kilovolt-ampere ratings of the necessary transformers?

Tube	PIV	I_{pk}, amperes	I_{avg}, amperes	Cost
A	5000	15	5	$ 15
B	5000	25	5	25
C	2000	40	7	40
D	3000	40	7	55
E	1500	75	20	150

24–12. Three identical transformers, each having two 220-volt windings and one 440-volt winding, have their high-voltage windings connected in delta to a 440-volt three-phase line and their other windings connected in zigzag to supply a three-phase half-wave rectifier. For a load current of 200 amperes, compute:

PROBLEMS

(a) D-c power to the load. *Ans.* 89 kw.
(b) Average tube current. *Ans.* 67 amperes.
(c) Rms current in transformer secondaries, and the total kilovolt-amperes of the transformer secondary windings. *Ans.* 116 amperes; 153 kva.

24–13. The three transformers of Prob. 24–12 are connected to supply a six-phase rectifier that delivers the same 89 kw of d-c power. Determine:
(a) Average tube current. *Ans.* 50 amperes.
(b) Rms current in transformer secondaries. *Ans.* 122 amperes.
(c) Total kilovolt-amperes of the transformer secondaries.
Ans. 161 kva.

24–14. How could one experimentally determine (a) the actual ripple factor, and (b) the d-c power output of a polyphase rectifier?

24–15. Draw the circuit diagram for a three-phase half-wave rectifier that uses three thyratrons controlled by the bias-phase method. Sketch one cycle of the load-voltage wave form that would be obtained when the grid voltage is adjusted so that each tube starts to conduct when its source voltage is at the most positive value.

24–16. Figure 24–9 gives the load-voltage wave form for a three-phase half-wave grid-controlled rectifier for one particular setting of the grid-bias

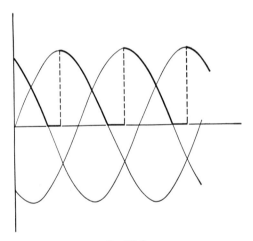

Fig. 24–9.

control. The corresponding peak inverse voltage across one of the three rectifier tubes is measured and found to be approximately equal to the peak inverse voltage rating for that tube. Is there any possibility of exceeding the rated peak inverse voltage as a result of adjustment of the grid-bias control? Explain.

■ 25 Electronic Amplifiers

25-1. INTRODUCTION

By far, the most important and the most frequent use of electronic devices is for amplification. Almost every piece of equipment that employs tubes or transistors does so, at least in part, for the purpose of amplification. Most of the tubes in a conventional radio receiver are straightforward amplifiers; all the tubes in a phonograph amplifier are amplifiers; and TV transmitters, TV receivers, various electronic-control circuits, electronic computers, and radars all make use of amplifying circuits of one type or another. As different as these various amplifiers may appear, they all have the basic features illustrated in the circuits of Fig. 25–1. In both these circuits, and any more complicated circuit, an input signal is impressed on one of the electrodes to cause corresponding

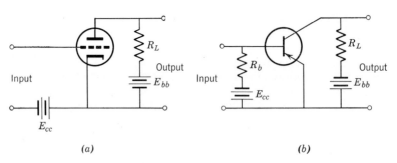

Fig. 25–1. Basic amplifier circuits.

INTRODUCTION

changes in the current through the plate or collector circuit. These changes may be employed directly as the output of the amplifier, or they produce corresponding changes in the voltage across a load impedance through which this plate or collector current flows. These changes in voltage are taken to be the output of the circuits shown in Fig. 25–1.

In addition to the basic elements mentioned in the preceding paragraph, certain practical items must be included in the circuit in order that the tube or transistor function properly. First, for there to be any appreciable plate or collector current, a battery must be included in series with the plate or collector; this is the supply battery labeled E_{bb} in the diagrams. Another item almost as important is a bias in the input circuit so that changes in the operating point of the tube or transistor that result from the application of an input signal take place in the most desirable range. Thus, in a vacuum-tube circuit, the grid-bias battery E_{cc} serves to insure that no grid current flows by keeping the net grid-cathode voltage always negative; that is, E_{cc} is made larger than the most positive input voltage expected, and its polarity is such as to make the grid negative. Under these conditions, the grid-cathode voltage, which is equal to the algebraic sum of the input voltage and the bias voltage, is assured of being negative.

In the case of the transistor, the bias voltage E_{cc} provides a d-c bias current through the base-emitter junction in the forward direction and of such a magnitude that the addition of an input current does not reverse the net current in the junction. Such a reversal would make the base-emitter junction a high impedance, and as a result a large input power would be required to obtain a desired value of input current. The situation is quite comparable to that in the vacuum tube where a positive grid voltage would require considerable input power to develop the desired input voltage. In both cases, the circuit acts best as an amplifier if the input power is kept small.

In many amplifier circuits, the external load to which the output terminals of the amplifier are connected affects the operation of the amplifier. If the output terminals are simply connected to the grid of another tube that draws no current, little effect is noted, and an analysis of the circuit alone is useful. If the load draws current that is comparable to the current through R_L, then consideration of that load must be given in any complete analysis of the circuit. If the load is the input to another transistor, the opposite extreme is reached; the load is virtually a short circuit. But, in the case of the transistor, the near short circuit is not detrimental. In order to obtain a large output from the second transistor, we should have as large a current as possible in its input, and the first circuit can supply the largest current to a low impedance or short circuit.

25–2. D-C AMPLIFIERS

The amplifiers of Fig. 25–1 amplify changes of voltage; that is, changes in the input voltage are reflected by proportional changes in the output voltage. However, if the total output voltage is considered, there is no such proportionality between instantaneous values of input and output voltages. For instance, at zero input voltage there is generally some plate current and an output voltage. If an amplifier must amplify direct voltages, the total output voltage must be made proportional to the input voltage. For a simple one-tube amplifier, this is accomplished by introducing into the output circuit a direct voltage of such magnitude and polarity that it cancels the zero-signal output voltage. The circuit of Fig. 25–2 illustrates a means of introducing this voltage. The rheostat P is adjusted until the output voltage is equal to zero when the signal is zero. If the output circuit draws no current, the output voltage is equal to the change in plate voltage from its zero-signal value, and the change in plate voltage equals the input voltage times the gain of the amplifier.

If a single stage of amplification does not provide adequate gain, two similar stages may be connected in cascade with an over-all gain equal to the product of the gains of the two individual stages. Cascading consists in applying the output voltage of one stage to the input terminals of the other, as illustrated in Fig. 25–3. This particular arrangement for coupling the two amplifier stages in Fig. 25–3 is called **direct coupling**.

Theoretically, any number of stages may be connected in cascade to

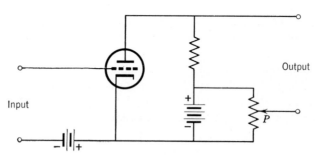

Fig. 25–2. D-c amplifier.

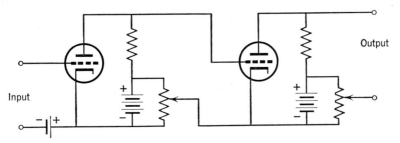

Fig. 25-3. Two-stage direct-coupled amplifier.

achieve a desired gain. Practically, it becomes difficult, even with two stages, to obtain a satisfactory amplifier by this means. When a signal is small enough to require several stages of amplification, variations in the first stage due to changes in temperature, changes in the supply voltages, etc., are comparable to the signal and may entirely mask the desired signal variations. In general, some sort of balancing circuit must be used in direct-coupled amplifiers to minimize the effects of these undesired variations. Even with balanced circuits, the design of a satisfactory d-c amplifier having a gain of more than one thousand presents many difficulties.

25-3. A-C AMPLIFIERS

The most common use of vacuum-tube amplifiers is the amplification of alternating voltages, or continuously varying voltages, such as the output voltage of a microphone or phonograph pick-up, or the small voltage developed in the antenna of a radio receiver. A d-c amplifier could be used for such purposes, but, because of the difficulties mentioned in the previous article, it is much more satisfactory to filter out all but the desired a-c component between the stages of an amplifier. The term **a-c amplifier** refers to any amplifier in which the d-c component of the output voltage of one stage is blocked from the input to the following stage, or from the amplifier load. The necessary filtering is accomplished with little additional equipment and allows the use of a common power supply for the various stages of a cascaded amplifier.

The circuits of Fig. 25-1 are most easily analyzed by the load-line method, but it must be remembered that the analysis is directly useful only if the current drawn by any external load is negligible in comparison with the current through R_L. The procedure for an a-c amplifier is simply a point-by-point determination of the output voltage or current for the input voltage of interest. In Fig. 25–4 the load line is used to determine the plate voltage and current corresponding to a sinusoidal input voltage to the circuit of Fig. 25–1a. The net grid-cathode voltage is plotted to align with the corresponding plate characteristics so that, for any instant of time, the correct characteristic is indicated, and its intersection with the load line gives the instantaneous plate voltage and current directly. As an a-c amplifier, the output voltage would be the a-c component of the plate voltage; the rms value of the output voltage of this amplifier is E_{out}.

The analysis of an a-c transistor amplifier differs from that of a vacuum-tube amplifier only in that the input signal is in the form of a current instead of a voltage.

A consideration of Fig. 25–4 indicates that, if the plate characteristics

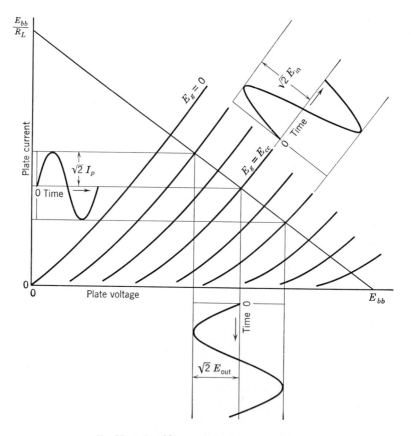

Fig. 25–4. Load-line analysis of an a-c amplifier.

are uniformly spaced along the load line, the plate-current and the plate-voltage variations are identical in shape with the input voltage; that is, the amplifier has faithfully reproduced the input voltage. The output voltage is 180 degrees out of phase with the input voltage, however, for, when the input goes more positive, the output goes less positive. The ratio of the output to the input voltage $E_{\text{out}}/E_{\text{in}}$ is known as the **voltage gain,** or simply **gain.**

For a given circuit, the load line allows one to determine the gain, or for a given tube, a consideration of various load lines indicates how best to select an operating point and plate load resistor. For example, it is evident from Fig. 25–4 that, for a fixed quiescent operating point (condition for zero input voltage), the larger R_L, the greater the voltage gain. But it also becomes evident that, with any increase in R_L, a larger plate supply voltage E_{bb} is required to maintain the same operating point. Thus, the design of an amplifier is seen to be a compromise between the gain obtainable and the supply voltage required.

A study of the characteristics of any tube or transistor indicates that the characteristics are never completely uniformly spaced along a load line. The gain for a small signal depends upon where along the load line the tube or transistor is operating. Maximum gain is obtainable by operating where the characteristics have the greatest spacing. A further consequence of the nonuniform spacing of the characteristics is that, for a large input signal, the gain of the amplifier is different for one portion of the signal than it is for another. The result is that the shape of the output voltage wave is no longer identical with that of the input; there has been some distortion of the signal, and this particular type is known as **nonlinear distortion** since it is a result of the nonlinear characteristic of the amplifier.

25–4. BIASING

Transistors require only small voltages and small currents to operate; so it is common to use batteries for the supply voltages as shown in the diagram of Fig. 25–1b. The two batteries in this circuit are of the same polarity, and so a single battery could and actually would be used in this circuit. Other arrangements are sometimes encountered in which the bias is made to depend upon the collector voltage; such an arrangement makes the circuit more adaptable to variations in different transistors that might be used or to changes in the characteristics of a given transistor with temperature. An example of such a self-biasing circuit is given in Prob. 26–18.

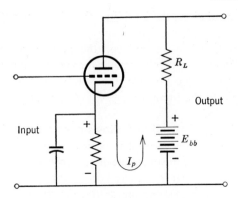

Fig. 25–5. Amplifier with cathode biasing.

Since the vacuum tube requires considerable power for cathode heating and a relatively high voltage for the plate supply, it is customary to do away with batteries entirely whenever commercial power is available. Thus, most amplifier tubes use indirectly heated cathodes with alternating current supplied to the heaters, and the plate supply consists of one of the single-phase rectifiers considered in Chapter 23.

The grid-bias battery could also be replaced with a rectifier, but the same effect can be achieved with a resistor in series with the cathode of the tube. Inspection of the circuit of Fig. 25–5 indicates that the plate current through this resistor causes a voltage drop across it, with the polarity as labeled. This method of developing the grid bias is known as **cathode biasing,** and the resistor across which the bias is developed is called the cathode bias resistor, or the **cathode resistor.**

For the cathode bias to be completely equivalent to the grid-bias battery, a capacitor must be connected in parallel with the cathode resistor. Without this capacitor, signal variations, which produce variations in the plate current, cause variations of the bias voltage. The capacitance must be of such magnitude that the capacitor voltage cannot change appreciably in the time that the signal goes through one cycle of its variation. (An approximate rule for the selection of the cathode capacitor is that the capacitive reactance at the lowest frequency to be encountered should be no more than one-tenth the resistance of the cathode resistor.)

The magnitude of the cathode bias is equal to the product of the average plate current and the cathode resistance; therefore, the bias can be altered by changing the cathode resistance. A frequent design problem requires the choice of a cathode resistor to produce a desired bias voltage for a particular value of average plate current. The required

RESISTANCE-CAPACITANCE COUPLING

cathode resistance is found by dividing the bias voltage by the plate current.

In the circuit of Fig. 25–5, the effective plate-supply voltage is less than the battery voltage by an amount equal to the cathode bias. In general, however, the bias is so small in comparison with the plate supply voltage that neglecting the reduction causes little error.

25–5. RESISTANCE–CAPACITANCE COUPLING

The object of coupling two stages of an a-c amplifier is (a) to transfer the a-c component of the output voltage of the first stage to the input of the following stage with as little loss as possible, and (b) to provide d-c isolation between the two stages.

Figure 25–6 and 25–7 indicate two methods of achieving a-c coupling: with a capacitor, and with a transformer. The significance of the d-c isolation of the two stages is that voltage variations due to temperature variations or changes in the supply voltages are not amplified in succeed-

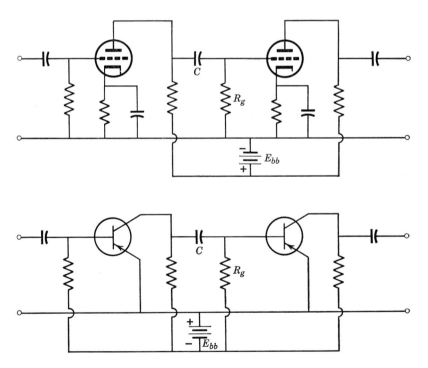

Fig. 25–6. Resistance-capacitance-coupled amplifiers.

ing stages. Thus, the a-c coupling overcomes the greatest difficulty of the direct-coupled amplifier, and a voltage gain of one million is common.

A further consequence of d-c isolation—by either of the two methods illustrated in Fig. 25–6 or 25–7—is that a single source may supply the plate circuits of all the stages of the amplifier. It is impossible to use the same plate supply for two stages of a direct-coupled amplifier.

In the resistance–capacitance coupling of Fig. 25–6, the output of the first stage is impressed across the series combination of a capacitor C and a resistor R_g. If the impedance of this combination is much larger than the load resistance of the first stage, the gain of the first stage is unaffected. The input voltage to the second stage, which is the voltage across R_g, is approximately equal to the a-c output voltage of the first stage if the reactance of the coupling capacitor at the frequency under consideration is much smaller than the resistance of R_g. Thus, with a sufficiently large coupling capacitor, d-c isolation of the two stages is accomplished with practically no reduction in the a-c signal. The only requirements for satisfactory resistance–capacitance coupling are that X_C be much smaller than R_g (an approximate rule is that R_g be at least 10 times X_C for all frequencies to be encountered), and that R_g be much greater than the load resistance of the preceding stage.

In the case of the vacuum-tube amplifier, the resistor R_g should have a large resistance, but it cannot be an open circuit, because any grid current in the second stage would charge the coupling capacitor. With no path for discharging, the capacitor charge increases until the grid is so negative that no plate current flows through the tube. In practice, the value of the grid resistor is generally limited to one or two megohms.

The resistor R_g in the transistor circuit serves a somewhat different function. Because R_g is shunted by the low impedance of the base-emitter junction, its value is not critical as far as coupling is concerned. It is desirable that the resistance be large enough so that not much of the signal current passes through it, but, more important, it still serves to determine the bias current through the base-emitter junction, and its value is chosen accordingly.

A measurement of the gain of any R–C coupled amplifier as a function of the signal frequency reveals a comparatively wide band of frequencies over which the gain is uniform. At both high and low extremes of frequency, however, there is a marked reduction in the gain of the amplifier. At low frequencies, the reduction in gain may be caused by either or both of two effects. First, the reactance of the cathode capacitor becomes so large that an appreciable alternating voltage is developed across it. This voltage is of such polarity that it subtracts from the input voltage and reduces the a-c component of the grid-to-cathode voltage. Second, the reactance of the coupling capacitor becomes so large that part of the a-c output of the first stage is lost across it. This voltage loss causes the gain to approach zero as the frequency approaches zero, since the capacitor is an open circuit at a frequency of zero.

At high frequencies, the reduction in gain is due to the presence of stray capacitance between the plate and the cathode. Much of this capacitance is the actual capacitance that exists between the plate and cathode in the tube; some is due to the capacitance between lead wires, and some may be due to the grid-to-cathode capacitance of the tube in the following stage. Whatever the source, the effect of the plate-to-cathode capacitance is that the load resistance is shunted by an a-c path; the higher the frequency, the lower the impedance of this shunting path. This capacitive shunting reduces the effective load impedance and, therefore, reduces the gain; part of the alternating current that would otherwise flow through the load resistor is shunted through this capacitance, resulting in a smaller current in the load resistor and therefore a smaller output voltage ($I_p R_L$) at higher frequencies.

25-6. TRANSFORMER COUPLING

When a transformer is employed as the coupling device, it may, and generally does, serve as the load impedance for the first amplifier stage. In the vacuum-tube circuit a single transformer replaces three circuit elements required for R–C coupling, but it is more expensive and, for some purposes, less satisfactory. The principal disadvantage to transformer coupling is that the frequency range is limited because of the transformer leakage reactance and the capacitance between turns. There are some advantages to transformer coupling, however, one being that the over-all gain of a stage is increased by the turns ratio of the transformer (assuming a perfect transformer). This advantage is limited, however, because the flux leakage becomes excessive for large turns ratios.

Interstage transformers in a transistor amplifier accomplish the same purpose as they do in a vacuum-tube amplifier except that in the transistor circuit one is interested in the possible current gain from the transformer instead of the voltage gain. Figure 25–7b illustrates a possible arrangement for obtaining the base bias for a transformer-coupled transistor amplifier. The bias current is supplied through the transformer, for otherwise much of it would be shunted past the junction by the low d-c impedance of the transformer secondary. The arrange-

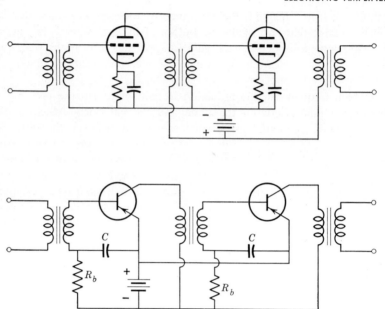

Fig. 25-7. Transformer coupled amplifiers.

ment is then completed by the addition of the capacitor C, which shunts R_b for the signal in much the same way that the cathode resistor in the vacuum-tube circuit is bypassed by the cathode capacitor.

A most important feature of transformer coupling is that the transformer may act as an impedance-matching device. Impedance matching is the transformation of the effective impedance of a given load as seen by an electric source to the optimum value for power transfer. This transformation can be accomplished by connecting a transformer of the proper turns ratio between the given load and source. For example, a practical amplifier having a 10-ohm loudspeaker as its load would exhibit almost no gain or power output; the amplifier might require a load impedance of 4000 ohms in order to develop maximum power output. A transformer with a turns ratio of 20 to 1 would make the 10-ohm load appear to have the desired impedance of 4000 ohms. This can be shown to be true by assuming an alternating voltage of 100 volts applied to the transformer primary. If voltage losses in the transformer are neglected, the secondary voltage would be 5 volts, which would produce a current of 0.5 ampere in the 10-ohm load. This secondary current would require a primary current of 25 milliamperes. Thus, viewed from the primary terminals of the transformer, the combination of the trans-

PUSH–PULL AMPLIFIER

former and load appears to have an impedance of the desired 4000 ohms. The transformer has made the load appear to have an impedance equal to the actual impedance times the square of the turns ratio.

25-7. PUSH–PULL AMPLIFIER

Any amplifier that supplies considerable a-c power output must necessarily have large current or voltage variations and is, therefore, subject to nonlinear distortion. The use of two tubes in parallel allows twice the current and hence twice the power output for the same amount of distortion. Similarly, a larger tube, which exhibits less nonlinearity for a given voltage or current variation, can be used to obtain increased power without increasing the distortion. The most effective means of increasing the power output without increasing the distortion, however, is through the use of the push–pull circuit, illustrated in Fig. 25–8. The output of the balanced push–pull amplifier is twice the output of a one-tube amplifier using one of the same tubes and operating under the same conditions of grid and plate voltage. More important, however, is that under these conditions, the distortion in the push–pull amplifier is actually less than in the one-tube amplifier. The final stage in most power amplifiers employs the push–pull arrangement.

The reduced distortion of the push-pull amplifier can be attributed to the fact that the two tubes in this circuit operate 180 degrees out of phase. The input transformer of Fig. 25–8 provides the two tubes with signals of equal magnitude but of opposite phase ($e_{g_1 m} = -e_{g_2 m}$), and these signals produce alternating components of plate currents

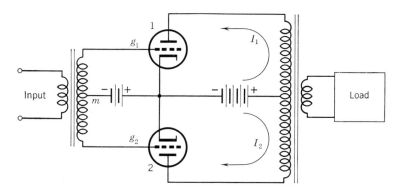

Fig. 25–8. Push–pull amplifier circuit.

that are also of equal magnitude and opposite phase ($i_1 = -i_2$). For sinusoidal signal voltages, i_1 and i_2 contain, in addition to the fundamental components, certain harmonics as a result of nonlinear distortion. Since i_1 and i_2 have the same wave shape, they must contain harmonics having the same amplitude and bearing the same phase relationship to their respective fundamentals. Thus, with the fundamentals 180 degrees out of phase, the two second-harmonic components are actually in phase with each other.

The current components mentioned in the preceding paragraph flow in the two halves of the output-transformer primary winding and may therefore produce corresponding voltages and currents in the load. A consideration of the fundamental components indicates that, since they are of opposite phase and since they link the transformer core in opposite directions, their effects on the load are additive; therefore, as would be expected, the two tubes produce twice the output that can be obtained from one of the tubes.

The second-harmonic components of the two plate currents are in phase, but they link the transformer core in opposite directions. Thus, the magnetizing effects of the second harmonics cancel, producing no corresponding component in the load. Further analysis would indicate that all even harmonics resulting from nonlinear distortion in a balanced push–pull amplifier are canceled, and that the relationship between the

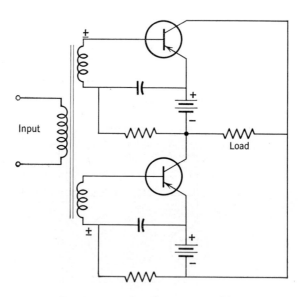

Fig. 25–9. A push–pull transistor amplifier.

odd harmonics and the fundamental is the same in the load voltage as in the individual plate currents.

The coupling of a push–pull amplifier to its load is generally accomplished by a transformer. The direct current that flows in the two halves of the primary produces no net d-c magnetizing component since the transformer core is linked in opposite directions. For this reason, a smaller transformer may be used in a push–pull amplifier than in a single-tube amplifier having the same power capacity.

Figure 25–8 also indicates the connection of a transformer to obtain the 180-degree phase difference between the two grid voltages. Instead of a transformer, another amplifier tube may be employed to produce the necessary phase shift.

The advantages of the push–pull circuit are equally applicable to a transistor amplifier. The same circuit may be used with transistors replacing the tubes, and there are other possibilities that are either not convenient or not available when tubes are used. For example, the circuit of Fig. 25–9 shows how two transistors may be arranged in push–pull to supply power to a common load without the use of an output transformer. Splitting the power supply and operating the emitters at different potentials as required by this circuit would not be practical in a vacuum-tube amplifier.

25–8. EFFICIENCY OF THE CLASS A AMPLIFIER

Amplifiers are classified according to the fraction of time that the tube or transistor conducts. Thus, a **class** A amplifier is one in which current flows continuously; a **class** B, one that conducts for about half the time; and a **class** C, one that conducts substantially less than half the time. Obviously, a class B or class C amplifier produces considerable nonlinear distortion of the current wave form, but this need not mean distortion of the output voltage, and only with the class B or class C amplifier can high efficiency be obtained.

The class A amplifier can be made to operate with little distortion, but its efficiency is never high. The maximum theoretical efficiency is 50 per cent, but this is twice as high as the actual efficiency of most practical class A amplifiers.

If it is assumed that the plate characteristics of an amplifier tube are linear over the operating range, the average value of plate current is unaffected by the application of an a-c signal to the grid of the tube. Thus, the power supplied by the plate-supply battery is independent of the magnitude of the input signal. The output of the amplifier, however,

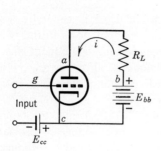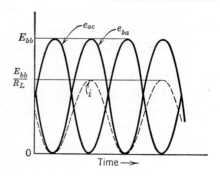

Fig. 25–10. Wave forms for resistance-loaded amplifier operating at maximum class A efficiency.

is proportional to the magnitude of the input signal; therefore, the larger the signal, the higher the efficiency of the amplifier. Maximum efficiency of a class A amplifier is therefore obtained by the application of the largest input signal consistent with class A operation. Thus, the signal and grid bias must be such that when the grid is most negative the plate current just goes to zero, and when the grid is most positive the plate voltage is negligible.

The conditions prescribed in the preceding paragraph can be applied to the circuit of Fig. 25–10 to determine the maximum theoretical efficiency of that circuit operating class A. The wave forms of Fig. 25–10 illustrate these conditions for a sinusoidal signal voltage. The plate voltage varies between zero and E_{bb}, the supply battery voltage; the current, between zero and the value obtained with the full battery voltage across the load resistor (assuming no voltage drop across the tube).

If no power is supplied by the input circuit (the plate-supply battery is the only source of power) and if the useful power output of the amplifier is that due to the a-c components of current and voltage in the load resistor,

$$\text{Power input} = E_{bb} I_{\text{avg}} = E_{bb} \frac{E_{bb}}{2R_L} = \frac{E_{bb}^2}{2R_L} \qquad (25\text{--}1)$$

and

$$\text{Power output} = \frac{E_{bb}}{2\sqrt{2}} \frac{E_{bb}}{2R_L \sqrt{2}} = \frac{E_{bb}^2}{8R_L} \qquad (25\text{--}2)$$

The maximum efficiency of this circuit is therefore 25 per cent. Much of the loss in this circuit is the d-c power dissipated in the load resistor. The maximum theoretical efficiency is raised to 50 per cent with trans-

PUSH–PULL CLASS B AMPLIFIER

former coupling, or some other device that prevents direct current from flowing in the load, because the d-c power loss is eliminated.

Even with transformer coupling, the efficiency of a practical class-A amplifier using tubes is seldom over 25 per cent because of transformer losses and because the input signal is always less than that assumed in obtaining Eqs. 25–1 and 25–2. With transistors, it is possible to more nearly approach the theoretical maximum efficiency, for the collector voltage can be driven almost to zero on positive peaks of the input signal.

25–9. PUSH–PULL CLASS B AMPLIFIER

Most of the effects of plate-current distortion in a class B amplifier are eliminated in the push–pull circuit. If the grid bias in a push–pull amplifier is proper for class B operation, one tube conducts for positive half-cycles of the signal voltage, and the other conducts for negative half-cycles. Thus, for a sinusoidal signal and linear tubes, the net magnetizing effect in the output transformer is sinusoidal.

Maximum efficiency of a class B amplifier can be determined by a consideration of the circuit and wave forms of Fig. 25–11. For zero input signal, no plate current flows in either tube, and for each tube the plate voltage equals E_{bb}, the voltage of the plate-supply battery. As for the class A amplifier, a signal is assumed that is large enough to reduce the plate voltage to a negligible value when the grid is most positive. Under these conditions, the plate voltage for each of the tubes varies between zero and $2E_{bb}$. (The plate voltage does not remain at E_{bb} when a tube is not conducting because of the induced voltage resulting from the current of the other tube flowing in the other half of the transformer primary.) The wave forms corresponding to these conditions are shown in Fig. 25–11.

The output of this amplifier may be found by assuming that all the transformer primary current flows in one half of the winding. This is a valid assumption, since the two halves are identical, and current in either of them produces the same effect in the output circuit. Thus, the net primary current is the difference between i_a and i_b; it is sinusoidal and has a peak value of I_m. The voltage across one half of the transformer primary equals the a-c component of the plate voltage. The power output is therefore the product of the rms values of this voltage and the net primary current. Thus,

$$P_{\text{out}} = \frac{I_m}{\sqrt{2}} \frac{E_{bb}}{\sqrt{2}} = \frac{I_m E_{bb}}{2} \qquad (25\text{–}3)$$

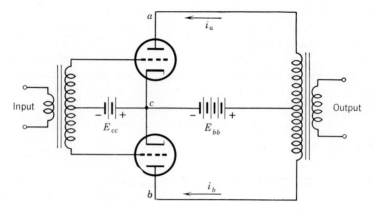

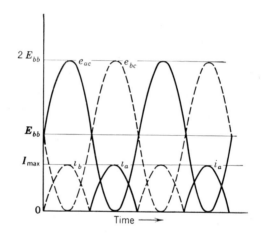

Fig. 25-11. Maximum efficiency for a push–pull class B amplifier.

The current in the plate-supply battery is the sum of i_a and i_b, the average value of which is $2/\pi$ times the maximum value of the current. The power supplied by the battery, the product of its terminal voltage and the average current through it, is

$$P_{in} = \frac{2}{\pi} I_m E_{bb} \qquad (25\text{-}4)$$

The efficiency is

$$\text{Efficiency} = \frac{P_{out}}{P_{in}} = \frac{\pi}{4} = 78.5\% \qquad (25\text{-}5)$$

TUNED AMPLIFIERS

if the power supplied by the signal is neglected. This value of efficiency is a theoretical maximum value; efficiencies of 50 to 60 per cent are practical with tubes, and almost the full theoretical efficiency can be realized with transistors.

25–10. TUNED AMPLIFIERS

The discussion in the previous articles is limited to resistance-loaded amplifiers—amplifiers in which the plate load is essentially resistive in nature. These amplifiers have the desirable characteristic of a uniform gain over a wide band of frequencies; but the frequency range for satisfactory operation is restricted by the effect of stray capacitance to frequencies within, or not far outside, the audio spectrum. Another limitation of the resistance-loaded amplifier is that distortion is excessive if, in order to obtain higher efficiency, it is operated class B or C. (The push–pull class B amplifier is an exception.) Under certain conditions, the tuned amplifier is capable of overcoming these limitations.

The tuned amplifier, several examples of which are shown in Fig. 25–12, has a parallel resonant impedance as its plate load. The gain

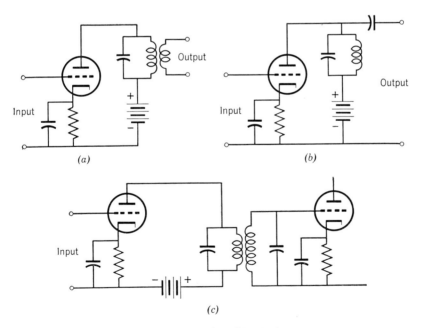

Fig. 25–12. Tuned-amplifier circuits.

of such an amplifier varies with frequency in approximately the same manner as the impedance of its plate load. Thus, at frequencies near the resonant frequency of the load, the gain of a tuned amplifier may be appreciable; at frequencies far from resonance, there is little or no gain. If a signal of many frequencies is applied to the input terminals of a tuned amplifier, only those components having frequencies equal to, or nearly equal to, the resonant frequency of the load are amplified so as to appear in the output. Signal components of other frequencies are effectively eliminated. Because of this characteristic, the tuned amplifier is said to be frequency-selective.

The ability of a radio receiver to select a single signal (a signal from a single station) from the many that are present in the antenna is a result of the frequency selectivity of one more tuned-amplifier stages in the receiver. To tune to a station of different frequency, it is necessary merely to change the resonant frequency of the amplifier load impedances. Ordinarily, this tuning is effected by varying the capacitance in the resonant circuits.

The stray capacitance, which is responsible for the upper-frequency limit of the resistance-loaded amplifier, also shunts the load impedance of the tuned amplifier. In the tuned amplifier, however, the shunt capacitance may be considered as part of the load. As such, it is a factor in determining the resonant frequency of the load impedance, but otherwise it offers no frequency limitation. Tuned-amplifier circuits, such as those shown in Fig. 25–12 are successfully employed at frequencies as high as 100 megacycles. (The inductors and transformers shown in the circuits of Fig. 25–12 have no iron cores. Iron cores may be used, but the higher frequencies at which tuned amplifiers are most commonly employed cause excessive hysteresis and eddy-current losses in iron cores similar to those of power transformers.)

The resonant load impedance of a tuned amplifier may be provided by the parallel combination of an inductor and a capacitor connected directly in the plate circuit as shown in Figs. 25–12a and b; it may be a parallel resonant circuit that is inductively coupled to the plate circuit; or it may be a "double-tuned circuit" such as that of Fig. 25–12c. The characteristic of the double-tuned circuit that often makes its use desirable is that its impedance may be made fairly constant over a band of frequencies that is several per cent of the mid-frequency, or resonant frequency.

If the operation of a tuned amplifier is limited to a single frequency, the resonant frequency of the plate load impedance, then distortion of the output voltage may be negligible, even though plate current flows for only a portion of each cycle. A single-frequency input voltage, that is, a sinusoidal input, produces a distorted plate-current wave form when the amplifier is operating class B or C. This distorted current consists of a d-c component, an a-c component of the input signal frequency, and harmonics of the signal frequency, and flows through a plate load impedance that is resonant to the signal frequency. Thus, the output voltage, which is the sum of the IZ drops across the load impedance due to the various current components flowing through that impedance, is essentially sinusoidal; the load impedance is high enough to result in an appreciable output voltage at the signal frequency only.

The ability of the resonant load impedance to select a particular frequency when the plate current contains components of many different frequencies can also be applied to frequency conversion. If the grid bias is large enough to produce a considerable harmonic content in the plate current, and if the load impedance is made resonant to the frequency of a particular harmonic, only that harmonic appears in the output voltage across the load impedance. Thus, the tuned amplifier may act as a frequency doubler.

25-11. OSCILLATORS

An electronic oscillator is any electronic device that produces an alternating voltage, the frequency of which is independent of external excitation. That is, the frequency of an oscillator output voltage depends only upon the circuit constants of the oscillator. An amplifier in which the input voltage is supplied from the output may act as an oscillator. The only requirement for oscillations is that the magnitude and phase of the voltage fed back to the input of the amplifier be such that the output voltage is reproduced.

It might appear that an oscillator, as described, would function only if first started by some external signal. However, there are always small variations in the amplifier plate current, even with no signal applied, and these variations may be thought of as consisting of components of all frequencies. Those frequencies for which the gain of the amplifier is sufficiently great are amplified, fed back, and amplified again and again until equilibrium is reached. This equilibrium results because the nonlinearity of the tube characteristics reduces the amplifier gain for large signals.

The most common vacuum-tube oscillators are tuned amplifiers in which a feedback path is provided between the amplifier output and input, that is, between the plate and grid circuits of the amplifier tube; two examples are shown in Fig. 25-13. The plate load impedance in each of these circuits is a parallel combination of inductance and capacitance (L and C); therefore, the gain of either of these circuits is appreciable only at the resonant frequency of the load impedance. The feedback path essential to the production of oscillations is provided in the tuned-plate oscillator by magnetic coupling between two coils: one in the grid circuit, one in the plate circuit. In the Hartley oscillator, the voltage across a portion of the plate load impedance is the grid voltage. Both circuits oscillate at the resonant frequency of the plate load impedance if the feedback path provides a large enough voltage of the proper polarity.

It is often difficult to prevent an amplifier from oscillating. This is especially true in high-gain amplifiers in which only a small part of the output voltage needs to be fed back to the input to cause oscillations. The proximity of the various component parts of the amplifier may provide sufficient feedback; a common plate supply for the various stages of an amplifier may provide adequate coupling between the input and output to cause oscillations; or, at high frequencies, the capacitance between the tube electrodes may provide the coupling. In general, oscillations in an amplifier are overcome by proper shielding of the

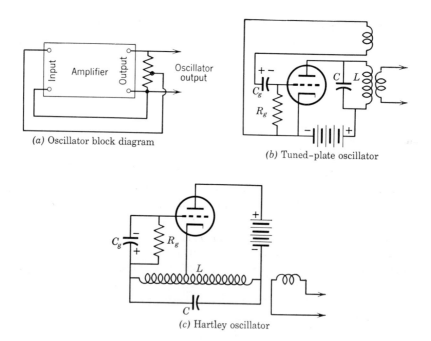

(a) Oscillator block diagram

(b) Tuned-plate oscillator

(c) Hartley oscillator

Fig. 25–13. Oscillator circuits.

amplifier components, by adequate filtering in the plate-supply rectifier, by isolation of the plate circuits of the various stages of an amplifier, and by pentode amplifier tubes. In low-power high-frequency applications, pentodes are used because the low interelectrode capacitance eliminates one source of undesirable feedback.

To obtain high efficiency, a power oscillator—one that supplies a-c power—must operate class C. However, the grid bias for class C operation is such that no plate current flows when the signal input is zero. Thus, with the proper value of grid bias supplied by a d-c battery, no plate current flows when the circuit is energized, and therefore no oscillations can build up. Grid bias for a class C oscillator must be provided by some method that allows an initial flow of plate current. Cathode biasing (Art. 25–4) is one method that allows an initial flow of plate current and also allows a sufficient bias voltage to build up for class C operation. Most oscillator circuits, including those shown in Fig. 25–13, employ **grid-leak biasing**. Grid-leak biasing is applicable only to circuits in which there is appreciable grid current. Most oscillators meet this requirement since oscillations build up until limited by some nonlinearity, usually in the positive-grid region of the tube characteristic.

In both the circuits of Fig. 25–13, the grid-to-cathode voltage consists of two components: the a-c feedback voltage developed across an inductor, and the voltage across the capacitor C_g. The capacitance of C_g is large enough so that there is no alternating voltage across it, only a d-c component. When grid current flows, C_g is charged with the polarity indicated. The resistor R_g provides a path through which the average, or d-c, component of the grid current may flow. When the circuit is first energized, there is no grid current and hence no grid bias; therefore, plate current flows, and, if the feedback path is adequate, oscillations build up. As soon as there are oscillations, however, an alternating voltage is impressed on the grid, making it positive for part of the time. During the time that the grid is positive, grid current flows to charge C_g with the polarity indicated in Fig. 25–13; while it is negative, some of the charge "leaks" off through R_g. The resulting average voltage across C_g is the grid bias.

The exact value of the grid bias obtained across the grid capacitor depends upon the values of R_g and C_g and also upon the magnitude of the voltage that is fed back. Since it depends upon the magnitude of the feedback voltage, the grid bias acts as a regulator on the oscillator. If the oscillations tend to become greater, the grid current increases, thereby increasing the bias voltage. The greater the negative grid bias, the smaller the plate current, and the smaller the amplitude of the oscillations. Because of the variability of the grid-leak voltage, and because of the dependence upon grid current, grid-leak biasing is seldom employed in ordinary voltage amplifiers.

Vacuum-tube oscillators are common sources of high-frequency power for induction and dielectric heating. Operating class C, an oscillator is an efficient source of energy for these purposes. For the high-frequency heating of conducting materials (induction heating), the material to be heated is placed in the magnetic field produced by the inductor of the oscillator resonant circuit, or in the field produced by another inductor that is coupled to the oscillator output. The high-frequency magnetic field produces eddy-current and, sometimes, hysteresis losses that serve to heat the material. At the high frequencies commonly employed for induction heating, the eddy currents tend to concentrate on the surface of the material; therefore, induction heating actually develops heat only on the surface. Normally, heat is conducted to the interior of the material to produce a uniform temperature. However, it is possible to take advantage of the "skin effect" for the case hardening of steel. In this application, energy is supplied at a rate high enough to heat the surface to the desired temperature before appreciable heat has been conducted to the interior of the material. The material being case-hardened can then be quenched when only the surface is at a high temperature.

For the high-frequency heating of insulating materials (dielectric heating), the material to be heated is placed in the electric field between the plates of the capacitor of the oscillator resonant circuit. The continuous reorientation of the molecules of the insulating material resulting from the high-frequency electric field heats the material. If the material is homogeneous and the electric field is uniform, the heat is produced at a uniform rate throughout the material. A few applications of dielectric heating are the bonding of plastics, the setting of glue, and the drying of wood.

PROBLEMS

25–1. A type-6J5 triode is used as an a-c amplifier with a plate-supply voltage of 300 volts, a load resistance of 30,000 ohms, and a grid-bias battery of 4.0 volts.

(*a*) Using the load-line method, determine the gain of this amplifier.

(*b*) Assume a sinusoidal input signal of 2.0 volts, and sketch curves of input voltage, plate current, and plate voltage showing the proper phase relationship and the approximate shape.

(*c*) Repeat (*b*) for a grid bias of 18 volts.

PROBLEMS

25-2. What value of cathode resistance could provide the 4 volts bias required in the amplifier of Prob. 25-1? What capacitance is needed to provide reasonable bypassing of this resistance at 200 cps? At 200 kc?
Ans. 770 ohms; 10 μf; 0.01 μf.

25-3. A certain a-c amplifier consists of a type-6SJ7 pentode, a plate supply of 300 volts, and a load resistance of either 30,000 ohms or 100,000 ohms.

(a) For each value of load resistance, specify a value of grid bias that will allow the application of the largest signal with a minimum of distortion. Approximately what magnitude of signal voltage may be applied in each case?

(b) What is the gain in each case?

25-4. A type-6L6 beam-power tube is connected in a simple amplifier circuit with a d-c plate supply of 300 volts and a plate load resistor of 3000 ohms. Plot one cycle of the plate-current and voltage wave forms for a negative grid bias of 10 volts and a sinusoidal input signal that has a peak value of 10 volts.

25-5. A transistor whose collector characteristic is given by Fig. 22-6 is used in the circuit of Fig. 25-1b with a single 12-volt battery and a collector load resistance of 2000 ohms.

(a) Select a value of resistance for R_b that will result in a no-signal collector voltage of about 6 volts. *Ans.* 43 kilohms.

(b) What is the available voltage gain of this amplifier; that is, what is the ratio of alternating voltage across the collector load resistor to the input source voltage? Assume that the source has an internal impedance of 10 kilohms and that there is negligible voltage drop between the base and emitter. *Ans.* 2.

(c) What is the current gain, assuming that the output current is the current through the collector load resistance? *Ans.* 10.

25-6. What is the efficiency of the amplifier of Prob. 25-5 with the largest possible signal that does not cause pronounced distortion? Assume that the power loss in the input circuit is negligible and that the output power is the a-c power developed in the collector load resistance. What is the power dissipated in the transistor under these conditions? *Ans.* 20%; 11 mw.

25-7. A type-6J5 triode is used in an amplifier with a pure resistance load of 30,000 ohms. Determine the efficiency of this amplifier when the d-c grid bias is -6.0 volts, the plate-supply voltage is 300 volts, and the input signal is a sinusoidal voltage having a peak value of 6.0 volts. Assume that the average plate current equals the current that would exist with no signal. Would the calculated efficiency be greater or less if this assumption were not made? Account for all the losses. *Ans.* 9.4%.

25-8. The d-c supply voltages in a class A amplifier are adjusted so that the plate dissipation just equals the rated value for no applied signal. Explain why the rated plate dissipation will not be exceeded when an input signal is applied to the amplifier.

25-9. Two transistors of the type used in Prob. 25-5 are connected in a common-emitter push–pull class B amplifier. The collector supply voltage is 12 volts, and the common load resistance is 2000 ohms. What is the collector-circuit efficiency when the signal is as large as possible without appreciable distortion?   *Ans.* 73%.

25-10. What is the smallest supply voltage that could be used with the amplifier of Prob. 25-9 to allow an output power of 50 mw without excessive distortion.   *Ans.* Approx. 16 volts.

25-11. The load impedance of a single-tube class B amplifier is equivalent to the parallel combination of a 100,000-ohm resistance, a 15.9-$\mu\mu$f capacitance, and a 1.59-millihenry inductance. What is the percentage of the second harmonic in the output voltage when the signal frequency is 1.0 megacycle? Assume a linear amplifier tube such that the current wave consists of the positive halves of a sine wave. The Fourier series expressing this current wave shape is

$$i = K\left(1 + \frac{\pi}{2}\cos x + \frac{2}{3}\cos 2x - \frac{2}{15}\cos 4x \cdots\right)$$

Ans. 2.8% of fundamental.

26 Analysis of Linear Electronic Circuits

26-1. INTRODUCTION

The load-line analysis of the a-c amplifier as employed in the preceding chapter is very useful but it also has some severe limitations. For the basic amplifier circuit it provides all the information that could be desired; the d-c, or quiescent, operating point was found, and for any given input voltage or current, the corresponding output voltage was found. If the circuit is made more complicated, however, it is not always convenient or even possible to apply the load-line method.

In many electronic circuits, the voltage and current variations resulting from an applied signal are so small that the tube or transistor may be considered a linear device, in which case all the methods of linear circuit analysis are available. No general rule can be made as to just how small the variations must be in order that the device may properly be called linear; it depends upon the device and the accuracy required of the analysis. It is sometimes useful to accept the inaccuracy of a linear analysis in order to obtain an understanding of the operation of a circuit that cannot be broached by any more accurate method.

The load line is still the only convenient means for determining the operating point of a tube or transistor, or for determining anything about nonlinear distortion that may be present. Furthermore, the application of a linear analysis to a specific circuit requires a knowledge of the characteristics of the device at the operating point, and this operating point is obtained from the load-line construction.

In order to apply any linear analysis to an electronic circuit, it is first necessary to replace these elements with their linear equivalents. The object of this chapter is first to develop equivalent circuits for the vacuum tube and the transistor that represent these devices as far as a-c components are concerned, and then to apply these circuits to some practical electronic circuits.

26–2. A-C EQUIVALENT CIRCUIT FOR THE VACUUM TUBE

From a consideration of the characteristics of a high-vacuum triode, it is evident that the plate current i_b, is a function of two other quantities: the plate voltage e_b, and the grid voltage e_c. Furthermore, changes in these quantities can be related by

$$\Delta i_b = \Delta e_c \frac{\partial i_b}{\partial e_c} + \Delta e_b \frac{\partial i_b}{\partial e_b} \qquad (26\text{–}1)$$

if the changes are small enough so that the characteristics may be considered linear. Equation 26–1 comes from the Taylor expression for one variable in terms of two independent variables and the partial derivatives, with all derivatives except the first equated to zero for this linear case. The two partial derivatives are dimensionally conductances and are named accordingly: $\partial i_b/\partial e_c$, is known as the **plate conductance,** and its reciprocal the **plate resistance;** the other, $\partial i_b/\partial e_c$, is known as the **mutual conductance** since it relates the current in one circuit to the voltage in another. The mutual conductance and plate resistance are symbolized by g_m and r_p, respectively.

Equation 26–1 provides an a-c equivalent circuit directly if the changes

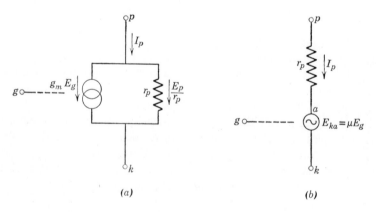

Fig. 26–1. A-c equivalent circuits for a high-vacuum triode.

A-C EQUIVALENT CIRCUIT FOR THE VACUUM TUBE

in current and voltage are interpreted as a-c components of current and voltage—which, in fact, they are. Thus, the a-c component of plate current may be expressed as the sum of two parts, one equal to g_m times the a-c component of the grid voltage, and the other equal to the a-c component of the plate voltage divided by the plate resistance. As far as a-c components are concerned, the plate-to-cathode path through the tube may then be represented by two parallel paths, one for each of the two parts of the plate current, that is, by the parallel combination of r_p and a current source as illustrated in Fig. 26–1a. In this diagram the grid terminal is shown for completeness—it is maintained negative with respect to the cathode. (The symbols I_p, E_p, and E_g are used for the a-c components of plate current, plate voltage, and grid voltage, respectively.)

The equivalent circuit could just as well have been developed starting with the plate voltage expressed as a function of plate current and grid voltage. Thus,

$$E_p = I_p \frac{\partial e_b}{\partial i_b} + E_g \frac{\partial e_b}{\partial e_c} \tag{26-2}$$

The new partial derivative, $\partial e_b/\partial e_c$, is by definition the negative of the **amplificaton factor**. This derivative is always negative because an increase in the potential of one of the electrodes must be accompanied by a decrease in the potential of the other to maintain constant plate current. As defined, the amplification factor is then always positive. Equation 26–2 may be rewritten as

$$E_p = I_p r_p - \mu E_g \tag{26-3}$$

where $\mu = -\partial e_b/\partial e_c$ = the amplification factor.

Equation 26–3, which equates the plate-to-cathode voltage to the sum of two voltages, may be used as the basis for another equivalent circuit; that is, two series elements can be chosen that are equivalent to the plate-to-cathode path through the tube. A resistance r_p provides for the IR drop, and an equivalent generator for the remaining term μE_g. This circuit is illustrated in Fig. 26–1b, which also indicates the proper polarity for the equivalent generator in order that Eq. 26–3 be satisfied.

The two circuits of Fig. 26–1 must be equivalent to each other since they were derived to be equivalent to the same thing. Therefore, the open-circuit plate voltage of these two circuits may be equated, and this provides a relationship among the three tube coefficients, as μ, g_m, and r_p are called. Thus,

$$g_m E_g r_p = \mu E_g$$

or

$$g_m r_p = \mu \tag{26-4}$$

Equation 26–4 could also have been derived quite apart from any consideration of equivalent circuits by applying the mathematical relationship that exists among the three partial derivatives represented by the tube coefficients.

Although the equivalent circuits were developed for the triode, there is nothing about the derivation that makes it peculiar to the triode; it is equally applicable to any situation in which a current is the function of two other variables. In particular, a pentode in which the screen and suppressor grids are maintained at fixed potentials can be represented by either of the two circuits of Fig. 26–1. In the case of the pentode, the current-equivalent form, Fig. 26–1a, is usually preferred because of the possibility of simplification. The plate resistance, which is by definition the reciprocal of the slope of the plate characteristic, is much larger for a pentode than for a similar triode. When it is much larger than the impedance shunting it in the external circuit, or, more precisely, when the current through it is negligible in comparison with the other currents in the circuit, the plate-resistance branch can be eliminated from the circuit with no sacrifice to the accuracy of any analysis of this circuit.

Before the equivalent circuits developed for the vacuum-tube are employed, the significance and limitations of these circuits must be considered. First, they are a-c equivalent circuits and apply only to a-c components of the various voltages and currents. Second, they are only as good as the tube (or other device) is linear over the operating range imposed by the input signal. And last, when these circuits are used to obtain numerical solutions to a problem, the values of the tube coefficients cannot be taken from a handbook; they must be appropriate for the specified d-c operating conditions of the problem.

26–3. VOLTAGE GAIN OF THE BASIC VACUUM-TUBE AMPLIFIER

Figure 26–2 shows the basic vacuum-tube amplifier circuit together with one of its possible a-c equivalent circuits in which the actual circuit elements are replaced by their a-c equivalents. The tube is replaced by one of the equivalent circuits developed in the preceding section; the batteries by a short circuit, since no alternating voltage can be developed across a perfect battery; and the resistors are left unchanged. If the batteries had not been perfect, it would have been necessary to replace them by their internal impedances.

To find the a-c gain of this amplifier, one needs only to find the

VOLTAGE GAIN OF THE BASIC VACUUM-TUBE AMPLIFIER

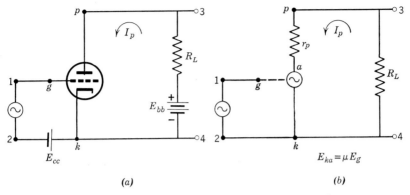

Fig. 26-2. Basic vacuum-tube amplifier circuit and a-c equivalent.

voltage E_{34} that results from an applied voltage E_{12}. Thus,

$$\text{Gain} = \frac{E_{34}}{E_{12}} = \frac{-I_p R_L}{E_{12}} = -\frac{E_{ka} R_L}{(r_p + R_L) E_{12}} = -\frac{\mu R_L}{r_p + R_L} \quad (26\text{-}5)$$

The minus sign indicates that the voltage E_{34} is 180 degrees out of phase with the input voltage E_{12}, which is in agreement with the results of the load-line analysis.

All the information contained in Eq. 26-5 is available from the load-line analysis, and the equation may also be derived from the load-line construction. Figure 26-3 shows a load line on a section of a tube's

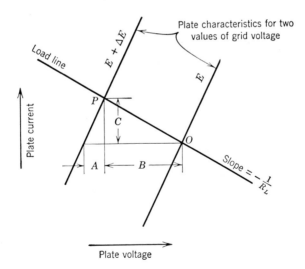

Fig. 26-3. Magnified section of load line on triode characteristic.

characteristic about the operating point. For a grid voltage E, the tube operates at the plate current and voltage given by point O on the diagram. If the grid voltage is changed by an amount ΔE, the operating point changes to P; that is, the plate voltage changes by an amount B, and the plate current by an amount C. By definition, the gain is

$$G = \frac{B}{\Delta E} \qquad (26\text{--}6)$$

and the amplification factor is

$$\mu = \frac{A + B}{\Delta E} \qquad (26\text{--}7)$$

Solving Eq. 26–7 for ΔE and substituting in Eq. 26–6 yields

$$G = \mu \frac{B}{A + B} = \mu \frac{B/C}{A/C + B/C} \qquad (26\text{--}8)$$

The ratio A/C is the plate resistance and B/C, the load resistance; therefore

$$G = \mu \frac{R_L}{r_p + R_L} \qquad (26\text{--}9)$$

Except for the sign, this expression for gain is the same as Eq. 26–5, which was developed from the equivalent circuit. A minus sign should also be included with Eq. 26–9 since inspection of the characteristic reveals that a positive increase in the plate voltage is accompanied by a negative increase in the grid voltage.

26–4. GAIN OF AN R-C COUPLED AMPLIFIER

In Art. 25–5, the reactive effects in an R-C coupled amplifier are qualitatively discussed for operation at both high and low extremes in frequency. By applying the a-c equivalent circuit for the amplifier, it is a relatively simple matter to determine these effects quantitatively. Because it is possible to make different simplifications in the circuit for operation at different frequencies, no attempt is made to develop a perfectly general solution for the gain of this amplifier; instead, an appropriate equivalent circuit is developed for each of three different frequency ranges.

(a) **Mid-Frequency Gain.** At frequencies for which the amplifier is designed to operate, that is, the mid-frequency range, reactive effects are presumably negligible, and it is therefore possible to derive a special load line that is valid for this case. However, the equivalent circuit is so

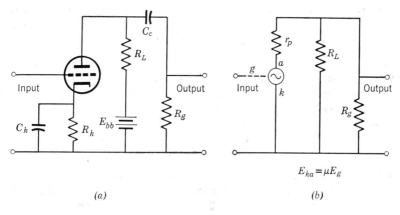

Fig. 26–4. One stage of R–C coupled amplifier and its mid-frequency a-c equivalent.

simple that for small signals there is really no justification for deriving the special load line.

Following the rules used for drawing the equivalent circuit for the basic amplifier and assuming the amplifier is properly designed so that in the mid-frequency range both the capacitors are effective short circuits results in the equivalent circuit shown in Fig. 26–4b. It differs from that for the basic circuit only in the addition of the resistance R_g shunting the load resistance. The effect then is simply to reduce the effective load resistance of the circuit, which immediately gives the gain of this circuit as

$$G = \frac{\mu R_L'}{r_p + R_L'} \qquad (26\text{--}10)$$

Where R_L' is the parallel resistance of R_L and R_g. Often, R_g is enough larger than the plate-load resistance that its presence can be ignored, but in any case its effect is to reduce the voltage gain below the value expected without the coupling.

(b) Low-Frequency Gain. At low frequencies, the assumption that the two capacitors are short circuits is no longer valid, and one or both of them must be included in the equivalent circuit. In Fig. 26–5a, the low-frequency equivalent circuit is drawn including only the coupling capacitor. To make the problem simpler the realistic assumption is made that, as the frequency is decreased, the effect of the coupling capacitor is felt before that of the cathode capacitor; for this analysis, the cathode capacitor is assumed to be still an effective short circuit.

The gain of this circuit is most easily found by replacing that portion of the circuit to the left of the dashed line in Fig. 26–5a by its Thévenin equivalent generator, resulting in the circuit of Fig. 26–5b. This is a simple series circuit that can be solved for any specific set of circuit parameters or in general terms to obtain an expression for gain.

Almost as much information is conveyed about this, or any amplifier, by stating the frequency at which the gain drops to a certain fraction of its mid-frequency value as by writing a complicated expression for the gain in terms of the frequency and circuit parameters. A convenient and commonly used frequency is known as the lower or upper half-power frequency, depending on whether reference is made to the reduction in gain at the low or high end of the spectrum. For the output power to change to one half, the voltage gain must be reduced to 0.707 times its mid-frequency value, assuming a constant input and a constant load resistance. In the circuit of Fig. 26–5b, the lower half-power frequency is reached when the current I is 0.707 of its mid-frequency value, or when the reactance of C_c is equal to the total series resistance in the circuit. Thus the lower half-power frequency is

$$f' = \frac{1}{2\pi C_c \left[R_g + \dfrac{r_p R_L}{r_p + R_L} \right]} \qquad (26\text{–}11)$$

(c) High-Frequency Gain. At the high-frequency end of the operating range of an amplifier, there is no problem with an alternating voltage drop across either the coupling or the cathode bypass capacitors. There is, however, current in the capacitance that inevitably shunts the load impedance. This current becomes larger as the frequency increases, and at some point becomes large enough so it no longer may be neglected. Figure 26–6 shows the high-frequency a-c equivalent circuit for the R–C

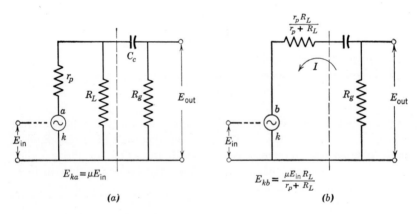

Fig. 26–5. Low-frequency equivalent circuits for R–C coupled amplifier.

DISTORTION IN AMPLIFIERS

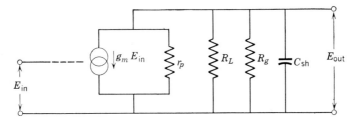

Fig. 26–6. High-frequency equivalent circuit for R–C coupled amplifier.

coupled amplifier stage of Fig. 26–4a with the capacitance C_{sh} representing the combined effects of all shunting capacitances. In this diagram, the current-source equivalent circuit is used since it results in the simpler circuit; with this equivalent circuit, all impedance elements are in parallel.

In the circuit of Fig. 26–6, current from the source divides among four parallel branches, and the output voltage may be thought of as the product of the total current through the resistive branches and the net resistance of that parallel combination. At mid-frequency, all the current flows through the resistive elements, but at higher frequencies some is shunted through the capacitance to reduce the output and thereby the gain of the amplifier. Again, the effect of this reactance can most conveniently be described by giving the frequency at which its effect is to reduce the power in the load resistor to one-half its mid-frequency value. When the reactance of C_{sh} equals the parallel resistance of r_p, R_g, and R_L, the resistive and reactive currents are each equal to 0.707 times the total current; that is, the output voltage equals 0.707 times its mid-frequency value, and the power dissipated in R_L (or R_g) is one half its mid-frequency value. Thus equating resistance to reactance gives an upper half-power frequency of

$$f'' = \frac{1}{2\pi R''_L C_{sh}} \qquad (26\text{–}12)$$

where R''_L is the combined resistance of the parallel combination of r_p, R_L, and R_g.

26–5. DISTORTION IN AMPLIFIERS

The possibility of distortion from the combination of large signals and a nonlinear characteristic is discussed in Art. 25–3. That particular type is known as nonlinear distortion. With complex signals, which contain

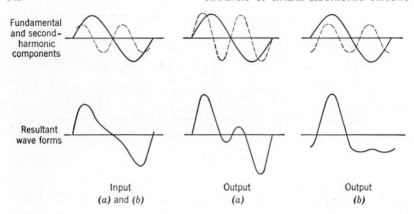

Fig. 26–7. Distortion due to (a) nonuniform amplification of different frequency components; (b) nonuniform phase shift of different frequency components.

components of more than one frequency, there are two additional possible sources of distortion: **amplitude** and **phase distortion.**

Amplitude distortion results from the fact that an amplifier does not equally amplify signals of different frequencies. If a signal has components of different frequencies, and if the amplification is different for the different frequencies, then the wave shape of the output voltage differs from that of the input. Wave forms are plotted in Fig. 26–7a to illustrate this type of distortion.

In general, the effect of amplitude distortion of an audio signal is a change in the quality of the sound. Thus, amplitude distortion of a signal corresponding to the tone of a French horn (which differs from the tone of other instruments in the number and strength of the overtones, or harmonics, that are present) reduces the relative strength of the harmonics, with the result that the output of the loudspeaker might sound more like a trumpet, or, in the extreme, if the harmonics were completely removed, a tone similar to that of a tuning fork.

The capacitances that affect the gain of an amplifier at different frequencies also affect the phase shift between the input and output voltages. Components of different frequencies are shifted different amounts, which may cause distortion even when amplitude distortion is negligible. The wave forms of Fig. 26–7b illustrate phase distortion. Actually, the human ear is unable to distinguish phase differences; therefore, phase distortion is of no significance in audio amplifiers. However, for many applications, such as television, the signal wave shape must be faithfully reproduced, and undue phase distortion cannot be tolerated.

26–6. A-C EQUIVALENT CIRCUIT OF A TRANSISTOR

The results of the derivation of the equivalent circuits for the vacuum tube can be applied directly to the transistor. The collector-emitter path through the transistor is quite analogous to the plate-cathode path through the vacuum tube, and, therefore, it is possible to represent the one by either of the two circuits developed for the other. A minor difference is that the equivalent generators in the collector-to-emitter branch are more conveniently expressed as functions of, or proportional to, the base current rather than to a voltage. In a like manner, the base-to-emitter path, since it conducts current, should also be capable of representation by similar circuits. As pointed out in Art. 22–10, the dependence of the base current upon the voltage or current in the collector is not strong, and is often ignored. However, for completeness, a generator to represent this dependence should be included in the base branch of the equivalent circuit, and this generator is more conveniently expressed as being proportional to the collector-emitter voltage.

One of the more popular of the many possible arrangements for a transistor equivalent circuit is shown in Fig. 26–8. It is called a hybrid circuit because a voltage source is used in one branch and a current source in the other. All parameters in this circuit are labeled h to indicate that they refer to this hybrid circuit, and the second subscript e is used to indicate the common-emitter configuration. (An equivalent circuit that is equally valid could have been developed with either of the two other electrodes common to the two branches. Since the transistor is more commonly used with the emitter common to the input and output circuits, only this equivalent circuit is considered here.) The first subscript associated with each of the h parameters refers to "input" or

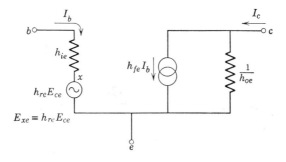

Fig. 26–8. Common-emitter hybrid equivalent circuit for a transistor.

"output" for the two impedances and to "forward" or "reverse" for the current and voltage ratios, respectively. Thus,

h_{ie} = input impedance = $\partial e_b/\partial i_b$ with constant e_{ce}
h_{oe} = output admittance = $\partial i_c/e_{ce}$ with constant i_b
h_{fe} = forward current transfer ratio = $\partial i_c/\partial i_b$ with constant e_{ce}
h_{re} = reverse voltage transfer ratio = $\partial e_{be}/\partial e_{ce}$ with constant i_b.

Average values of the common-emitter h parameters for a typical low-power transistor, the type 2N45, are as follows:

$$h_{ie} = 500 \text{ ohms}$$
$$h_{oe} = 12.5 \text{ } \mu\text{mhos}$$
$$h_{fe} = 11.5$$
$$h_{re} = 2.5 \times 10^{-4}$$

When the input to a transistor amplifier is supplied from the output of another, the effective impedance of the source is so much larger than the base-to-emitter impedance that the accuracy is not greatly affected by considering the base-emitter to be short-circuited. If this simplification is made, the transistor equivalent circuit is no more complicated than that for the vacuum tube.

26–7. GAIN OF TRANSISTOR AMPLIFIER

Figure 26–9 shows the circuit diagram of a simple grounded-emitter transistor amplifier with an a-c equivalent circuit that is applicable when the frequency of the supply E_s is high enough so that the coupling capacitor C may be considered a short circuit but low enough so that no shunting capacitances need be considered. Except that this circuit is a bit more complicated than the corresponding vacuum-tube circuit, there is no particular difficulty in calculating the gain of this amplifier. It is worth while, however, first to consider how the circuit is employed and then to decide for those conditions just what is meant by gain, or what gain is the most useful.

The vacuum tube is inherently a voltage amplifier in that it responds to an input voltage; therefore, the only gain that has meaning for a vacuum-tube amplifier that has for its load another vacuum tube is voltage gain. Voltage gain is a measure of how much the input to the second amplifier is improved by the insertion of the stage of amplification under consideration. The situation is just reversed for a transistor

GAIN OF TRANSISTOR AMPLIFIER

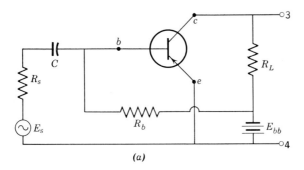

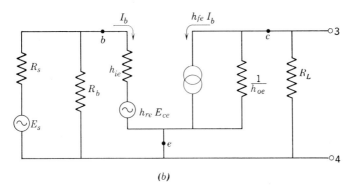

Fig. 26-9. Common-emitter amplifier and its mid-frequency equivalent circuit.

amplifier that has for its load another transistor. Here, the gain of interest is current gain or how much the base or input current to a second transistor is increased by the amplifier under consideration. Furthermore, assuming that the load for the amplifier is another transistor makes possible a considerable simplification of the circuit. The input impedance of a grounded-emitter transistor is so much smaller than its output impedance that, by comparison, the base-emitter circuit of both the transistor under consideration and the one that constitutes its load can usually be considered short circuits. Applying these simplifications to the circuit of Fig. 26-9 means that in ordinary usage the output current is approximately equal to the short-circuit current between terminals 3 and 4, or $h_{fe}I_b$, and that the base current is approximately

$$I_b = E_s/R_s$$

The improvement, or current gain, of this amplifier is the ratio of the current it can supply to another transistor input to the current that can be supplied by the source E_s directly to the transistor input. Thus, the available current gain is

$$G = \frac{h_{fe}I_b}{E_s/R_s} = h_{fe} \tag{26-13}$$

Equation 26–13 gives the approximate current gain of the transistor amplifier reduced to its simplest form for the simplest possible application of the amplifier. But this solution is not always adequate. If the amplifier of Fig. 26–9 is to be used to supply the input or grid voltage to a vacuum-tube amplifier, the terminals 3 and 4 would be connected to an open circuit, and the effectiveness of the amplifier would be measured as a voltage gain, the ratio of the open-circuit output voltage E_{34} to the available input voltage E_s. (If the supply were connected directly to the load, which in this case is a vacuum-tube grid, the full voltage E_s would be available for there would be no IR drop across R_s.) If the base-emitter voltage is assumed to be negligible and if negligible current flows through R_b, the base current is

$$I_b = E_s/R_s \tag{26-14}$$

and the output voltage equals the product of the current in the equivalent generator and the parallel resistance of R_L and $1/h_{oe}$. Thus

$$E_{34} = \frac{h_{fe}I_b R_L}{h_{oe}R_L + 1} \tag{26-15}$$

Combining Eqs. 26–14 and 15 gives the approximate voltage gain for the amplifier of Fig. 26–9 as

$$G = \frac{E_{34}}{E_s} = \frac{h_{fe}R_L}{R_s(h_{oe}R_L + 1)} \tag{26-16}$$

One might be tempted to call the ratio of E_{34} to E_{12} the voltage gain, in which case the gain would be much larger than that given by Eq. 26–16. For this amplifier employed as described, the larger value of gain would be misleading; it is not a true measure of the improvement offered by the amplifier. It is the gain as given by Eq. 26–16 that indicates the quality of this amplifier.

If one is not satisfied with the approximation of zero base-emitter impedance that was made in the development of the gain formula of Eq. 26–16, the complete circuit of Fig. 26–9b can be analyzed. The circuit can be simplified somewhat by replacing the supply voltage, its impedance, and the resistance R_b by a Thévenin equivalent generator as shown in Fig. 26–10. The circuit is still rather complicated, and so the approach used is to assume that the equivalent voltage source in the base

GAIN OF TRANSISTOR AMPLIFIER

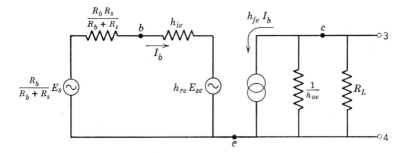

Fig. 26-10. Equivalent circuit of Fig. 26-9 with Thévenin generator replacing the source and R_b.

branch is negligible, and then to calculate for a specific case the approximate error introduced by this assumption. With this assumption, the base current is

$$I_b = \frac{R_b E_s}{(R_b + R_s)\left[h_{ie} + \dfrac{R_b R_s}{R_b + R_s}\right]} = \frac{R_b E_s}{h_{ie}(R_b + R_s) + R_b R_s} \quad (26\text{-}17)$$

and the output voltage is the product of h_{fe}, this base current, and the net resistance between the output terminals. Thus,

$$E_{34} = \frac{R_L I_b h_{fe}}{(h_{oe} R_L + 1)} \quad (26\text{-}18)$$

and the voltage gain is

$$G = \frac{h_{fe} R_b R_L}{(h_{oe} R_L + 1)[h_{ie}(R_b + R_s) + R_b R_s]} \quad (26\text{-}19)$$

To check the validity of the various assumptions made in the derivation of Eqs. 26–16 and 19, calculations can be made for a type-2N45 transistor having the h parameters given in Art. 26–6 and with $R_b = 10{,}000$ ohms, and $R_L = R_s = 2000$ ohms. The gain for this amplifier as given by Eq. 26–16, in which the base-emitter path is assumed to be a short circuit, is 11.2. The use of Eq. 26–19, which neglects only the equivalent voltage generator $h_{re}E_{ce}$, results in a calculated gain of 8.6. The error introduced in the first calculation would have been smaller had the source impedance R_s been very much larger than h_{ie}. The results of the more accurate of the two calculations can now be used to determine the error caused by ignoring $h_{re}E_{ce}$. For this particular situation, $E_{ce} = 8.6 E_s$ and $h_{re}H_{ce} = 2.5 \times 10^{-4} \times 8.6 E_s = 2.15 \times 10^{-3} \times E_s$. Thus, in this example one is quite justified in neglecting the equivalent generator in the base circuit, but not the base resistance h_{ie}.

PROBLEMS

26–1. Using the curve of Fig. 20–9, determine the three tube coefficients for the type-6J5 tube at the following operating points:
 (a) Plate voltage = 120, plate current = 8.0 ma.
 (b) Plate voltage = 120, plate current = 0.6 ma.

26–2. From the curves of Fig. 20–11, determine the approximate tube coefficients for the type-6SJ7 pentode operating at a plate voltage of 150 volts and a plate current of 7.0 ma.

26–3. A type-6L6 beam-power tube is used with a plate supply voltage of 300 volts, a plate load resistance of 1000 ohms, and a negative grid bias of 5 volts. What are the g_m and r_p at the quiescent operating point? What value of cathode resistance could be used to supply the negative grid bias?
Ans. 9000 μmhos; 9000; 37 ohms.

26–4. Determine α, the current-amplification factor, for the transistor whose common-base characteristics are given in Fig. 22–5. *Ans.* 0.93.

26–5. For the transistor whose characteristics are given by Fig. 22–6, determine h_{oe} and h_{fe}, the common-emitter hybrid parameters, at a collector voltage and current of 20 volts and 3.0 ma, respectively.
Ans. 36 μmhos; 11.

26–6. Draw a circuit diagram of an *NPN* transistor in a common-emitter connection employing a supply battery of 30 volts, a collector load resistor of 3000 ohms, and a base bias resistor of 33 kilohms.
 (a) What would have to be changed to accommodate a *PNP* transistor?
 (b) Assuming that the collector characteristic of this transistor is given by Fig. 22–6, determine the operating point.
Ans. E_{ce} = 3.5 volts; I_c = 8.8 ma.
 (c) Determine h_{oe}, and h_{fe} about this operating point. Compare with answers to Prob. 26–5. *Ans.* 100 μmhos, 6.7.

26–7. Draw an a-c equivalent circuit for the amplifier of Fig. 25–5. Explain why neither R_k nor C_k need to be considered in the determination of the gain of this amplifier operating in the frequency range for which it was designed.

26–8. Determine the gain of the amplifier in Prob. 26–7 if the cathode bypass capacitor is removed.

$$Ans. \quad \frac{\mu R_L}{R_L + r_p + R_K(1 + \mu)}.$$

26–9. Figure 26–11 shows the circuit diagram for a cathode follower. Although the cathode follower always has a voltage gain less than unity, it is useful because of the low current required at the input terminals. In analyzing this circuit for mid-frequency operation, one can neglect the current through R_g in comparison with the plate current as well as neglecting any reactance effects.

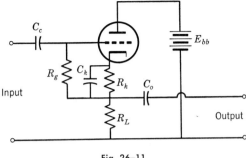

Fig. 26-11.

(a) Determine the voltage gain of this circuit.

$$Ans. \quad \frac{\mu R_L}{r_p + R_L(1 + \mu)}.$$

(b) What is the ratio of input voltage to input current, or input impedance, of this circuit?

$$Ans. \quad \frac{R_g[r_p + R_L(1 + \mu)]}{r_p + R_L}.$$

(c) What are the voltage gain and input impedance of this circuit using a type-6J5 triode, $R_L = 10$ kilohms, $R_g = 1.0$ megohm, and a bias resistance and plate supply voltage that produces no-signal operation at a plate current of 4.0 ma and grid-cathode voltage of -6 volts?

$$Ans. \quad 0.905; \quad 10.54 \text{ megohm}.$$

26-10. The transistor equivalent to the cathode follower is the common-collector amplifier shown in Fig. 26-12. Draw an a-c equivalent circuit for

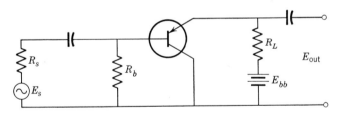

Fig. 26-12. Common-collector amplifier.

this amplifier, and determine both voltage and current gains, assuming negligible current in R_b, $R_L << 1/h_{oe}$, and $R_S >> h_{ie}$.

$$Ans. \quad \frac{R_L(1 + h_{fe})}{R_s + R_L(1 + h_{fe})}; \quad 1 + h_{fe}.$$

26-11. Draw an a-c equivalent for the circuit shown in Fig. 26-13. Show

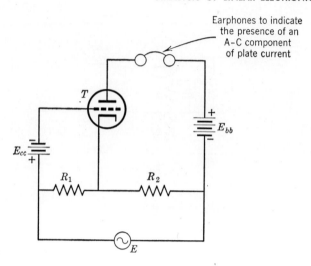

Fig. 26–13.

that, if R_1 and R_2 are so adjusted that there is no sound in the earphones, the amplification factor of the tube T is given by $\mu = R_2/R_1$.

26–12. Show that for the circuit of Fig. 26–14 the condition for no sound in the earphones is $g_m = 1/R$. (Both this circuit and the one of Fig. 26–13 are useful in determining tube coefficients experimentally.)

26–13. Figure 26–15 shows a circuit that might be employed for the experimental determination of h_{fe} of a transistor. For the conditions $R_b \gg h_{ie}$, $R \gg h_{ie}$, and $R_c \ll 1/h_{oe}$, show that h_{fe} is equal to $V_2 R/V_1 R_c$.

26–14. Draw vector diagrams for the amplifier of Fig. 26–4 for both upper and lower half-power frequencies. Include sufficient information on the diagrams so that the relationship between input and output voltages is clear.

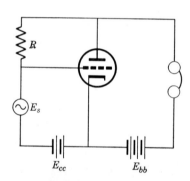

Fig. 26–14. G_m bridge.

26–15. The two stages of a certain amplifier are coupled by means of a 0.05-μf coupling capacitor and a 100,000-ohm grid resistor. The voltage gain of this amplifier is 1000 at a frequency of 2500 cps. To what value is the gain reduced for a frequency of 20 cps? Assume that the gain of the first stage is unaffected by changes in the reactance of the coupling capacitor.
Ans. 532.

26–16. Repeat Prob. 26–15 for two coupling networks in the amplifier, each identical with the single one described. *Ans.* 283.

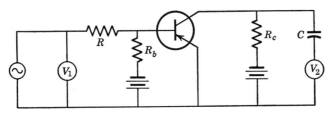

Fig. 26-15.

26-17. One stage of an R–C coupled amplifier employs a triode having a μ of 20 and an r_p of 10 kilohms, a plate load impedance of 5 kilohms, coupling capacitance of 0.01 μf, a grid resistance of 500 kilohms, and an effective shunting capacitance of 63.6 $\mu\mu$f.

(a) What are the lower and upper half-power frequencies?
Ans. 31.8 cps; 0.5 mc.

(b) What is the gain at 16 cps, and at 1.0 mc? *Ans.* 3.0, 3.0.

26-18. Figure 26-16 illustrates a possible self-biasing scheme whereby the biasing current depends somewhat on the characteristics of the particular

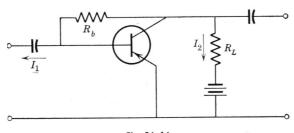

Fig. 26-16.

transistor. Show that the current gain for this amplifier is approximately

$$\frac{I_2}{I_1} = \frac{h_{fe}R_b}{R_b + R_L(1 + h_{fe})}$$

How does this compare with the current gain when the bias is supplied from a separate battery through R_b?

■ 27 Electronics in Communications

27-1. MODULATION

Under proper conditions, it is possible to radiate electric energy into space and to receive a portion of that radiated energy at another location. This process is practical only at relatively high frequencies—those well beyond the audible range. Because of this frequency limitation, audio signals cannot be transmitted directly by radio; the information to be transmitted must be impressed upon a high-frequency signal, or carrier, which is capable of being transmitted through space. The process whereby information is impressed upon a **carrier** is called **modulation**.

The general expression for a sinusoidal carrier voltage,

$$e_c = E \sin (2\pi f_c t + \phi) \qquad (27\text{-}1)$$

indicates that there are three parameters that might be varied according to the desired audio signal in order to modulate the carrier. The amplitude E, the frequency f_c, or the phase angle ϕ may be varied to accomplish **amplitude modulation (AM), frequency modulation (FM),** or **phase modulation,** respectively. Amplitude modulation and frequency modulation are illustrated in Fig. 27–1. (For simplicity, only a sinusoidal audio signal is shown in this illustration.) The effect of varying the phase angle of the carrier is so similar to that of varying the frequency that the difference would not be apparent on a plot such as that of Fig. 27–1. With slight modification, phase modulation can be, and is, used to produce a frequency-modulated signal.

Although other forms of modulation are possible and find application

AMPLITUDE MODULATION

in special situations, all commercial broadcasting and most other radio communication is by an amplitude- or frequency-modulated sinusoidal carrier signal. All stations in the standard broadcast band (550 to 1500 kc) use amplitude modulation, the FM band between 88 and 108 megacycles is for frequency modulation, and the standard television broadcasts employ both—amplitude modulation for the video, or picture, information, and frequency modulation for the audio.

27-2. AMPLITUDE MODULATION

For a sinusoidal audio or modulating signal, the amplitude of the amplitude-modulated carrier varies sinusoidally about its unmodulated

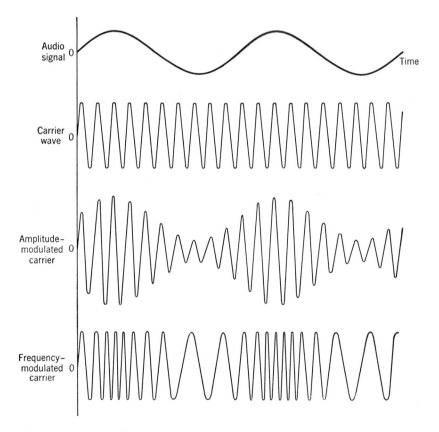

Fig. 27-1. Wave forms illustrating amplitude and frequency modulation.

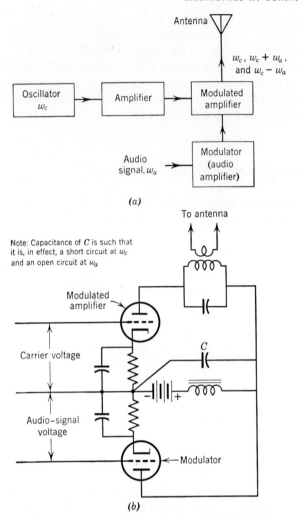

Fig. 27–2. Block diagram for an amplitude-modulation transmitter and a circuit for accomplishing amplitude modulation.

value. Thus, a general expression for an amplitude-modulated carrier is

$$e = E(1 + A \sin 2\pi f_a t) \sin 2\pi f_c t \qquad (27\text{--}2)$$

where f_a is the frequency of the modulating signal, and the factor A is a measure of the degree of modulation. When the value of A is unity, the amplitude of the modulated carrier varies between zero and twice its unmodulated value, and the modulation is said to be 100 per cent. For the example in Fig. 27–1, A is about 0.6.

Equation 27–2 can be expanded into the following:

$$e = E\left[\sin 2\pi f_c t + \frac{A}{2}\cos 2\pi(f_c - f_a)t - \frac{A}{2}\cos 2\pi(f_c + f_a)t\right] \quad (27\text{–}3)$$

For every frequency component f_a of the modulating signal, the modulated carrier contains two frequency components in addition to that of the original carrier. These two components, which have frequencies equal to the sum and difference of the carrier and modulating frequencies, are called **upper** and **lower side frequencies,** respectively. Any antenna, transmission line, amplifier, etc., that is to transmit an amplitude-modulated wave must be capable of transmitting these side frequencies without undue loss.

Amplitude modulation of the output of a high-frequency oscillator can be accomplished by varying the grid bias or the plate-supply voltage according to the instantaneous value of the modulating signal. It is more common, however, to use one or more stages of amplification following the oscillator, and to impress the modulating signal on one of the amplifier tubes. This arrangement prevents interaction between the oscillator and modulating circuits, thereby eliminating a source of frequency variation in the oscillator. Figure 27–2 shows a block diagram of a radio transmitter employing this arrangement and a circuit diagram to illustrate one of the many possible means of actually accomplishing the modulation.

The plate-supply voltage for the modulated amplifier of Fig. 27–2 is the output of an audio amplifier, which is the modulator. This voltage consists of a d-c component, and an a-c component that corresponds to the audio signal. Since the amplitude of the output of the modulated amplifier depends upon the plate-supply voltage to that amplifier, the amplitude of the output varies according to the instantaneous value of the audio signal.

27–3. DETECTION OF AMPLITUDE-MODULATED SIGNALS

To make any use of the amplitude modulation, it must be possible to regain from the modulated carrier the original modulating signal. The process of obtaining the audio signal from a modulated carrier is called **demodulation,** or **detection.** Detection of an amplitude-mod-

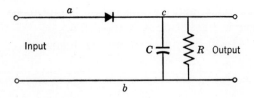

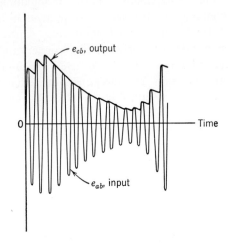

Fig. 27-3. Amplitude-modulation diode detector.

ulated signal can be accomplished in a number of ways. The most common method is also the simplest, and only that method is discussed here.

The diode detector of Fig. 27-3 is, in reality, nothing more than a single-phase half-wave rectifier with a capacitor filter. The action of this circuit as a detector depends upon the proper choice of values for the load resistance R and the filter capacitance C. The product RC must be large in comparison to the period of the carrier, but small in comparison to the period of the highest audio frequency to be encountered. With values of R and C that meet this condition, the capacitor cannot discharge appreciably between successive positive peaks of the carrier, but it can discharge so that the capacitor voltage follows the variations of the peak value of the modulated carrier, as shown by the wave forms of Fig. 27-3. In practice, the ratio of carrier frequency to modulating frequency is much greater than can be shown with any degree of clarity on a diagram such as that of Fig. 27-3 (for the standard radio frequencies, this ratio is greater than 100); therefore, the steps indicated on the output-voltage wave form of Fig. 27-3 are actually so

AMPLITUDE-MODULATION RECEIVERS

small that they may be neglected. In fact, they are entirely eliminated in any conventional audio amplifier that follows the detector.

27-4. AMPLITUDE-MODULATION RECEIVERS

Theoretically, an amplitude-modulation receiver requires only a detector, a suitable antenna for receiving the radiated signal, and earphones or a loudspeaker to transform the audio signal into sound; practically, several stages of amplification are required to make the receiver valuable for any but extremely strong signals. (Ordinary signals picked up on a radio antenna are measured in microvolts.) The tuned-radio-frequency receiver uses one to four stages of amplification between the antenna and detector, in addition to one or two stages of audio amplification between the detector and loudspeaker. Those stages preceding the detector are tuned amplifiers adjusted for maximum sensitivity, or gain, at the frequency of the signal to be received. If the receiver is to be used over a broad band of frequencies, such as the standard broadcast band, the tuning is accomplished by adjusting the capacitance in each of the tuned circuits. Commonly, the several variable capacitors in a receiver with more than one tuned stage are mechanically ganged together so that tuning is accomplished by means of a single control. Accurate tuning of the various stages over the full range of the frequency band is difficult to achieve with a single control.

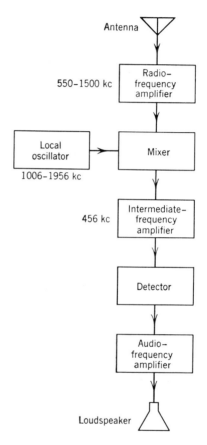

Fig. 27-4. Block diagram of standard broadcast receiver.

Practically all radio receivers employ the **superheterodyne** circuit, rather than the tuned-radio-frequency circuit. A block diagram for a superheterodyne receiver is shown in Fig. 27-4.

The essential feature of the superheterodyne circuit is that, regardless of the signal frequency, most of the amplification takes place at a single frequency, or, to be more exact, over a band of frequencies only wide enough to include the necessary side frequencies. This allows the use of amplifiers with double-tuned circuits, which may be carefully tuned to provide high and uniform gain. In order to transform all signals to a single frequency, the incoming signal is mixed with the signal from a local oscillator whose frequency is so adjusted that the resulting frequency has the desired constant value. This requires that the local oscillator and the input circuit be tuned simultaneously so that the difference between their frequencies is constant. This difference frequency is the intermediate frequency that is applied to the high-gain intermediate-frequency amplifier.

There are many possible variations in the components of a superheterodyne receiver. For example, the radio-frequency amplifier may or may not be present, and the functions of the mixer and the local oscillator may be combined in a single tube called a **converter**.

27–5. FREQUENCY CONVERSION

When voltages of different frequencies are impressed in series across a nonlinear device, the resulting current contains, in addition to components of the original frequencies, components having frequencies equal to the sum and difference of the original frequencies. This principle can be applied directly to form a superheterodyne mixer. Furthermore, modulation and detection may also be looked at as frequency-conversion operations, and both may be accomplished by the mixing process in a nonlinear device. The differentiation among a mixer, a modulator, and a detector circuit may be simply the selection of the component or components that are to be considered as the output of the device.

To describe the mixing process it may be assumed that a voltage consisting of components of two different frequencies is impressed across a device having the current–voltage characteristic similar to that given in Fig. 27–5. If the impressed voltage, expressed by

$$e = E_1 \cos \omega_1 t + E_2 \cos \omega_2 t \tag{27-4}$$

has the amplitudes E_1 and E_2 small enough so that the nonlinear characteristic may be acurately expressed by the first three terms of a power series, then

$$i = I_0 + a_1 e + a_2 e^2 \tag{27-5}$$

FREQUENCY CONVERSION

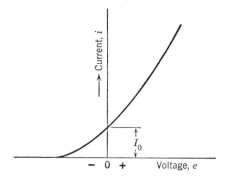

Fig. 27–5. Nonlinear characteristic for a diode mixer.

Substituting in Eq. 27–5 the assumed value for e from Eq. 27–4, we get

$$i = I_0 + a_1 E_1 \cos \omega_1 t + a_1 E_2 \cos \omega_2 t + a_2 E_1^2 \cos^2 \omega_1 t \\ + a_2 E_1 E_2 \cos \omega_1 t \cos \omega_2 t + a_2 E_2^2 \cos^2 \omega_2 t \quad (27\text{–}6)$$

The last three terms of Eq. 27–6 can be expanded by the use of well-known trigonometric identities to give

$$i = I_0 + a_1 E_1 \cos \omega_1 t + a_1 E_2 \cos \omega_2 t + \frac{a_2 E_1^2}{2}(1 + \cos 2\omega_1 t) \\ + \frac{a_2 E_1 E_2}{2}[\cos(\omega_1 - \omega_2)t + \cos(\omega_1 + \omega_2)t] + \frac{a_2 E_2^2}{2}(1 + \cos 2\omega_2 t) \quad (27\text{–}7)$$

Equation 27–7 indicates that, when the sum of two voltages of frequency ω_1 and ω_2 is impressed across the nonlinear device, the current through that device contains components of the following frequencies: 0(d-c), ω_1, ω_2, $\omega_1 - \omega_2$, $\omega_1 + \omega_2$, $2\omega_1$, and $2\omega_2$. (If higher-order terms had been assumed in the power series of Eq. 27–5, even more frequency components would be present). The frequency conversion as required by the superheterodyne mixer is completed by employing this current to supply a circuit that is sensitive only to $\omega_1 - \omega_2$ or to $\omega_1 + \omega_2$.

In the block diagram of Fig. 27–4, the intermediate-frequency amplifier is tuned to $\omega_1 - \omega_2$, the difference between the frequencies of the incoming signal and the local oscillator. Since the incoming signal has many components if it is modulated, the output also contains many components whose frequencies equal the difference between each of the input frequencies and the local oscillator frequency. Thus, the intermediate-frequency amplifier must be capable of amplifying over a broad enough band to include all the side frequencies associated with the incoming signal. For the standard broadcast band, an ideal intermediate-frequency amplifier should amplify equally well over 10 kilocycles and reject everything outside of that band.

A consideration of the frequency components present in Eq. 27–7 indicates that the mixing process produces all the components required for amplitude modulation. If ω_1 represents the carrier frequency and ω_2 the modulating frequency, then the three components of the amplitude-modulated wave as given in Eq. 27–3 are present: ω_1, $\omega_1 - \omega_2$, and $\omega_1 + \omega_2$. Amplitude modulation is achieved if those three and only those three components are selected as the output.

Similarly the necessary components for the output of a detector are produced in this mixing process. The input to an amplitude-modulation detector is the sum of a carrier and two side frequencies differing from the carrier by the frequency of the modulating signal. Thus, the output component representing the difference between the input frequencies is the desired modulating signal. For very small signals, the operation of the detector shown in Fig. 27–3 can be described in terms of the mixing principle. Under this condition it is not accurate to consider the diode as a perfect rectifier; it is a nonlinear device having a current–voltage characteristic of the type illustrated by Fig. 27–5.

At carrier frequencies less than 100 megacycles, it is convenient to use a more efficient mixer than the simple two-terminal nonlinear device. The only useful output component as far as the superheterodyne mixer is concerned results from the term in Eq. 27–6 that is the product of the two input voltages. Thus, if the two voltages can be multiplied directly, many of the unwanted components can be eliminated. Special mixer and converter tubes have been developed to approximate this multiplication. The two voltages are applied to two different grids of such a tube so that the g_m of the tube between one of the grids and the plate varies approximately linearly with the voltage of the other grid, and the plate-current variations are proportional to the product of the two voltages.

27–6. FREQUENCY MODULATION

A frequency-modulated signal is one in which the frequency of the carrier is varied about a mean value by an amount proportional to the instantaneous value of the modulating signal. For an audio-modulating signal, frequency modulation is most easily accomplished with a condenser microphone—two parallel plates, one rigid, the other flexible so as to vibrate according to sound pressure waves that impinge upon it—connected in parallel with, or acting as the capacitor portion of, the resonant circuit of an oscillator. Since the frequency of the oscillator is a function of the capacitance of its resonant circuit, changes in the capacitance of the condenser microphone due to sound waves cause corresponding changes in the oscillator frequency. Thus, intelligence corresponding to an audio signal is impressed upon a high-frequency carrier signal by varying the frequency of that signal.

Although the scheme just described for providing frequency modulation is impractical as such, it is the basis for a practical system. In the practical system, the condenser microphone is replaced by a pentode that is connected to act as a reactance. The audio signal, which is applied to the grid of this tube in the form of varying voltage, varies the effective reactance of the tube, thereby varying the oscillator frequency.

27-7. FREQUENCY-MODULATION RECEIVER

A typical frequency-modulation receiver employs a superheterodyne circuit with a detector that is capable of recovering the modulating signal from a frequency-modulated carrier. In most frequency-modulation receivers, a **limiter** tube is connected between the intermediate-frequency amplifier and the detector to limit all voltage variations to a fixed amplitude. This action of the limiter eliminates the effects of atmospheric conditions, fading, and other disturbances, leaving a pure frequency-modulated signal to be demodulated by the detector.

The diagram for a frequency-modulation detector circuit is shown in Fig. 27-6. This circuit consists of two half-wave rectifiers with capacitor filters of such magnitude that the output voltage of each follows the peak value of the source voltage of the corresponding rectifier. Thus, e_{cm} equals the peak value of e_{am}, and changes as that value changes. Likewise, e_{dm} equals the peak value of e_{bm}, and changes accordingly.

The input to the frequency-modulation detector is a constant-amplitude frequency-modulated voltage. However, the amplitudes of the voltages e_{am} and e_{bm}, which supply the two half-wave rectifiers, are not constant; these voltages are developed across parallel resonant circuits and, therefore, depend upon the frequency of the signal. The parallel combination of L_1 and C_1 is adjusted to be resonant at a frequency slightly above the carrier frequency, so that the amplitude of e_{am} and the instantaneous value of e_{cm} vary with frequency as indicated in Fig. 27-6. The combination of L_2 and C_2 is resonant at a frequency slightly

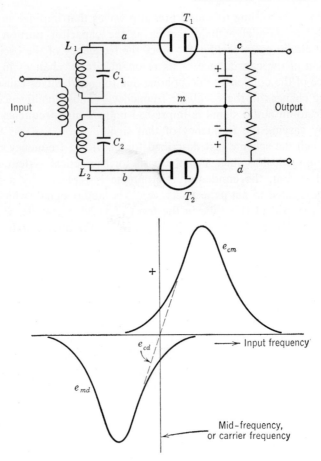

Fig. 27-6. Frequency-modulation detector.

below the carrier frequency, and the polarity of the second rectifier is reversed. Thus, on the diagram of Fig. 27-6, the curve of e_{md} is shifted to the left and inverted. The output voltage of this detector is the sum of the voltages e_{cm} and e_{md} and, as indicated, is approximately proportional to the deviation in the frequency of the signal from the carrier frequency. Since the deviation in the frequency was made proportional to the instantaneous value of the modulating signal, this circuit is a successful detector.

Proper operation of the frequency-modulation detector requires that its input voltage be of constant amplitude. The function of the limiter stage is to limit all signals to this amplitude. As long as the signals are equal to, or greater than, this limiting value, a constant-amplitude signal is applied to the detector. The limiting action is the source of an important advantage of frequency modulation over amplitude modulation; unless the signal voltage at the limiter input drops below the limiting value, all effects of fading, atmospheric disturbances, etc., are eliminated by the limiter. It is for this reason that frequency-modulation reception is relatively static free.

27-8. OTHER APPLICATIONS OF MODULATION

The discussion of modulation and demodulation in this chapter applies equally to systems in which the connection between the transmitter and receiver is by wire or cable instead of by radio waves in space. In telephony, a single cable can simultaneously transmit several telephone messages if the different messages are transmitted as modulated carriers of different frequencies. For amplitude modulation, it is necessary only that the difference between adjacent carrier frequencies be twice the frequency of the highest audio frequency to be transmitted and that there be some filters, or tuned circuits, that are capable of sorting out the different modulated signals at the receiving end of the cable.

Telephone conversations or signaling information may be transmitted by carrier on a 60-cycle power line. Fig. 27–7 illustrates the simple connections necessary for such dual use of the power line. The inductors in series with the line are of sufficiently low inductance to offer no objectionable impedance to 60-cycle current, but at the carrier frequency of the communication system the impedance is large enough to confine the communication signals to the desired section of the line. Conversely, the capacitors in series with the transmitter and receiver offer no imped-

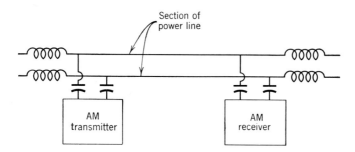

Fig. 27–7. Connection of communication system to a power line.

ance at the carrier frequency but effectively bar the flow of 60-cycle current into the communication equipment.

PROBLEMS

27-1. Over what range should the local oscillator of an FM receiver be tuned in order to cover the entire FM band when the IF amplifier is tuned to 10.7 Mc?

27-2. Why are there two possible answers to Prob. 27-1, and why might this provide difficulty if the intermediate frequency was less than 10 Mc?

27-3. An AM signal on a carrier of 456 kc is to be detected using the circuit of Fig. 27-3. If the detector is to be effective for audio signals between 100 cps and 5 kc, and if $R = 1$ megohm, what is a reasonable value for C?

27-4. The input signal to the receiver of Fig. 27-4 is a 1000-kc carrier amplitude modulated at 5 kc. What frequency or frequencies are present in the RF amplifier, local oscillator, IF amplifier, and AF amplifier?

■ 28 *Electronic Control*

28–1. INTRODUCTION

Electronic circuits are widely used in control systems, whether the system is basically electric or not. Vacuum-tube amplifiers and grid-controlled rectifiers are perhaps the most common electronic components of such systems. Some closed-loop systems are entirely electronic in nature; a voltage regulator that fits this category is discussed in the following article.

The use of electronic control is not limited to closed-loop systems. Electronic circuits find wide application in one-way, or open, systems as well. Such systems are, in effect, relays that respond to a certain stimulus to cause a desired reaction. For example, in a rolling mill an electronic relay may indicate flaws in the sheet steel as it passes a certain point, or actually initiate cutting of the sheet wherever flaws appear.

Most of the electronic components encountered in control systems are discussed in earlier chapters. The greater part of this chapter, therefore, is devoted to the study of some examples of electronic control.

28–2. VOLTAGE REGULATOR

For low-power loads that require a more constant voltage than is available from a given line or rectifier, the load voltage may be automatically regulated by grid control of a vacuum tube connected in series with the load. A simple circuit that provides regulation by this means

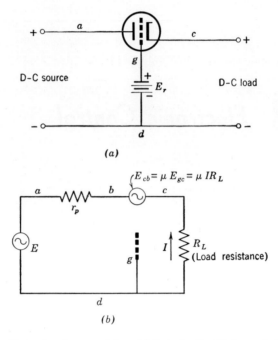

Fig. 28–1. Electronic voltage regulator. (a) Basic circuit. (b) A-c equivalent circuit.

is shown in Fig. 28–1a. There is no amplifier in this circuit, but otherwise all the components of a complicated closed-loop system are present. The d-c load voltage, the quantity under control, is compared with a fixed direct voltage, and the difference in voltage is applied between the grid and cathode of a tube that acts as the error-correcting device. An increase in the source voltage tends to send more current through the load, thereby increasing the load voltage, but this makes the grid more negative with respect to the cathode. This reaction opposes an increase in the current through the circuit, or increases the effective impedance of the series tube. Thus, an increase in the supply voltage does not result in a proportionate increase in the load voltage. There must be some change in the load voltage, however; otherwise there would be no change in the grid-to-cathode voltage of the series tube to effect the correction. The circuit can only reduce load-voltage variations; it cannot eliminate them entirely.

An analysis of the degree to which this regulator reduces the load-voltage variations is best approached by considering that variations in the source voltage are produced by a small a-c generator in series with the d-c source. The a-c equivalent circuit method can then be applied to determine the load-voltage variation (a-c component) that results from a given supply-voltage variation (a-c generator voltage). If the regulator is effective, the a-c component of the load voltage should be much smaller than the a-c component in the supply.

The current I that flows in the a-c equivalent circuit of Fig. 28–1b equals the net voltage of the two sources divided by the total resistance in the circuit. Thus,

$$I = \frac{E_{da} - E_{cb}}{r_p + R_L} = \frac{E - \mu I R_L}{r_p + R_L} \qquad (28\text{–}1)$$

Solving Eq. 28–1 for I and multiplying by R_L gives the a-c component of the load voltage,

$$E_{dc} = IR_L = \frac{ER_L}{r_p + R_L(1 + \mu)} \qquad (28\text{–}2)$$

The ratio of the a-c load voltage to the a-c source voltage is

$$\frac{E_{dc}}{E_{da}} = \frac{R_L}{r_p + R_L(1 + \mu)} \qquad (28\text{–}3)$$

The ratio given by Eq. 28–3 is a figure of merit for the regulator. It is an indication of the degree to which the regulator has succeeded in reducing the load-voltage variations. As indicated in Eq. 28–3, this ratio is less than unity (which is essential if the regulator is to be of any value at all), and the exact value of the ratio depends upon the tube characteristics and upon the value of the load resistance.

A considerable reduction in the ratio of a-c load to a-c supply voltage can be effected by amplifying the error voltage before applying it to the grid of the series tube. The circuit for a practical voltage regulator that employs an amplifier is shown in Fig. 28–2.

The constant d-c reference voltage in Fig. 28–2 is provided by the cold-cathode glow tube T_3 instead of by a battery. The difference between this reference voltage and the voltage E_{gh} (proportional to the load voltage) is applied to the grid of the amplifier tube T_2. Thus, a change in the load voltage produces an amplified voltage change appearing across the amplifier load resistor R_s. The amplified voltage change reacts on the series regulator tube just as in the basic circuit of Fig. 28–1. The control P offers a means of adjusting the grid bias on the amplifier tube, thereby varying the d-c output voltage.

Regulating load voltage by means of a series tube is practical only for low-current loads. Its principal application is to the rectifiers that supply

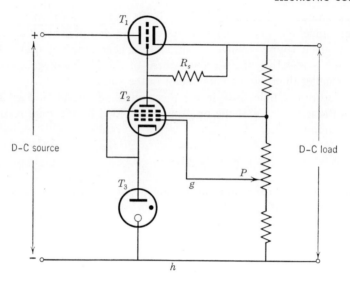

Fig. 28–2. A practical electronic voltage regulator.

direct voltages to precision instruments and other low-power electronic devices.

28–3. GENERATOR VOLTAGE REGULATOR

If d-c power is supplied by a generator, the load voltage may be regulated by applying the correction to the field current of that generator. This method of regulation is applicable to any d-c generator, regardless of its size. A simplified diagram of a circuit that accomplishes regulation by control of a generator field current is shown in Fig. 28–3.

The terminal voltage of the generator in Fig. 28–3 is compared with a reference voltage E_1, and the difference between the two is applied to the grid of the tube T_3 for amplification. The series combination of the amplifier output and the alternating voltage e_{ab} (which lags the supply voltage e_{db} by approximately 90 degrees) constitutes the grid-to-cathode voltage of the thyratron T_2 that controls the field current to the d-c generator. (See Art. 23–12 for bias-phase rectifier control.)

An increase in the generator terminal voltage, caused by a decrease in the generator armature current or an increase in the generator speed, causes the current through R_1 to decrease and therefore makes the grid of T_3 less negative. As in any amplifier, an increase in the grid voltage (less negative) is accompanied by a decrease in the plate voltage. This change in the potential of the plate of T_3 is transferred directly to the grid of T_2, causing T_2 to start conducting later in the cycle; the later T_2 starts to conduct, the smaller the average current through T_2 and the generator field. Thus, any increase in the generator terminal voltage is accompanied by a decrease in the generator field current, and therefore a decrease in the generated voltage. Perfect regulation with this circuit is impossible for the same reason that it is impossible with the circuit of Fig. 28–1; there must be some error in the generator terminal voltage to produce the correcting tendency.

The terminal voltage of the generator of Fig. 28–3 can be varied by adjustment of the magnitude of the reference voltage E_1.

The generator field in the circuit of Fig. 28–3 is supplied by a grid-controlled half-wave rectifier. Because of the highly inductive nature of the field winding, the field current tends to continue after the end of the positive half-cycle of the rectifier supply voltage. Tube T_1 provides a low-impedance path through which the current continues to flow during that part of each cycle when the thyratron does not conduct. Since the current "coasts" with no current in the supply for half of each cycle, this circuit is frequently called a free-wheeling rectifier.

For a highly inductive load, the free-wheeling circuit provides the same load voltage that would be obtained from a full-wave rectifier with twice the transformer voltage. However, the fact that it draws an unbalanced current from the a-c supply limits its application to low-current loads.

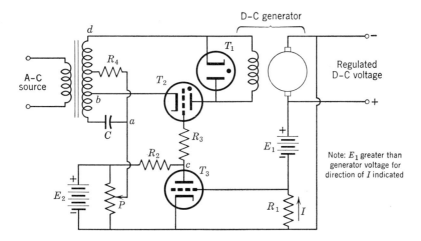

Fig. 28–3. D-c generator voltage regulator.

28-4. SPEED REGULATION OF A D-C MOTOR

The same circuit that is used for the regulation of the terminal voltage of a d-c generator can be applied to a d-c motor to regulate its speed. The only differences in the application of the circuit are that the grid-controlled rectifier in the speed regulator supplies the armature of the motor instead of its field, and the controlled quantity, which is compared to a reference voltage, is the output voltage of a mechanically coupled tachometer. Except for possible changes in the ratings of the various components required, nothing else needs to be done to the voltage-regulator circuit to transform it into a motor-speed regulator. In effect, the speed regulator is just a voltage regulator in which the voltage being regulated is the tachometer voltage. The tachometer voltage is proportional to its speed, and the regulator affects the tachometer voltage by changing the motor speed.

Most commercial speed regulators include a means for manual control of the speed. When a tachometer voltage is compared with a d-c reference voltage, the control is obtained by adjusting the reference voltage.

28-5. ANTI-HUNTING

In general, closed-loop systems such as the regulators discussed in the preceding articles are subject to hunting. Any disturbance to these systems produces a correcting force that tends to bring the controlled quantity back to the steady-state value. If the system has any inertia (mass in a mechanical system or inductance in an electric system), the correcting force may cause the controlled quantity to overshoot the steady-state value, and oscillations of the controlled quantity are likely to result. These oscillations may be damped out rapidly, they may persist, or they may build up so that the system loses all control. An analysis of these oscillations requires a knowledge of all the electrical and mechanical characteristics of the system.

It is desirable that a disturbance to a regulator-type system be corrected rapidly. The greater the restoring force, however, the greater the likelihood of oscillations. Most regulator-type systems or closed-loop systems include "anti-hunting" circuits to oppose the oscillations. Although the anti-hunting circuit must necessarily reduce the restoring force, it reduces the time required to correct for a disturbance by eliminating the oscillations, or hunting.

To provide anti-hunting in an electronic system, it is necessary to so connect a reactor—either a capacitor or an inductor—in the circuit that the current through it, or the voltage across it, is proportional to the rate at which the controlled quantity is changing. This reaction is then fed into the amplifier input to oppose the voltage that produces the restoring force. Thus, the effect of the anti-hunting circuit is the prevention of the build-up of rapid changes in the controlled quantity; without rapid changes, the possibility of hunting is greatly reduced.

28–6. ELECTRIC RESPONSE FROM NONELECTRICAL QUANTITIES

The regulator circuits discussed in the previous articles control electric potential. In order to employ electronic components in systems that control other than electrical quantities, there must be some means of producing voltage changes corresponding to the changes in the controlled quantity. The most direct method of accomplishing this transformation is by the use of a unit that generates an emf that is proportional to, or is some function of, the quantity to be controlled. A few examples of such units are: the photovoltaic cell, which develops an emf when it is illuminated; the piezoelectric crystal, which develops an emf proportional to the pressure exerted upon it; and the thermocouple, which develops an emf proportional to the difference in temperature between its two junctions.

Electric responses to changes in nonelectrical quantities can also be obtained by means of circuit elements whose volt-ampere characteristics or resistance depends upon some nonelectrical quantity. The phototube (Arts. 20–7 and 21–5) is such an element; its volt-ampere characteristic depends upon the amount of illumination on its cathode. Changes in voltage corresponding to changes in illumination can be obtained with a phototube connected in series with a resistor and a d-c source. An increase in the illumination on the phototube causes an increase in the current, and hence an increase in the voltage across the resistor and a corresponding decrease in the voltage across the phototube. Actually, this circuit is more satisfactory than the photovoltaic cell in obtaining an electric response to a change in illumination; for a given change in illumination, a greater change in voltage results.

A thermistor is a resistor whose electrical resistance depends upon temperature—an increase in the temperature causes a decrease in the resistance. The same circuit that is employed for the phototube is applicable to the thermistor to produce voltage variations corresponding to changes in temperature.

The strain gage makes use of a conducting wire firmly attached to the surface of the material in which the strain is of interest. For any deformation of the material, there is a corresponding deformation of the conducting wire and therefore a change in its electrical resistance.

The Pirani tube makes use of a conducting wire exposed to a low-pressure gas, the pressure of which is to be measured or controlled. The temperature of the wire, and hence its resistance, depends upon the ability of the gas to conduct away the heat developed in the wire. The lower the pressure, the lower the efficiency of the gas in conducting away the heat, and therefore the higher the temperature of the wire.

The changes in voltage that would be developed across either the strain-gage resistor or the Pirani tube when connected in the simple series circuit used for the phototube and the thermistor are so small as to make this circuit impractical. Therefore, the indicating elements are generally connected as arms of a balanced bridge circuit, and the degree of unbalance is measured as an indication of strain or pressure.

28–7. PHOTO-CONTROLLED RELAY

The circuit of Fig. 28–4 indicates a means whereby a relay can be actuated as the result of a reduction in the illumination on a phototube. The phototube current flows through R_1, and, as long as the illumination is sufficient, the resulting voltage drop across R_1 maintains the grid of the thyratron T_2 so negative that it cannot conduct. If the illumination is interrupted, however, the negative grid voltage on the thyratron is reduced, allowing the thyratron to conduct and current to flow through the coil of the relay CR. The capacitor filter C_1 serves to eliminate relay-contact chatter. (If current flows through the relay coil only during positive half-cycles, the relay contacts chatter badly, and operation would be unsatisfactory.) When the illumination on the phototube is restored, the thyratron ceases to conduct, and the relay is de-energized.

The circuit of Fig. 28–4 might be applied to turn on lights when the illumination becomes too low, or to turn on a fan when smoke density becomes so great that it interferes with the illumination of the phototube. For most applications, however, this circuit is too insensitive and is not independent of changes in the general illumination.

The circuit of Fig. 28–5 is more sensitive than that of Fig. 28–4, but only to sudden changes in illumination (for example, the change

PHOTO-CONTROLLED RELAY

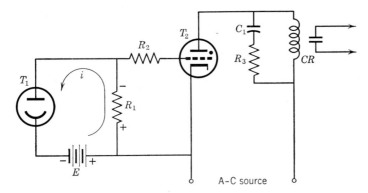

Fig. 28-4. Photo-controlled relay.

in reflected light that might result from a black mark on a rapidly moving sheet of paper). The increased sensitivity of this circuit is obtained by the addition of an amplifier connected between the phototube and the thyratron; independence of general illumination is achieved by the use of capacitive coupling between the different components. The capacitor coupling makes the circuit completely insensitive to gradual changes in illumination; only a sudden change is transmitted through the capacitor coupling to trigger the thyratron.

The three tubes in the circuit of Fig. 28-5 are supplied by a common d-c source, such as the output of a rectifier. Resistor R_1 is the load resistor for the phototube T_1; R_3 for the amplifier tube T_2; and the coil of the relay CR for the thyratron T_3. Under normal conditions, the phototube is illuminated, resulting in current through R_1 and a voltage drop across R_1 having the indicated polarity. The grid of T_2 is at

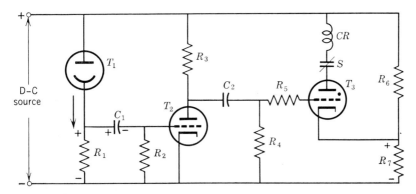

Fig. 28-5. Sensitive photo-controlled relay.

cathode potential; therefore, a relatively large current flows, and the plate of T_2 is at a low positive potential. The voltage across R_7 maintains the grid of T_3 so negative with respect to the cathode that T_3 does not conduct.

When the illumination on the phototube is changed, the voltage across R_1 changes, but, if these changes are slow, the current that flows through R_2 to charge C_1 is so small that the remainder of the circuit is unaffected. Thus, slow changes in the phototube illumination, such as those resulting from an accumulation of dirt on the light source, aging of the light source, or changes in general background illumination, cause no response in the relay.

The effect of a sudden reduction in the illumination on the phototube is much different. In addition to the decrease in the voltage across R_1, there is an equal change in the voltage across R_2 because the voltage across C_1 cannot change instantaneously. This change in the voltage across R_2 makes the grid of T_2 negative with respect to the cathode, and, as in other amplifiers, the more negative grid produces a more positive plate. The change in plate potential of T_2 is transmitted through the capacitor coupling (the charge on a capacitor cannot change instantaneously) to the grid of the thyratron T_3; therefore, T_3 conducts and the relay coil is energized.

Since the thyratron anode circuit is supplied from a d-c source, the thyratron continues to conduct after the initiating change in illumination has passed. The anode circuit of the thyratron must be opened to de-energize the relay coil; the normally closed switch S is provided for this purpose. The switch may be manually operated, or, it may be a limit switch operated by the equipment that is actuated by the closing of the relay contacts. For example, if the relay is used to reject black objects from a mixture of black and white objects passing the phototube, the relay may energize a rejector that operates a switch to open the thyratron anode circuit.

28–8. TIMING CIRCUITS

The time required for a capacitor to charge or discharge to a definite voltage makes a convenient basis for timing, or for the introduction of a time delay in an electronic circuit. The charging rate for a capacitor depends upon its capacitance, the resistance through which the charging current flows, and the voltage of the source. Variation of any of these quantities provides a means of controlling the time required for the voltage across a capacitor to change to a given value.

TIMING CIRCUITS

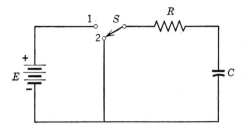

Fig. 28-6. Basic timing circuit.

For the simple circuit of Fig. 28-6, the charging current may be expressed by

$$i = \frac{E}{R} \epsilon^{-t/RC} \text{ amperes} \tag{28-4}$$

where t is the time in seconds after switch S is closed to position 1, and E, R, and C are expressed in volts, ohms, and farads, respectively. The voltage across the series resistor is, therefore,

$$e_R = iR = E\epsilon^{-t/RC} \tag{28-5}$$

and the voltage across the capacitor is

$$e_C = E - e_R = E(1 - \epsilon^{-t/RC}) \tag{28-6}$$

The voltage across the capacitor starts at zero and builds up, approaching the value E as a limit; the voltage across the resistor starts at E and approaches zero.

If S is switched to position 2 after the capacitor voltage has reached a definite value, then both the resistor and the capacitor voltages start at that value and approach zero exponentially. Since the rate of change of any of these voltages is easily controlled, any of them may be applied as the basis of a timing circuit.

An inductive circuit might also be used for timing, since time is required to change the current through an inductance. The R–C circuit is preferred, however, since capacitors are less expensive and require current flow only during the charging or discharging period.

A timing circuit may introduce a definite delay after the closing of one circuit before a second circuit is energized, or it may act to energize a particular circuit for a definite period. An example of a circuit that energizes a relay coil for a definite period is shown in Fig. 28-7. The relay coil CR is energized when the line switch S is closed, and it remains energized for a period that depends upon the constants of the circuit.

The relay coil in Fig. 28-7 is in the anode circuit of a thyratron that

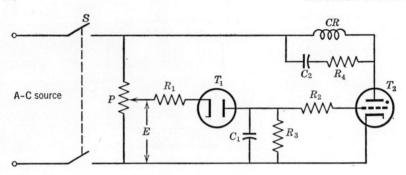

Fig. 28–7. Electronic timer circuit.

is supplied from an a-c source. When S is first closed, there is no charge on C_1; therefore, the grid of the thyratron T_2 is at cathode potential, and the thyratron starts conducting on positive half-cycles of the supply voltage. The thyratron continues to conduct (during alternate half-cycles) until its grid becomes sufficiently negative to prevent conduction. While CR is energized, current also flows through R_1 and T_1 to charge C_1. The rectifier T_1 is so oriented that the resulting charge on C_1 makes the thyratron grid negative. The time required to stop conduction therefore depends upon the rate at which C_1 charges, and this in turn depends upon the time constant R_1C_1 and the voltage E; E, and therefore the timing, can be varied by adjustment of potentiometer P. Since the voltage E is alternating, T_1 allows charging current to flow for only a part of each cycle, and the exact time that the thyratron conducts cannot be calculated by direct application of Eqs. 28–4 through 28–6. However, the equations do provide an indication of the effects of the various circuit elements on the timing.

The resistance of R_3, which shunts C_1, is so large that it does not interfere with the charging of C_1. The function of this resistor is to discharge C_1 after the timing period.

The capacitor C_2 allows the relay coil to remain energized during negative half-cycles of the supply voltage; resistor R_4 limits the peak thyratron current required for charging C_2.

28–9. RESISTANCE-WELDING CIRCUITS

A common method of welding metals consists in passing an electric current through the pieces to be joined; the heat developed in the resistance of the contact fuses the metals. Often, the currents involved

RESISTANCE-WELDING CIRCUITS

are so great that the mechanical opening and closing of contacts in the current circuit is impractical, and the precision of the timing needed may be greater than can be expected from a mechanical timer. Electronic circuits are quite adaptable for both making and breaking the circuit, and for timing the duration of the current flow.

In the basic electronic welding control, two ignitrons act as the contactors, as shown in Fig. 28–8. They are connected "back-to-back" so that one may conduct during positive half-cycles of the supply voltage and the other during negative half-cycles. Thus, a means is provided for the control of the current in both directions through the primary of the welding transformer.

While switch S of Fig. 28–8 is closed, the two ignitrons conduct on alternate half-cycles. When the supply voltage e_{ab} is positive, current flows through the transformer primary, X_2, S, X_4, and the ignitor of T_2. (X_1, X_2, X_3, and X_4 are copper-oxide rectifier units, and the symbols representing them in Fig. 28–8 indicate the current direction for low resistance.) The current flow through the ignitor initiates conduction in T_2 since the anode of T_2 is positive with respect to its cathode; T_2 continues to conduct until the supply voltage reverses. When e_{ab} becomes negative, current flows through X_3, S, X_1, the ignitor of T_1, and the transformer primary, causing conduction in T_1. Thus, while S is closed, an alternating current of approximately sinusoidal wave form flows in the transformer primary.

Switch S may be operated manually or, for more accurate control of the welding time, by an electronic timer similar to that of Fig. 28–7.

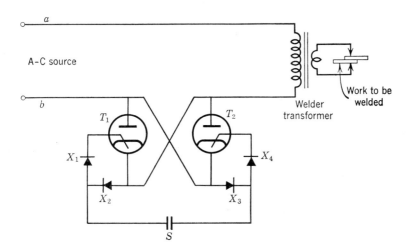

Fig. 28–8. Basic circuit for electronic control of resistance welding.

Since S does not carry the entire transformer current—only that needed for the ignitors—its contacts may be much smaller than those capable of directly making and breaking the transformer circuit.

The timing that can be achieved with the circuit of Fig. 28–8 is limited to the control of the number of half-cycles of current flow; for most purposes, a finer adjustment is unnecessary. Resistance welding of stainless steel or the light metals, however, requires both accurate and short timing. In welding these materials, control of the number of half-cycles of current flow is generally much too coarse to achieve satisfactory welding; for example, four half-cycles may be too long for a particular welding job, and three half-cycles—the next shortest available welding time—may be too short.

For finer welding control, the switch S of Fig. 28–8 is replaced by a pair of thyratrons connected "back-to-back" in the same manner as the ignitrons. These thyratrons are controlled for the number of half-cycles of conduction (by a timer) and for the time in each cycle at which conduction starts (by any of the methods of rectifier grid control). Such a welder control has two manual adjustments: a timing control, which varies the resistance, capacitance, or voltage in the timer circuit; and a "heat" control, which changes the time in the cycle at which conduction starts (usually a variable resistance in a phase-shifter circuit).

The entire sequence of a repetitive welding operation may be electronically controlled. One timer controls the time that the welder jaws are open between welds; another, the time that the jaws squeeze the work before the weld is made; another, the duration of the flow of welding current; and a fourth, the time that the welder jaws are held together before releasing the work. Each timer delays the start of the following operation by a prescribed amount, and, at the end of the delay, sets off the next timer.

PROBLEMS

28–1. A type 6J5 is to be used in the circuit of Fig. 28–1 to supply 8.0 ma. from a 360-volt source to a 200-volt load. What ratio of load-voltage ripple to supply-voltage ripple should be expected? What should E_r be to achieve these operating conditions? *Ans.* 0.047; 196 volts.

28–2. The reference voltage E_r of Fig. 28–1 is to be provided by a voltage-regulator tube, and the path for current through this tube is completed by a resistor connected between the grid and plate of the series tube or between the grid and cathode of the series tube. Show that, with either connection, the circuit will provide the desired voltage regulation, but that the regulation may be improved with the resistor connected between grid and cathode.

PROBLEMS

28-3. Draw an a-c equivalent circuit for the voltage regulator of Fig. 28-2, and write circuit equations from which the ratio of load-voltage ripple to supply-voltage ripple could be determined.

28-4. What would be the effect on the operation of the regulator of Fig. 28-3 of removing T_3?

28-5. How could the circuits of Figs. 28-4 and 28-5 be altered so as to respond to an increase in illumination?

28-6. An increase of 2 volts is required on the grid of the thyratron of Fig. 28-5 in order to fire the tubes and energize the relay coil. What is the corresponding change in voltage at the grid of T_2, if T_2 is a 6J5, $R_3 = 25$ kilohms, $R_4 = 50$ kilohms, and the d-c supply voltage is 300? Assume instantaneous changes so that capacitive reactances may be neglected.

Ans. 0.14 volt.

28-7. The relay CR in the timer of Fig. 28-7 is used to control the duration of exposure for a photographic printing process. What elements in the circuit might be adjusted to double the exposure time? Where possible, indicate how much the change should be.

28-8. Draw a diagram to indicate how two thyratrons might be added to the circuit of Fig. 28-8 to allow control of the time in the cycle at which conduction starts in the ignitrons.

■ 29 Miscellaneous Electronic Circuits

29–1. INTRODUCTION

The possible variations in electronic circuits and their application are so nearly limitless that only some of the basic circuits and components can be treated with any degree of thoroughness in this book. However, to provide a broader picture of the application of these basic elements, in this chapter examples are presented of a few circuits and concepts typical of the fields of electronic instrumentation, wave shaping as required in many TV and radar circuits, and electronic computers.

29–2. VACUUM-TUBE VOLTMETERS

Vacuum-tube voltmeters provide an interesting application of electronic circuits and an almost indispensable tool for the experimental study of electronic circuits. Their principal virtues are that (1) they may be operated over a wide range of frequencies, from direct current to 100 or more megacycles, and (2) they have a high input impedance; voltages can be measured with a vacuum-tube voltmeter without taking so much current that conditions in the circuit under observation are altered.

Some vacuum-tube voltmeters, or VTVM's as the name is often abbreviated, are designed to give an indication that is proportional to the average value of the voltage being measured, others to the average value of the rectified voltage, and still others to the peak value of the voltage.

VACUUM-TUBE VOLTMETERS

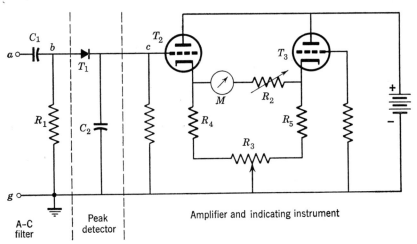

Fig. 29-1. Schematic diagram of a simple vacuum-tube voltmeter.

A common arrangement is one that responds to, or gives an indication proportional to, the peak value of the a-c component of the voltage under test; the circuit diagram of a simple VTVM of this type is shown in Fig. 29-1.

To illustrate the behavior of the VTVM circuit of Fig. 29-1, its reaction to a typical nonsinusoidal voltage is described. For this purpose, it is assumed that the input voltage e_{ag} is the output of a full-wave rectifier as shown in Fig. 29-2a. The input voltage is first applied to an R-C coupling network that is identical, in both circuit arrangement and function, to the coupling used between stages of an amplifier; the d-c component of e_{ag} appears across the capacitor, and the a-c component appears across the resistor as e_{bg}. (R_1C_1 must be much larger

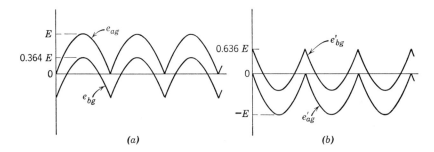

Fig. 29-2. Sample wave forms for vacuum-tube voltmeter.

than the period of the voltage variations, but a large R is desired anyway to maintain the high input impedance that makes this instrument so valuable.) This voltage e_{bg}, which is also shown in Fig. 29–2a, then serves as the input voltage to a simple half-wave rectifier that has a capacitor filter. The R–C time constant of the rectifier load is large enough so that the capacitor charges to the positive peak value of the rectifier input voltage. Thus, the capacitor C_2 is charged to the positive peak value of the a-c component of the voltage being measured. For this example, the capacitor charges to $0.364E$, where E is the peak value of the input voltage as shown in Fig. 29–2a. (If the connections between the voltmeter and the voltage under test had been reversed, the wave forms would be as shown in Fig. 29–2b, and the voltage across the capacitor would be $0.636E$.)

The amplifier that connects the capacitor voltage e_{cg} to the indicating instrument is operated as a linear amplifier so that the indication is also proportional to the positive peak of the a-c component of the voltage being measured. Since the instrument is used so frequently to measure the rms value of a sine wave, it is calibrated to read that value directly; that is, the meter dial is calibrated to read 0.707 times the capacitor voltage. In general, it does not read the rms value, however, but 0.707 times the positive peak value of the a-c component of the voltage applied to the terminals of the VTVM.

The amplifier shown in Fig. 29–1 is a cathode follower whose output voltage is compared by means of the indicating meter to the cathode of a second tube that has no signal applied to its grid. This arrangement provides the high impedance necessary across the rectifier output; if offers an easy method of setting the indicator to zero for zero applied voltage— accomplished by adjustment of the potentiometer R_3; and the use of two tubes provides some compensation for changes in the tube characteristics as it heats, for identical changes in both T_2 and T_3 do not affect the meter reading. Calibration of the instrument may be accomplished by adjustment of resistor R_2 that is connected in series with the indicating meter.

The sensitivity of the vacuum-tube voltmeter can be increased considerably by inserting a voltage amplifier between the peak detector and the cathode follower shown in Fig. 29–1. Other possible variations are that the voltage to be measured might be applied directly to terminals b and g to provide an instrument that responds to the positive peak voltage, or to terminals c and g to provide one that responds to the average value. In the latter case, the meter M, which is an ordinary d-c meter, does the averaging.

29-3. CATHODE-RAY OSCILLOSCOPE

The cathode-ray oscilloscope is another electric instrument that is both an interesting application of electronic circuits and an invaluable aid to the study of other electronic circuits. Its basic function is the display of voltage or current wave forms, that is, plots of voltage or current versus time, but it has many other uses. The screen on which the wave forms are displayed may be calibrated so that instantaneous values of alternating voltage or current, or direct voltages or current, can be measured. (The oscilloscope is essentially a voltage-indicating device; to measure currents, it is necessary only to observe the voltage across a small resistor through which the current in question flows.) By proper application of the cathode-ray oscilloscope, it is also possible to compare two voltages for relative phase or relative frequency.

The heart of the cathode-ray oscilloscope is a cathode-ray tube, the essential components of which are shown in Fig. 29-3. As indicated in this diagram, these components are: (*a*) a screen, made of material that fluoresces when bombarded by electrons, (*b*) an electron gun, which produces a sharply focused beam of electrons directed toward the screen, and (*c*) two pairs of deflecting plates between which the electron beam must pass. The deflecting plates are so mounted that a voltage applied between the two plates of one pair causes a horizontal deflection of the electron beam, and a voltage applied to the other pair causes a vertical deflection. In each case, the deflection is proportional to the voltage applied to corresponding deflecting plates.

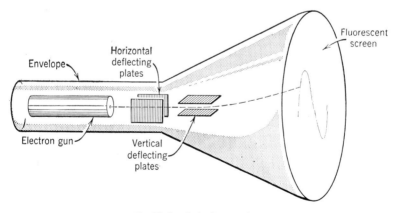

Fig. 29-3. Cathode-ray tube.

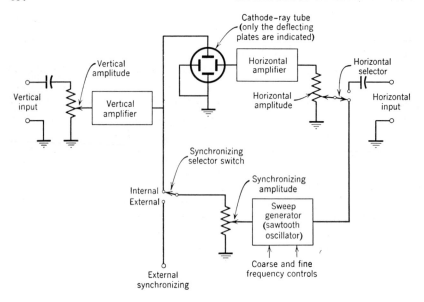

Fig. 29–4. Block diagram of cathode-ray oscilloscope.

The cathode-ray tube requires a low-voltage filament supply, and a low-power source of direct voltage—of at least 1000 to 2000 volts—which is generally furnished by a half-wave rectifier with a capacitor filter. These voltage sources, as well as some associated controls, are contained in the oscilloscope. The **focus** and **intensity controls** are manually operated rheostats that affect the direct voltages applied to the electron gun. As the names imply, they provide an adjustment of the focus and intensity of the electron beam and, therefore, provide the same adjustment of the fluorescent spot on the screen. **Horizontal** and **vertical positioning controls,** also manually operated rheostats, serve as adjustments of the direct components of voltage applied to the respective pairs of deflecting plates. These controls provide a means of positioning, or centering, whatever pattern happens to be under observation on the screen.

In addition to the cathode-ray tube and its associated equipment, an oscilloscope includes (*a*) provision for a linear time base for the observation of wave forms, (*b*) amplifiers to make the instrument applicable to small voltages, and (*c*) various controls and switching arrangements. The block diagram of Fig. 29–4 shows the interconnection of these components as found in most oscilloscopes.

As indicated in Fig. 29–4, means are provided for amplifying both

CATHODE-RAY OSCILLOSCOPE

the horizontal and vertical deflecting voltages before application to the deflecting plates of the cathode-ray tube. Controls are also provided for the adjustment of the amplitude of deflection in either direction.

Resistance–capacitance coupling to both amplifiers is employed so that, at any time, the deflection on the screen is proportional to the instantaneous value of the a-c component of an applied signal. To obtain a deflection corresponding to a direct voltage, that voltage must be applied directly to the deflecting plates, and most oscilloscopes provide for this connection, however, some employ direct coupling to the amplifier, so that the d-c component of a voltage may be observed while the amplifier is still retained.

To determine the relative phase or relative frequency of two voltages, the two voltages are applied to the horizontal and vertical input terminals of the oscilloscope, amplified by the amplifiers, and applied to the deflecting plates. The desired information can be obtained by an analysis of the resulting pattern on the screen of the cathode-ray tube. A more common use of the cathode-ray oscilloscope, however, is for the viewing of voltage wave forms. For this purpose, the vertical deflection is made proportional to the instantaneous value of the voltage under observation by connection to the vertical input terminals, but the horizontal deflection must be proportional to time. This linear time base is supplied by a **sweep generator,** and, as shown in the block diagram, a selector switch allows for obtaining the horizontal deflecting voltage from either the horizontal input terminals or from this sweep generator.

The sweep generator is a self-contained oscillator that produces an output voltage having a sawtooth wave shape, as shown in Fig. 29–5. **Coarse** and **fine frequency controls** are provided for the adjustment of the sweep frequency to the desired value. The application of this sawtooth sweep voltage to the horizontal deflecting plates causes the luminous spot to sweep from left to right across the screen at a constant rate, and then to snap back to the starting position in negligible time and repeat the sweep. Thus, if a signal voltage is applied to the vertical input terminals while the sweep voltage is applied to the horizontal

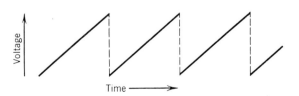

Fig. 29–5. Sawtooth sweep voltage.

deflecting plates, the luminous spot on the screen traces out a plot of the signal voltage versus time. Furthermore, if the sweep frequency is equal to the frequency of the signal, or equal to a submultiple of the signal frequency, the spot retraces itself for each sweep, producing a stationary pattern on the screen. For equal sweep and signal frequencies, one cycle of the signal-voltage wave form appears on the screen; for a sweep frequency of one-half the signal frequency, two cycles appear; etc. Thus, by adjustment of the sweep frequency controls, it is possible to obtain a stationary picture of any number of cycles of the signal voltage.

If the sweep frequency is not equal to the signal frequency, or to one of its submultiples, the pattern on the screen appears to move to the left or right. Since it is impossible to maintain the frequency of an independently operated sweep generator exactly at a fixed frequency, there is always a tendency for the picture to drift unless the sweep generator is synchronized to the signal frequency. If the natural frequency of the sweep generator is approximately the desired frequency, the generator may be synchronized by the injection into the sweep generator of a small voltage of the desired frequency. Such synchronization causes the sweep generator to "lock in" with a fixed phase relationship to the injected synchronizing voltage and causes the pattern on the screen to become absolutely stationary.

Since the frequency of the synchronizing voltage should be identical with that of the signal, the signal itself is the logical synchronizing voltage. Accordingly, with the **synchronizing selector switch** in the *internal* position, the synchronizing voltage is obtained from the output of the vertical amplifier. This is the most convenient method of synchronizing.

A **synchronizing amplitude control** is provided for the adjustment of the magnitude of the synchronizing voltage. This control should be left at zero until adjustment of the frequency controls has been made. When the sweep frequency is approximately the desired value, the synchronizing amplitude control is advanced only far enough to cause the picture to "lock" in place. Further advancing may distort the picture.

With the synchronizing selector in the *external* position, the synchronizing voltage is whatever is connected to the **external synchronizing terminals**. It is convenient, when the different voltages to be observed are all derived from the power line, to apply a small voltage of power-line frequency as an external synchronizing voltage. There are certain advantages to external synchronizing: The magnitude of the synchronizing voltage is unaffected by changing the signal; therefore, the synchronizing amplitude control need not be readjusted every time a different signal is applied. Also, the phase of the synchronizing voltage

29-4. THYRATRON RELAXATION OSCILLATOR

One of the simplest means of obtaining the sawtooth wave form required for the horizontal sweep in an oscilloscope is the thyratron relaxation oscillator, which is illustrated in Fig. 29–6a. (A relaxation oscillator is one whose period of oscillation is determined by the charging or discharging time of a capacitor through a resistance.) In this circuit, the anode supply voltage for the thyratron is the voltage across capacitor C, which is charged from E_{bb} through R_1. The voltage across C would build up along an exponential curve, as indicated in Fig. 29–6b, if there were no other circuit elements present. However, the thyratron breaks down when the capacitor voltage reaches E_{bkd}, which is related to the grid voltage by the thyratron control characteristic. As soon as the thyratron breaks down, its anode voltage drops to the normal tube drop, E_{TD}, and C discharges rapidly through R_2. (Resistor R_2 is present only to limit the peak anode current through the thyratron). If C dis-

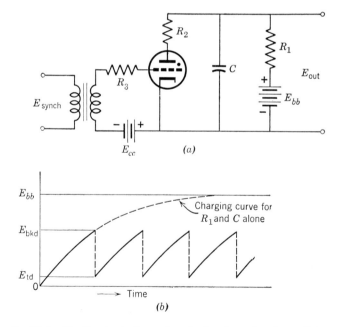

Fig. 29–6. Circuit and output wave form for thyratron relaxation oscillator.

charges much more rapidly than it can be recharged through R_1, the thyratron is extinguished when the voltage across C becomes insufficient to sustain the arc. When this occurs, C is recharged through R_1, and the cycle repeats. The resulting wave form for the output voltage is shown in Fig. 29–3b.

A consideration of the curves of Fig. 29–6b indicates that the period of oscillation of this sweep generator can be increased or decreased by increasing or decreasing R_1, C, or E_{bkd}; either R_1 or C can be adjusted directly, and E_{bkd} is controlled through adjustment of the grid voltage. Common practice is to employ a continuously variable R_1 to provide a fine frequency control and to switch to different fixed values of C for coarse changes in frequency.

Thyratron grid control of oscillation frequency is not employed so that the grid can be left free for synchronization; that is, so that a synchronizing voltage may be conveniently connected to the grid. When the grid is made more positive as a result of an applied synchronizing voltage, E_{bkd} is reduced, and thyratron breakdown (which coincides with the start of a new sweep) is most likely to occur when the synchronizing voltage is going positive. Actually, after a few cycles, the operation stabilizes with all sweeps starting at the same phase of the synchronizing voltage if it is large enough to have any effect at all. Under this condition, the duration of the sweep is equal to an integral number of periods of the synchronizing voltage. If the synchronizing voltage amplitude is increased, it is possible to make the thyratron breakdown one or more cycles earlier; that is, by control of the amplitude of the synchronizing voltage, the period of the sawtooth sweep can be made equal to any integral number of periods of the synchronizing voltage, so long as the total period is less than the period of the relaxation oscillator running free with no synchronizing voltage.

One disadvantage of the thyratron sweep generator is the frequency limitation imposed by the gas-discharge tube. Because a finite time is required for the gas in the thyratron to deionize after each firing, these circuits do not operate satisfactorily at frequencies much greater than 50 kilocycles. For higher-frequency operation, vacuum tubes must be used. A vacuum-tube circuit capable of supplying a sawtooth output wave form is discussed in Art. 29–6.

29–5. DIODE WAVE SHAPING

The circuits considered in this section are simply half-wave rectifiers, or half-wave rectifiers with slight modifications, looked at from the point of view of their effects on the wave shape of the applied voltages. Thus, the resistance-loaded half-wave rectifier of Fig. 29–7a is a negative clipper. The output voltage e_{bc} is equal to the input voltage e_{ac} with its negative half removed, or "clipped" off. If the polarity of the rectifier is reversed, or if the positions of R and Y are interchanged, the circuit becomes a positive clipper as shown in Fig. 29–7b and c, since the positive half of the sine-wave input has been clipped off to form the output voltage.

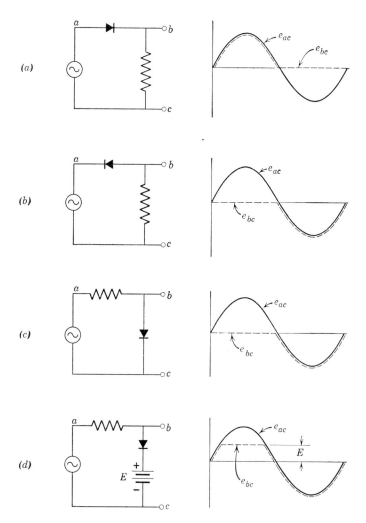

Fig. 29–7. Clipping circuits and wave forms.

A slight modification of the simple rectifier can make it a much more versatile clipper; a d-c potential introduced in series with the *b*-to-*c* branch alters the level at which clipping occurs. Thus, if a battery is placed in series with the rectifier of Fig. 29–7c, as illustrated in Fig. 29–7d, only that portion of the input voltage that exceeds the battery voltage is clipped. When e_{ac} is less than E there is no current, and e_{bc} must equal e_{ac}; when e_{ac} exceeds E, current flows, and, if the drop across Y is negligible, e_{bc} equals E. (This is precisely the circuit that was investigated as a battery charger in Art. 21–9.) The clipping level can

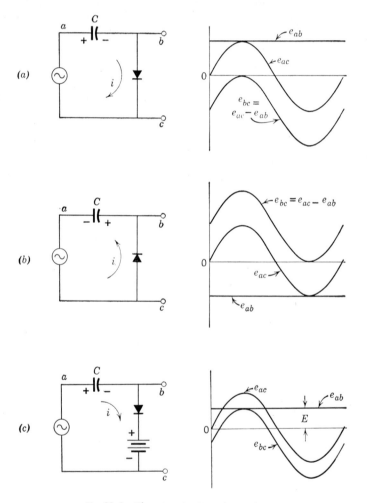

Fig. 29–8. Clamping circuits and wave forms.

DIODE WAVE SHAPING

equally well be adjusted to negative values by reversing the polarity of the battery.

Sinusoidal wave forms have been used to illustrate the behavior of these simple clipping circuits, but there is obviously nothing about them or about their operation that limits their use to any particular wave form. A clipper might be employed to square off the top of an imperfect square wave, or to clip a series of pulses of variable amplitude to make them uniform.

The clipping circuits of Fig. 29–7c and 29–7d can be converted to clamping circuits by replacing the resistors with capacitors, as shown in Fig. 29–8. The resulting circuits are half-wave rectifiers with large capacitor filters identical with those analyzed in Art. 23–4. Thus, in the circuit of Fig. 29–8a, current flows in the direction indicated until C is charged to the positive peak value of e_{ac}. After this initial charging, the output voltage, which equals the input voltage less the capacitor voltage, is of the same shape as the input voltage and differs from it only by being vertically displaced so that the positive peak value is zero. The positive peak is said to be "clamped" at zero.

If the polarity of the rectifying element of the clamping circuit discussed in the preceding paragraph is reversed, the negative peak is clamped at zero as shown in Fig. 29–8b. Furthermore, the level at which the peak is clamped can be controlled in the same manner that the clipping level is controlled. That is, if a d-c potential is inserted in the b-to-c branch as shown in Fig. 29–8c, the peak is clamped at the battery voltage instead of at zero.

A possible application of a clamping circuit is to provide a bias at the input of a vacuum-tube amplifier so that the grid cannot go positive with respect to the cathode. As illustrated in Fig. 29–9, the grid–cathode portion of the tube acts as the rectifier element; grid current flows on positive peaks to charge the capacitor until, when it is fully

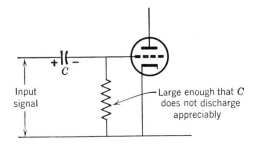

Fig. 29–9. Clamping circuit applied to grid biasing.

charged, the positive peaks of the input voltage are clamped at the cathode potential. The clamping action automatically provides the proper bias for the tube to operate in its most favorable range.

29-6. MULTIVIBRATORS

A multivibrator is a two-stage amplifier with the output of the second stage fed back to provide the input signal for the first. Such an arrangement, for which a circuit diagram is given in Fig. 29–10, differs from the ordinary feedback oscillator only in the lack of a frequency-selective network and in the use of R–C coupling instead of inductive coupling from the output to the input of the amplifier. The multivibrator is probably more accurately described as a relaxation oscillator whose period of oscillation is determined primarily by the time constants of the R–C coupling networks $R_{g1}C_1$ and $R_{g2}C_2$. It is possible to obtain a sawtooth wave form from the multivibrator similar to that obtained from the thyratron relaxation oscillator, but its most common use is as a source of square or rectangular waves. Wave forms for this circuit are illustrated in Fig. 29–11.

Because the output of a second stage of an R–C coupled amplifier is in phase with the input to the first, any disturbance in the current or voltage in a multivibrator is continuously amplified around the loop with one grid becoming more and more negative and the other less and less negative until one of the tubes is cut off. Disturbances are unavoidable—either as a result of differences in the heating of the two tubes, or because the current in the tubes is carried by finite charges—so that, as soon as the power is turned on, one tube is immediately cut off, and the other left

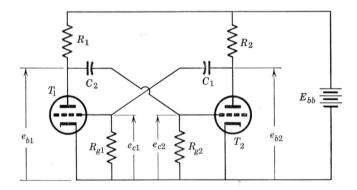

Fig. 29–10. Basic multivibrator circuit.

heavily conducting with its grid at or near cathode potential. This is not a stable condition, however, because the only d-c connection from the grid of the cut-off tube is to its cathode through a resistor. In order to maintain the grid at a highly negative potential, current must flow through this resistor, and this current can only charge or discharge the coupling capacitor: Thus, the cut-off condition cannot be maintained with steady voltages in the circuit. When the tube that had been cut off starts to conduct again, a new disturbance is created that is amplified to cut off the other tube. There are, then, two states depending upon which tube is cut off, and the multivibrator continuously switches back and forth between these two states. The actual switching time is normally but a fraction of a microsecond, but the period between successive switchings is a function of the circuit parameters and can be made to vary over a wide range.

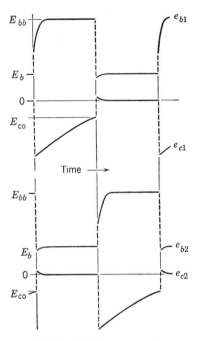

Fig. 29–11. Multivibrator wave forms.

To describe the operation of the multivibrator more specifically, it is assumed that the tubes have arrived at one of the states mentioned in the preceding paragraph: for example, T_1 conducting with its grid at cathode potential, and T_2 cut off. Since the cut-off condition is not stable, it is further assumed that the grid of T_2 is just approaching E_{co}, the voltage at which T_2 starts to conduct. These are the conditions illustrated just prior to $t = 0$ in the diagram of Fig. 29–11. The plate voltage of T_1 is E_b, the value fixed by the tube characteristic, plate supply voltage E_{bb}, and plate load resistance, R_1; the plate voltage of T_2 is E_{bb} since that tube is not conducting.

At $t = 0$, T_2 starts to conduct, and switching takes place; T_1 is cut off, and T_2 becomes heavily conducting. The exact nature of this switching is complicated; however, an understanding of the general behavior of the multivibrator can be obtained by neglecting the effects of interelectrode capacitances and other stray capacitances, and by assuming that the grid resistors are so much larger than the plate load resistors that the effects of the capacitor charging currents through the load resistors can be neglected. Accordingly, when T_1 is cut off at $t = 0$, its plate voltage e_{b1} rises immediately, and, since the charge on C_2 cannot change instantaneously, e_{c2} rises with it. Because of grid current, e_{c2} cannot go far positive, so the instantaneous change in both e_{c2} and e_{b1} is approximately equal to E_{co}. The time constant that governs the charging of C_2 when the grid of T_2 is conducting is very short; therefore, e_{c2} and e_{b1} soon reach their final values, zero and E_{bb}, respectively, as shown in the wave forms of Fig. 29–11. At the same time, the change in the grid voltage e_{c2} causes a corresponding change in the plate voltage e_{b2} as dictated by the tube characteristic and plate load resistor R_2. This change in e_{b2} is transmitted through C_1 (because the charge on this capacitor cannot change instantaneously) to make an equal change in e_{g1}, thereby driving T_1 well beyond cut-off.

Once switching has been completed, conditions would be stable except for the fact that the grid of the cut-off tube is connected to its cathode. The coupling capacitor C_1 discharges through R_{g1} to cause the grid voltage e_{g1} to vary from $(E_{bb} - E_b)$ immediately following switching along an exponential path toward zero, as shown in Fig. 29–11. However, as soon as e_{c1} becomes more positive than the cut-off value E_{co}, T_1 starts to conduct again, and switching takes place in the opposite direction. Wave forms for the complete cycle that results from this continuous switching back and forth are shown in Fig. 29–11. The plate-voltage wave forms are reasonably good square waves and are sometimes used directly as such. If a squarer wave is required, one of the clipping circuits discussed in Art. 29–5 might be employed to cut off the rounded tops.

There are many variations of the basic multivibrator circuit shown in Fig. 29–10. For example, it is not essential that the circuit be symmetrical as assumed in drawing the wave forms of Fig. 29–11; by using different time constants for the two coupling networks, the time that the circuit remains in one state can be made to differ from the time it remains in the other. If carried far enough, this nonsymmetry converts the multivibrator into a pulse generator; one plate voltage stays at E_{bb} for only a small fraction of the total period, while the other remains at E_b for the same small fraction of the time. Thus, one could obtain either a positive or a negative pulse, depending upon which plate was employed as the output terminal.

Another common variation in the multivibrator circuit is in the choice of the d-c potential to which the grid is connected. If the grid resistors are connected to a positive potential rather than to the cathode as shown in the circuit of Fig. 29–10, an improvement can be effected in the stability of the multivibrator operation. With the positive grid return, the slope of the grid-voltage curve as it intersects the E_{co} line is increased, thereby reducing the effects of tube aging or heating up that tend to alter the value of E_{co}.

Multivibrators may be synchronized in the same manner in which the thyratron relaxation oscillator is synchronized; that is, an alternating voltage may be superimposed on one of the grid voltages to cause switching to take place sooner than it otherwise would. If the frequency of the synchronizing voltage is greater than the free-running frequency of the multivibrator, the period of the multivibrator can be made equal to an integral number of periods of the synchronizing voltage.

29-7. MONOSTABLE AND BISTABLE MULTIVIBRATORS

The multivibrator can be made to serve other functions if it is arranged to be stable in either one or both states. When triggered by the application of a pulse, the **monostable,** or **one-shot, multivibrator** switches to the unstable state, remains there for a period depending on the time constant of the R–C coupling network, and then switches back to the stable state and remains there until triggered again. Such an arrangement is capable of delivering either positive or negative voltage pulses, or gates, whose width can be controlled by choice of RC elements. In the **bistable multivibrator,** (also called a **"flip-flop"** or **Eccles-Jordan** circuit) a trigger causes switching from one stable state to another that is equally stable, and so a second trigger is required to return the multivibrator to its initial state. Because the bistable multivibrator can then be arranged to provide one output pulse for every two input pulses, it is indispensable in electronic counters and digital computers. Like the

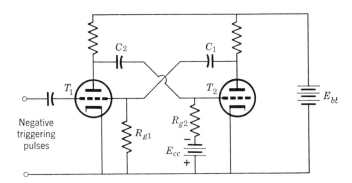

Fig. 29-12. Schematic diagram for a monostable Multivibrator.

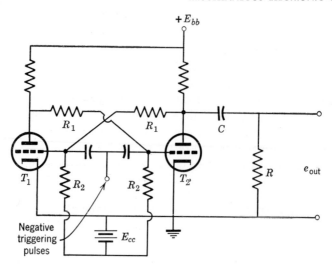

Fig. 29–13. Bistable multivibrator circuit.

free-running multivibrator, it can also provide a square-wave output, which, in this case, must be synchronized with a set of trigger pulses.

In the monostable multivibrator circuit of Fig. 29–12, a sufficiently negative bias is applied to the grid of T_2 so that, if undisturbed, T_2 is cut off. The grid of T_1, however, is returned to its cathode, and so the circuit is completely stable with T_1 conducting and T_2 cut off. Any disturbance or trigger that is large enough to bring T_2 into conduction

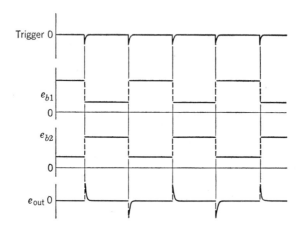

Fig. 29–14. Wave forms for bistable multivibrator.

causes switching to an unstable state and a subsequent return in the same manner as in the free-running multivibrator. The time in the unstable state and therefore the width of the output pulse is a function principally of the time constant $R_{g1}C_1$.

Triggering of the monostable multivibrator of Fig. 29–12 can be accomplished by the direct application of a positive pulse to the grid of T_2. A smaller trigger can be used, however, if applied as a negative pulse to the grid of T_1; T_1 amplifies and converts the pulse to a positive pulse before applying it to the grid of the cut-off tube T_2.

To achieve a practical bistable multivibrator, resistance coupling is employed between the two tubes as shown in the circuit of Fig. 29–13. The grid voltage for either tube is determined by the voltage divider connected between the plate of the other tube and a negative potential with respect to the cathode. A balance of resistance and voltage values is obtained such that, when one tube is cut off, the voltage at the grid of the other tube is near cathode potential causing it to conduct, and such that the low plate voltage of the conducting tube causes the grid voltage of the other to be sufficiently negative to cut off its plate current. Thus, either combination of one tube conducting and the other cut off is a stable condition with the plate voltage of each tube providing the proper grid potential to maintain the other tube in its conducting or nonconducting state. Switching occurs exactly as it does in the monostable circuit; a negative pulse applied to the grid of the conducting tube is amplified, and the resulting positive pulse at the grid of the other tube brings it into conduction.

In order for the bistable multivibrator to serve as a scale-of-2 counter, input pulses are applied to both grids, as shown in the circuit of Fig. 29–13. Alternate pulses act on one tube to produce switching in one direction; the other pulses, on the other tube to produce switching in the other direction. The reaction of this circuit to a train of negative pulses is illustrated by the plate-voltage wave forms of Fig. 29–14.

The square-wave plate voltage is converted to a series of alternating positive and negative pulses by the use of a **differentiating circuit,** which is nothing more than an R–C coupling network with a short time constant. With the short time constant, the capacitor cannot maintain its charge for the period between switchings, and so only the rapid changes in voltage are transmitted across the capacitor. The addition of the differentiating circuit has made the bistable multivibrator into a true scale-of-2 counter; one negative output pulse is produced for every two input. (Negative pulses are so much more effective in actuating another bistable circuit that the positive output pulses are ignored. If the positive pulses were large enough to cause triggering of the following

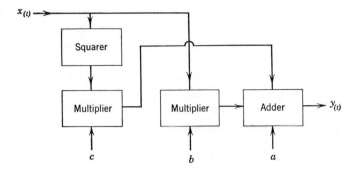

Fig. 29–15. Block diagram for computing $a + bx + cx^2$.

circuit, a simple clipping circuit could be arranged to eliminate them.) An electronic device that can count pulses requires only a series of bistable multivibrators and a means for indicating the state of each one. The first multivibrator switches its state for every input pulse, and, with every second pulse, provides a trigger for the second circuit. The output of the second triggers the third for every fourth input pulse, and so forth. The total number of pulses entering the counter can be calculated if the states of all the multivibrators are known.

Because a large electronic digital computer may use several thousand bistable multivibrators, it is an ideal application for the transistor with its smaller size, lower power requirements, and simpler wiring.

29–8. ANALOG COMPUTERS

An electronic analog of a mathematical expression can be set up so that the expression may be solved with all variables expressed as voltage. Thus, if it is desired to solve the equation

$$y = a + bx + cx^2 \qquad (29\text{--}1)$$

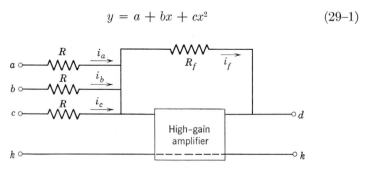

Fig. 29–16. Circuit for addition of voltages.

ANALOG COMPUTERS

and if electronic circuits are available to perform addition, multiplication, and squaring of voltages, a circuit analog can be devised that would solve the equation for any set of constants or for y as a function of time if x is known as a function of time and can be so expressed as a voltage. A circuit set up according to the block diagram of Fig. 29-15 would perform this operation.

Instead of attempting to describe analogs for all mathematical operations, the following discussion is limited to the application of feedback to a high-gain amplifier so that the addition of voltages as required by example of Eq. 29-1 can be achieved or so that it integrates with respect to time. The latter arrangement is essential to the solution of differential equations by an analog computer.

If an amplifier with high-voltage gain is arranged with resistance feedback, as shown in Fig. 29-16, the output voltage is proportional to the sum of any voltages that are connected to its input through equal and high-valued resistances. For the circuit shown

$$e_{dk} = k(e_{ak} + e_{bk} + e_{ck}) \qquad (29\text{-}2)$$

This is shown to be true for the case of the gain approaching infinity and for zero input current to the amplifier.

For e_{dk} much larger than e_{gk}, that is, for high gain,

$$i_f = \frac{e_{dk}}{R_f} \qquad (29\text{-}3)$$

and, with no current to the amplifier input,

$$i_a + i_b + i_c = i_f$$

or

$$\frac{e_{ak} - e_{gk}}{R} + \frac{e_{bk} - e_{gk}}{R} + \frac{e_{ck} - e_{gk}}{R} = \frac{e_{dk}}{R_f} \qquad (29\text{-}4)$$

Equation 29-4 can be rearranged to give

$$e_{dk} = \frac{R_f}{R}(e_{ak} + e_{bk} + e_{ck} - 3e_{gk}) \qquad (29\text{-}5)$$

and if again use is made of the fact that e_{gk} is negligible with respect to e_{dk} (high gain), the last term can be neglected. The output voltage is, for the conditions specified, proportional to the sum of the three input voltages. Thus,

$$e_{dk} = \frac{R_f}{R}(e_{ak} + e_{bk} + e_{ck}) \qquad (29\text{-}6)$$

If R_f is made equal to R in this circuit, it performs addition directly; otherwise it introduces a multiplying factor R_f/R that might conceivably be employed.

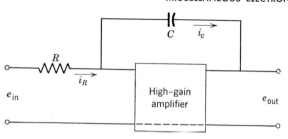

Fig. 29–17. Circuit to perform integration.

The same circuit with a capacitive feedback, as shown in Fig. 29–17, and with but a single input voltage performs integration. Making the same approximations as for the adder,

$$i_R = i_C \tag{29-7}$$

$$i_C = C \frac{d(e_{out})}{dt} \tag{29-8}$$

and

$$i_R = \frac{e_{in}}{R} \tag{29-9}$$

Combining Eqs. 29–7 thru 29–9 gives

$$e_{in} = RC \frac{d(e_{out})}{dt}$$

or

$$e_{out} = \frac{1}{RC} \int e_{in} \, dt \tag{29-10}$$

As in the case of the adder, this circuit does not perform the desired operation alone unless the feedback impedance is properly chosen. For the integrator the time constant RC should be made equal to unity if no multiplying factor is desired.

PROBLEMS

29–1. A properly calibrated vacuum-tube voltmeter of the type illustrated in Fig. 29–1 indicates 20 volts when connected across the output terminals of a thyratron relaxation oscillator. Assuming that the wave form is a perfect sawtooth and that the thyratron extinction potential is 10 volts, calculate the anode voltage at which the thyratron breaks down. Does the assumption of a perfect sawtooth wave form result in a calculated value of breakdown voltage that is higher or lower than the actual?   *Ans.* 66.6 volts.

PROBLEMS

29-2. When connected between terminals b and c of the circuit shown in Fig. 29-7a, the VTVM of Prob. 29-1 indicates 10 volts. Assuming that the diode is a perfect rectifier and that the supply voltage e_{ac} is a perfect sine wave, calculate the rms value of e_{ac}. Account for the possibility of two different answers. *Ans.* 14.7 or 31.4 volts.

29-3. The d-c grid bias for a thyratron relaxation oscillator is so adjusted that the thyratron breaks down at an anode voltage of 100 volts. If the anode supply is 250 volts, the time constant of the R–C circuit is 0.001 sec, and the thyratron extinction potential is negligible, what is the frequency of the oscillations? *Ans.* 1960 cps.

29-4. If the control characteristic of the thyratron in the oscillator of Prob. 29-3 is given by Fig. 21-9, determine the grid voltage required to produce an oscillation frequency of 4500 cps. *Ans.* −0.2 volts.

29-5. The input voltage e_{ab} to the circuit of Fig. 29-18 is $100 \sin \omega t$

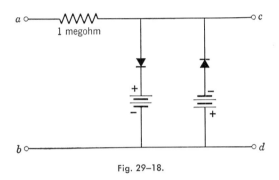

Fig. 29-18.

volts. Sketch the wave form of the output voltage e_{cd}, assuming perfect diode rectifiers.

29-6. Analyze the circuit of Fig. 29-19 to determine the waveshape of the output voltage.

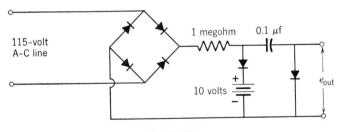

Fig. 29-19.

29-7. Draw the diagram for a circuit that will eliminate positive pulses from a train of alternately positive and negative pulses of 20 volts amplitude, and that will leave the tip of the negative pulse clamped at −10 volts.

29–8. The multivibrator of Fig. 29–10 is symmetrical and consists of two type-6J5 triodes (characteristics given by Fig. 20–8), a 300-volt plate supply, 50,000-ohm plate load resistors, 1.0-megohm grid resistors, and 0.001-μf coupling capacitors. Label the corresponding voltage magnitudes on the wave forms of Fig. 29–11, and calculate the approximate frequency of operation. *Ans.* 380 cps.

29–9. How could the multivibrator of the preceding problem be altered to provide a negative square pulse of 0.10 msec duration at a repetition rate of about 350 cps? Between what two points in the circuit would you observe this pulse?

29–10. A 1.0-megohm resistor is used as R_f in the circuit of Fig. 29–15. What values of input resistors might be used to accomplish the following operation?

$$e_{\text{out}} = 0.3e_1 + e_2 + 2e_3$$

Appendix

TABLE 1

COPPER WIRE TABLE, SOLID CONDUCTORS

Size A. W. Gage (B. & S.)	Diameter Mils	Area Circular Mils	Area Square Mils	Weight Pounds per 1000 ft	Breaking Strength Minimum Pounds for Hard Drawn	Maximum Resistance * Ohms per 1000 ft—20 C-68 F	
						Soft or Annealed †	Tinned—Soft or Annealed
0000	460	211 600	166 000	641	8143	0.04993	......
000	410	167 800	132 000	508	6722	0.06296	......
00	365	133 100	105 000	403	5519	0.07939	......
0	325	105 500	82 900	320	4517	0.1001	......
1	289	83 690	65 700	253	3688	0.1262	0.1275
2	258	66 370	52 100	201	3003	0.1592	0.1608
3	229	52 630	41 300	159	2439	0.2007	0.2028
4	204	41 740	32 800	126	1970	0.2531	0.2557
5	182	33 100	26 000	100	1591	0.3192	0.3225
6	162	26 250	20 600	79.5	1280	0.4025	0.4066
7	144	20 820	16 400	63	1030	0.5075	0.5127
8	129	16 510	13 000	50	826	0.6400	0.6465
9	114	13 090	10 300	39.6	661.2	0.8070	0.8153
10	102	10 380	8 160	31.4	529.2	1.018	1.039
11	91	8 234	6 470	24.9	422.9	1.283	1.310
12	81	6 530	5 130	19.8	337.0	1.618	1.652
13	72	5 178	4 070	15.7	268.0	2.040	2.083
14	64	4 107	3 230	12.4	213.5	2.573	2.626
15	57	3 257	2 560	9.86	169.8	3.244	3.812
16	51	2 583	2 030	7.82	135.1	4.091	4.176
17	45	2 048	1 610	6.20	107.5	5.158	5.266
18	40	1 624	1 280	4.92	85.47	6.505	6.640
19	36	1 288	1 010	3.90	67.99	8.202	8.373
20	32	1 022	802	3.09	54.08	10.34	10.56
21	28.5	810.1	636	2.45	43.07	13.04	13.31
22	25.4	642.4	505	1.95	34.26	16.45	16.79
23	22.6	509.5	400	1.54	27.25	20.74	21.17
24	20.1	404.0	317	1.22	21.67	26.15	27.26
25	17.9	320.4	252	0.97	17.26	32.97	34.37
26	15.9	254.1	200	0.769	13.73	41.58	43.34
27	14.2	201.5	158	0.610	10.92	52.43	54.66
28	12.6	159.8	126	0.484	8.698	66.11	68.92
29	11.3	126.7	99.5	0.384	6.908	83.37	87.85
30	10.0	100.5	78.9	0.304	5.502	105.1	110.8
31	8.9	79.70	62.6	0.241	4.376	132.6	139.7
32	8.0	63.21	49.6	0.191	3.485	167.2	176.1
33	7.1	50.13	39.4	0.152	2.772	210.8	222.1
34	6.3	39.75	31.2	0.120	2.204	265.8	280.1
35	5.6	31.52	24.8	0.0954	1.755	335.2	353.2
36	5.0	25.00	19.6	0.0757	1.396	422.6	445.4
37	4.5	19.83	15.6	0.0600	1.110	532.9	561.6
38	4.0	15.72	12.4	0.0476	.8829	672.0	708.1
39	3.5	12.47	9.8	0.0377	.7031	847.4	893.0
40	3.1	9.89	7.8	0.0299	.5592	1069	1126

* The resistances are maximum values at 20 C for nominal diameters based upon A.S.T.M. specifications of resistivities. Resistances for operating conditions should be corrected for operating temperature. Refer to footnote † on Table 3 for operating temperatures of insulated wires. Temperature coefficient 0.003 93 per degree Centigrade.
† Resistances of medium-hard-drawn wire will be approximately ½ per cent greater for sizes 0 to 0000 and 1½ per cent greater for sizes 1 to 40. Resistances of hard-drawn wire will be approximately 1 per cent greater for sizes 0 to 0000 and 2 per cent greater for sizes 1 to 40.

TABLE 2

COPPER WIRE TABLE, STRANDED CONDUCTORS

Size A. W. Gage or Circular Mils	Over-all Diameter Mils	Area Circular Mils	Number of Strands *	Diameter of Individual Strands Mils	Weight Pounds per 1000 ft	Maximum Resistance † D.C.—Ohms per 1000 ft—20 C–68 F		Multiplying Factor for 60-cycle A-c Resistance
						Soft or Annealed ‡	Tinned—Soft or Annealed	
4	232	41 700	7	77.2	129	0.258 2	0.263 5	1.000
3	260	52 600	7	86.7	163	0.204 7	0.209 0	1.000
2	292	66 400	7	97.4	205	0.162 4	0.165 7	1.000
1	332	83 700	19	66.4	258	0.128 8	0.131 4	1.000
0	373	106 000	19	74.5	326	0.102 1	0.104 2	1.000
00	418	133 000	19	83.7	411	0.080 97	0.082 66	1.000
000	470	168 000	19	94.0	518	0.064 22	0.065 56	1.000
0 000	528	212 000	19	105.5	653	0.050 93	0.051 45	1.000
250 000	575	250 000	37	82.2	772	0.043 11	0.044 00	1.005
300 000	630	300 000	37	90.0	926	0.035 92	0.036 67	1.006
350 000	681	350 000	37	97.3	1 080	0.030 79	0.031 43	1.009
400 000	728	400 000	37	104.0	1 240	0.026 84	0.027 22	1.011
450 000	772	450 000	37	110.3	1 390	0.023 95	0.024 19	1.014
500 000	814	500 000	37	116.2	1 540	0.021 55	0.021 78	1.018
550 000	855	550 000	61	95.0	1 700	....	....	1.021
600 000	893	600 000	61	99.2	1 850	0.017 96	0.018 33	1.025
650 000	929	650 000	61	103.2	2 010	....	....	1.029
700 000	964	700 000	61	107.1	2 160	0.015 40	0.015 55	1.034
750 000	998	750 000	61	110.9	2 320	0.014 37	0.014 52	1.039
800 000	1031	800 000	61	114.5	2 470	0.013 47	0.013 61	1.044
900 000	1093	900 000	61	121.5	2 779	0.011 97	0.012 10	1.055
1 000 000	1152	1 000 000	61	128.0	3 088	0.010 78	0.010 89	1.067
1 250 000	1289	1 250 000	91	117.2	3 859	0.008 622	0.008 710	1.102
1 500 000	1412	1 500 000	91	128.4	4 631	0.007 185	0.007 258	1.142
1 750 000	1526	1 750 000	127	117.4	5 403	0.006 158	0.006 221	1.185
2 000 000	1631	2 000 000	127	125.5	6 175	0.005 388	0.005 444	1.233
2 500 000	1824	2 500 000	127	140.3	7 794	0.004 353	....	1.326
3 000 000	1998	3 000 000	169	133.2	9 353	0.003 628	....	1.424
3 500 000	2159	3 500 000	169	143.8	11 020	0.003 139	....	1.513
4 000 000	2309	4 000 000	217	135.8	12 590	0.002 747	....	1.605
4 500 000	2448	4 500 000	217	144.0	14 300	0.002 465	....	1.685
5 000 000	2581	5 000 000	217	151.8	15 890	0.002 219	....	1.765

 * Standard concentric Class B stranding.
 † The resistances are maximum values at 20 C for nominal diameters based upon A.S.T.M. specifications of resistivities. Resistances for operating conditions should be corrected for operating temperature. Refer to footnote † on Table 3 for operating temperatures of insulated wires. Temperature coefficient 0.003 93 per degree Centigrade.
 ‡ Resistances of medium-hard-drawn wire will be approximately 1½ per cent greater. Resistances of hard-drawn wire will be approximately 2 per cent greater.

TABLE 3

CURRENT-CARRYING CAPACITIES OF COPPER CONDUCTORS*

Ampere Carrying Capacity †
(Based on Room Temperature of 30 C–86 F)

Size AWG MCM	Not More Than Three Conductors in Raceway or Cable ‡			Single Conductor in Free Air			
	Rubber Type R, Type RW, Type RU (14–6); Thermoplastic Type T (14–410), Type TW (14–410)	Rubber Type RH	Paper; Thermoplastic Asbestos Type TA; Var-Cam Type V; Asbestos Var-Cam Type AVB	Rubber Type R, Type RW, Type RU (14–6); Thermoplastic Type T, Type TW	Rubber Type RH	Thermoplastic Asbestos Type TA; Var-Cam Type V; Asbestos Var-Cam Type AVB	Slow-Burning Type SB; Weather-Proof Type WP, Type SBW
14	15	15	25	20	20	30	30
12	20	20	30	25	25	40	40
10	30	30	40	40	40	55	55
8	40	45	50	55	65	70	70
6	55	65	70	80	95	100	100
4	70	85	90	105	125	135	130
3	80	100	105	120	145	155	150
2	95	115	120	140	170	180	175
1	110	130	140	165	195	210	205
0	125	150	155	195	230	245	235
00	145	175	185	225	265	285	275
000	165	200	210	260	310	330	320
0000	195	230	235	300	360	385	370
250	215	255	270	340	405	425	410
300	240	285	300	375	445	480	460
350	260	310	325	420	505	530	510
400	280	335	360	455	545	575	555
500	320	380	405	515	620	660	630
600	355	420	455	575	690	740	710
700	385	460	490	630	755	815	780
750	400	475	500	655	785	845	810
800	410	490	515	680	815	880	845
900	435	520	555	730	870	940	905
1000	455	545	585	780	935	1000	965
1250	495	590	645	890	1065	1130	....
1500	520	625	700	980	1175	1260	1215
1750	545	650	735	1070	1280	1370	....
2000	560	665	775	1155	1385	1470	1405

CORRECTION FACTOR FOR ROOM TEMPERATURES OVER 30 C 86 F

C F							
40 104	.82	.88	.90	.82	.88	.90	....
45 113	.71	.82	.85	.71	.82	.85	....
50 122	.58	.75	.80	.58	.75	.80	....
55 131	.41	.67	.74	.41	.67	.74	....
60 140	...	.58	.67	...	.58	.67	....
70 158	...	.35	.52	...	.35	.52	....
75 167	...	...	.43	...	...	.43	....
80 176	...	...	.30	...	...	.30	....
90 194	...	...	...	...	...	...	....
100 212	...	...	...	...	...	...	....
120 248	...	...	...	...	...	...	....
140 284	...	...	...	...	...	...	....

* National Electrical Code allowable current-carrying capacities for copper conductors. For aluminum conductors the carrying capacities are 84 per cent of those given in the table.
† The current-carrying capacities are based on the following maximum allowable operating temperatures: Type R—60 C; Type RH—75 C; Type V—85 C; Type WP—80 C.
‡ If the number of conductors in a raceway or cable is from 4 to 6, the carrying capacities are 80 per cent of those given in the table. If the number of conductors in a raceway or cable is from 7 to 9, the carrying capacities are 70 per cent of those given in the table. A raceway is any enclosure for supporting the wires, such as metal conduit or wireways.

■ Index

Alternating current, 6, 122
A-c amplifier, 511
Alternation, 127
Alternator, 336
Ammeter, a-c, 202
   d-c, 198
   multirange, 199
Ammeter shunt, 198
Amortisseur winding, 353
Ampere, 4
Ampere turn, 57
Amplidyne, 388
   applications, 394
Amplification factor, 535
Amplifier, 508
   a-c, 511
   class A, 521
   class B, 521, 523
   class C, 521, 529
   common-collector, 549
   direct-coupled, 510
   d-c, 510
   distortion, 513, 541
   efficiency, 521, 523
   gain, 513, 544
   intermediate-frequency, 558
   magnetic, 374, 398
   push–pull, 519, 523
   radio-frequency, 557
   resistance–capacitance-coupled, 515, 538
   rotating, 374, 387, 394
   transformer-coupled, 517
   tuned, 525
Amplitude, 127
Analog computer, 598
Anode, 408, 411
Anode dissipation, 420
Antihunting, 570
Apparent power, 151
Arc discharge, 435
Armature, 221
Armature reaction, 243
Armature resistance, 231
Armature windings, 228
Atom, 1
Autotransformer, 299
Average value, 128
   of sine wave, 129

Battery-charging circuit, 442
Beam-power tube, 425
B–H curve, 61
Bias, cathode, 514
   grid, 485
   grid-leak, 529, 591
Blocking-layer rectifier, 458
Blowout, magnetic, 81, 277

Brake, friction, 283
Braking, dynamic, 283
Breakdown potential, 433
Bridge circuit, 48
Bridge rectifier, 475
Brushes, 223

Capacitance, 112, 117, 143
   effects of, 113, 143
Capacitor, 112
   connection of, 117
Capacitor motor, 362, 366
Carrier, 552
Carrier-current telephony, 563
Cathode, 408, 409
   mercury-pool, 490
   multicellular, 410
   oxide-coated, 409
Cathode bias, 514
Cathode follower, 548
Cathode-ray oscilloscope, 583
Charge, 1
Circuit, 4
   bridge, 48
   calculations, d-c, 33
   capacitive a-c, 143
   clamping, 590
   clipping, 589
   closed, 4
   differentiating, 598
   equivalent, 34
   inductive a-c, 141
   integrating, 602
   nonlinear, 39
   nonlinear a-c, 146
   open, 4
   parallel a-c, 153
   phase-shift, 485
   polyphase a-c, 174
   resistive a-c, 140
   series a-c, 147
   short, 11
   single-phase a-c, 139, 157, 174
   three-phase four-wire, 176
   three-wire, 45, 174, 297
Circular mil, 19
Clamping circuit, 590
Class A amplifier, 521
Class B amplifier, 521, 523
Class C amplifier, 521, 529

Clipping circuit, 589
Closed-loop system, 373
Commutating poles, 237
Commutating winding, 235, 237
Commutation, 232
Commutator, 223
Complex notation, 157
Compound machines, 237
Computers, 598
Conductance, 5
Conductivity, 18
Conductor, 2, 19, 603, 604
   current-carrying capacity of, 20, 605
   force on, 73
Contact rectifier, 458
Contacts, normally closed, 274
   normally open, 274
   push-button, 274
   thermal overload, 274
Control circuit, 273
Control grid, 422
Converter, 558
Copper oxide rectifier, 458
Coulomb, 112
Coupling, direct, 510
   resistance-capacitance, 515
   transformer, 517
Current, 3
   alternating, 6, 122
   conventional, 4
   direct, 6
   electron, 4
   pulsating, 6
Current-amplification factor, 462
Current regulator, d-c motor, 395
Current transformer, 306
Cycle, 127

D'Arsonval mechanism, 190
Delta connection, 183
Delta–delta connection, 304
Demodulation, 555
Detection, 555
   amplitude-modulation, 555
   frequency-modulation, 562
Dielectric, 112
Dielectric constant, 117
Dielectric heating, 530
Differentiating circuit, 598
Diode, 408

INDEX

Diode, double, 475
  high-vacuum, 411
  junction, 457
  point-contact, 457
  power loss in a, 419
  semiconductor, 458
Diode detector, 556
Diode wave shaping, 588
Direct coupling, 510
Direct current, 6
D-c amplifier, 510
D-c generator, 240
  build-up, shunt, 245
  construction, 226
  flat-compounded, 249
  operating characteristics, 248
  overcompounded, 249
  rating, 238, 240
  under-compounded, 251
  voltage regulator, 394, 568
D-c motor, 256
  braking, 282
  construction, 226
  control, 268
  current regulator, 395
  operation, 259
  power requirements, 260
  rating, 238, 256
  reversal, 282
  speed–load characteristics, 261
  speed regulator, 396, 570
  starting, 258, 268
  torque, 256
  torque–current characteristics, 263
Discharge resistor, 95
Distortion in amplifiers, 513, 541
Double diode, 475
Double-rotating-field theory, 361
Double-subscript notation, 123
Double-wye connection, 501

Eddy currents, 91
Effective value, 128
  of sine wave, 129
Electric field, 417
Electrodynamometer mechanism, 193
Electromagnetic forces, 72
Electromagnetic induction, 83
Electromotive force, 2
Electron, 1

Electron emission, 408
Electron theory, 1
Electron tubes, 407
Electronic control, 565
Emf, 2
Emission, electron, 408
Equivalent circuits, 34
  transistor, 543
  vacuum-tube, 534

Farad, 113
Field, electric, 417
  magnetic, 53
Field emission, 417, 491
Field windings, 235
Filament, 410
Filter, capacitor, 470
  rectifier, 477
Flip-flop circuit, 595
Fluorescent lamp, 439
Flux, leakage, 67
  magnetic, 54
  mutual, 109, 291
Flux cutting, 88
Flux density, 54
Flux linkage, 83
Force on a conductor, 73
Forces, electromagnetic, 72
Free-wheeling rectifier, 569
Frequency, 127
Frequency conversion, 558
Full load, 238

Gain, voltage, 513, 536
Gas discharge, 431
Generated voltage, 241, 242
Generator, electric, 220
  elementary, 221
  see also Amplidyne, D-c generator,
    Regulex, Rototrol, Synchronous
    generator
Generator action, 221
Glow discharge, 434
Glow tube, 437
Grid, 420
  control, 422
  screen, 422
  shield, 445
  suppressor, 422
Grid bias, 485

INDEX

Grid-controlled rectifier, 481
Grid-leak bias, 529, 591

Harmonics, 163
Henry, 101
High-frequency heating, 530
Hysteresis, 70

Ignitron, 491, 577
Impedance, 150
  matching, 518
  triangle, 150
Induced voltage, magnitude of, 85, 90, 100
  polarity of, 86, 89
  self-, 93, 100
Inductance, 100, 141
  mutual, 109
  physical aspects of, 101
  self-, 100
Induction, electromagnetic, 83
Induction heating, 530
Induction motor, breakdown torque, 325
  current, 327
  operation, 325
  pull-out torque, 325
  rating, 326
  rotating field, 315, 317
  rotor, 312, 314
  slip, 322
  speed control, 332
  squirrel-cage rotor, 314
  starting current, 328
  starting methods, 330
  starting torque, 328
  stator, 312, 313
  synchronous speed, 318
  three-phase, 312
  torque at standstill, 320
  torque–slip curve, 324
  wound rotor, 314
Inductor, 100
Instrument transformer, 306
Instruments, electrical, 189
Insulators, 2
Integrating circuit, 602
Intermediate-frequency amplifier, 558
Interphase transformer, 503
Ion, 2

Ionization, 431
Ionization potential, 433

Kilowatt-hour, 9
Kirchhoff's law, of current, 33
  of voltage, 31

Leakage flux, 67
Limiter, 561
Line-to-neutral voltage, 177
Line voltage, 177
Lines of force, 53
Load, balanced three-phase, 178
  balanced wye-connected, 179
  delta-connected, 183
  electric, 10
  full, 238
  unbalanced wye-connected, 182
Load-line construction, 41
  application to diode circuit, 413
  application to triode circuit, 426

Magnet, 52
  permanent, 56
  pull of, 72
Magnetic amplifier, 374, 398
  self-saturating, 403
Magnetic blowout, 81, 277
Magnetic circuit laws, 63
Magnetic circuit problem, solution of, 68
Magnetic field, 53
  about a conductor, 55
Magnetic relations, 60
Magnetics, 52
Magnetism, residual, 71, 245
Magnetization curve, 60
Magnetizing force, 60
Magnetomotive force, 57
Measurement of nonelectrical quantities, 212
Measurements, electrical, 189
Mechanism, electrodynamometer, 193
  instrument, 190
  moving-iron, 195
  permanent-magnet moving-coil, 190
Mercury-pool cathode, 490
Mho, 5
Microfarad, 113
Mixer, 483

INDEX 611

Mmf, 57
Modulation, 552
  amplitude, 552, 553
  frequency, 552, 560
  phase, 552
Motor, electric, 220
  elementary, 226
  see also D-c motor, Induction motor, Single-phase motor, Synchronous motor
Motor action, 225
Moving-iron mechanism, 195
Multicellular cathode, 410
Multiplier phototube, 416
Multiplier resistor, 198
Multivibrator, 592
  bistable, 595
  monostable, 595
  one-shot, 595
Mutual conductance, 534
Mutual flux, 109, 291
Mutual inductance, 109
  effects of, 110

Neon sign, 437
Neutral, 45, 177
Nonlinear circuits, 39
Nucleus, 1

Ohm, 5
Ohm-circular mils per foot, 18
Ohmmeter, 209
Ohm's law, 5
Open-delta connection, 305
Oscillator, 527
  Hartley, 528
  local, 557
  relaxation, 587, 592
  tuned-plate, 528
Oscilloscope, cathode-ray, 583
Oxide-coated cathode, 409

P–N junction, 455
Parallel connection, 10
Peak inverse voltage, 468
Pentode, 408
  characteristics, 424
Permanent-magnet moving-coil mechanism, 190
Permeability, 58, 62

Permeance, 58
Phanotron, 408, 440
Phase, 130, 177
Phase angle, 130
Phase difference, 131
Phase opposition, 131
Phase quadrature, 131
Phase sequence, 131, 177
Phase-shift circuit, 485
Phase voltage, 177
Photo-controlled relay, 572
Photoelectric emission, 408, 414
Phototransistor, 458
Phototube, 408, 414, 436
  applications, 572
  electrical characteristics, 415, 436
  multiplier, 416
Pirani gage, 572
Plate, 420
Plate characteristic, 421
Plate resistance, 534
Poles, commutating, 237
  field, 222
  magnet, 52
Polyphase circuits, 174
Potential difference, 2
Potential transformer, 306
Power, 7
  apparent, 151
  measurement of, 203
  measurement of three-phase, 205
  reactive, 152
Power factor, 151
  importance of, 152
Primary winding, 286
Prime mover, generator, 251
Push–pull amplifier, 519, 523

Radio-frequency amplifier, 557
Radio receiver, FM, 561
  superheterodyne, 557
  tuned-radio-frequency, 557
Reactance, capacitive, 144
  inductive, 142
  leakage, 293, 340
  synchronous, 340
Reactive factor, 153
Reactive power, 152
Receiver, electric, 10
Rectification efficiency, 503

Rectifier, 201, 467
  blocking-layer, 458
  bridge, 475
  contact, 458
  copper oxide, 458
  filter, 477
  free-wheeling, 569
  full-wave, 473
  grid-controlled, 481
    bias-phase control, 485
    phase-shift control, 484
  half-wave, 469
  multianode tank, 494
  power, 490
  selenium, 458
  six-phase half-wave, 498
  three-phase bridge, 500
  three-phase full-wave, 500
  three-phase half-wave, 495
  tube rating, 468
  vibrator, 479
Regulex, 388, 392
  applications, 394
Relaxation oscillator, 587, 592
Reluctance, 57
Reluctivity, 58
Repulsion motor, 369, 371
Residual magnetism, 71, 245
Resistance, 4, 15
  armature, 231
  effective, 146
  effect of temperature on, 21
  of resistors in parallel, 35
  of resistors in series, 35
  relationship to physical dimensions, 17
  temperature coefficient of, 22
Resistance–capacitance coupling, 515
Resistance–start motor, 362, 365
Resistance thermometry, 23
Resistance-welding circuit, 576
Resistivity, 17
Resistor, 5
  discharge, 95
  dynamic-braking, 283
  multiplier, 198
  rating of, 27
Resonance, parallel, 156
  series, 155
Rheostat, 26

Ripple factor, 503
Rototrol, 388, 392
  applications, 394

Saturable reactor, 398
Saturation, emission, 412
  magnetic, 61
Saturation curve, 244
Screen grid, 422
Secondary emission, 408, 416, 423
Secondary winding, 287
Selenium rectifier, 458
Self-induced voltage, 93, 100
Self-inductance, coefficient of, 100
  effects of, 102
Self-saturating magnetic amplifier, 403
Selsyns, 375
Semiconductor, 452
Separately excited field winding, 235
Series a-c motor, 368
Series connection, 10
Series field winding, 235
Shaded-pole motor, 362, 366
Shield grid, 445
Shunt, ammeter, 198
Shunt field winding, 235
Side frequencies, 555
Sine waves, addition of, 132
  effective and average values of, 129
  production of, 128
Single-phase motor, 359
  commutator-type, 368
    repulsion, 369
    repulsion–induction, 371
    repulsion–start induction-run, 371
    series, 368
  induction-type, 359
    capacitor, 362, 366
    capacitor-start, 362, 365
    resistance-start, 362, 365
    shaded-pole, 362, 366
    split-phase, 362, 364
Six-phase connection, 305
Slip, 322
Solenoid, 56
Source, electric, 10
Sparking potential, 433
Speed control, d-c motor, 281
Speed regulation, 263